Sliding Mask

Exam Tips:

1. Read each question carefully before looking at the possible answers.

2. After formulating an answer, determine which of the choices most nearly corresponds with that answer. It should completely answer the question.

3. Answer each question according to the latest regulations and procedures. You will receive credit if the regulations or procedures have changed. Computerized exams may be updated as regulations and procedures change.

4. There is only one answer that is correct and complete. The other answers are either incomplete or are derived from popular misconceptions.

5. If you do not know the answer to a question, try not to spend too much time on it. Continue with those you can answer. Then, return to the unanswered or difficult questions.

6. Unanswered questions will be counted as incorrect.

7. On calculator problems, select the answer nearest your solution. If you have solved it correctly, your answer will be closer to the correct answer than the other choices.

JEPPESEN
A Boeing Company

FAA AIRMAN KNOWLEDGE

PRIVATE PILOT
TEST GUIDE

The charts, tables, and graphs used in this publication are for illustration purposes only and cannot be used for navigation or to determine actual aircraft performance.

Cover Photo: Cirrus airplane in flight courtesy of Cirrus Aircraft

ISBN 978-0-88487-698-4

Jeppesen
55 Inverness Dr. East
Englewood, CO 80112-5498
Web Site: www.jeppesen.com
Email: Captain@jeppesen.com
Copyright © Jeppesen
10001387-025   All rights reserved. Published 1992-2009, 2014, 2015, 2018, 2019, 2024

Preface

Thank you for purchasing the *FAA Airman Knowledge — Private Pilot Test Guide*. This test guide helps you understand the learning objectives that apply to questions on the FAA Private Pilot Airplane — Airman Knowledge Test so you can take the test with confidence. This test guide contains examples of the types of questions that appear on the FAA knowledge test along with correct answers, explanations, and study references. Explanations of why the other choices are wrong are included where appropriate. Questions are organized by topic, with explanations located to the right side of each question. You can use the sliding mask to cover up the answers and test yourself. Full-color legends and figures identical to the figures on the FAA test are included in Appendix 1 and 2 in the back of the book. This test guide is a component of the Guided Flight Discovery Pilot Training System and is not intended as a stand-alone learning tool but as a supplement to your instructor-led flight and ground training.

GUIDED FLIGHT DISCOVERY PILOT TRAINING SYSTEM

The Guided Flight Discovery Pilot Training System provides the finest pilot training available. Rather than just teaching facts, Guided Flight Discovery concentrates on an application-oriented approach to pilot training. The comprehensive and complete system emphasizes the why and how of aeronautical concepts when they are presented. As you progress through your training, you will find that the revolutionary Guided Flight Discovery system leads you through essential aeronautical knowledge and exposes you to a variety of interesting and useful information that will enhance and expand your understanding of the world of aviation.

Although you can use each element of the Guided Flight Discovery Pilot Training System separately, the effectiveness of the materials is maximized by using all of the individual components in a systems approach. In addition to this test guide, the primary components of the Private Pilot Program are described below.

PRIVATE PILOT TEXTBOOK/E-BOOK

This *Private Pilot* textbook/e-book is your primary source for initial study and review. The text contains complete and concise explanations of the fundamental concepts and ideas that every private pilot needs to know. The subjects are organized in a logical manner to build upon previously introduced topics. Subjects are often expanded upon through the use of Discovery Insets, which are strategically placed throughout the chapters. The Summary Checklists, Key Terms, and Questions are designed to help you review and prepare for both the knowledge and practical tests.

Jeppesen e-books are electronic versions of traditional textbooks and reference materials that you can view on computers and other devices. Jeppesen e-books are available on iOS or Android devices and PC or Mac computers. Jeppesen e-books provide valuable features, including the ability to quickly jump to specific information, bookmark pages, and take notes. Direct linking to chapters in each book is provided through the table of contents.

PRIVATE PILOT MANEUVERS MANUAL

When used in conjunction with the other components of the Private Pilot Program, the *Private Pilot Maneuvers Manual* provides an effective, practical approach to your training. Maneuvers are numbered for ease of reference, are grouped into categories based on similar operational characteristics, and presented in the order in which they are typically introduced during training. This book uses colorful graphics and step-by-step procedure descriptions to help you visualize and understand each maneuver that you will perform in the airplane. Expanded instructional guidance, helpful hints, and explanations of common errors help you perform the maneuver more precisely the first time.

PRIVATE PILOT SYLLABUS

The syllabus is an outline of the Private Pilot Certification Course. The syllabus provides a basic framework for your training in a logical sequence and assigns appropriate study material for each lesson. Ground and flight lessons are coordinated to ensure that your training progresses smoothly and that you are consistently introduced to topics on the ground prior to being required to apply that knowledge in the airplane. The syllabus is available in print and e-book versions.

FAR/AIM

The Jeppesen FAR/AIM e-book includes the current Federal Aviation Regulations (FARs) and the Aeronautical Information Manual (AIM) in one publication. The FAR/AIM includes FAR Parts 1, 3, 11, 43, 48, 61, 67, 68, 71, 73, 91, 97, 103, 105, 107, 110, 119, 135, 136, 137, 141, 142, NTSB 830, and TSRs 1552 and 1562. The AIM is a reproduction of the FAA publication with full-color graphics and the Pilot/Controller Glossary. The AIM contains basic flight information and the ATC procedures to operate effectively in the U.S. National Airspace System.

PRIVATE PILOT PRACTICAL TEST STUDY GUIDE

The *Private Pilot FAA Practical Test Study Guide* provides guidance for you to pass your practical test with ease. The guide presents the information that you need to meet the knowledge, risk management, and skill requirements for each task in the Airman Certification Standards (ACS). An effective question and answer format helps you prepare for the oral portion of the test and step-by-step diagrams with helpful hints and common errors provides insight into performing the maneuvers proficiently during the flight.

PRIVATE PILOT ONLINE COURSE

Available from jeppdirect.com, the Private Pilot online course provides academic content in ground lessons with exams and interactive maneuvers lessons in a complete ground school, and outlines for every flight lesson. Ground school and maneuvers lessons are included using a combination of audio, video and graphics to clearly explain each topic. A Learning Management System (LMS) tracks your completions and test results specific to each question to assist you in identifying your strengths and weaknesses. You can use the online course together with the textbook and other Jeppesen products to enhance your learning experience.

Table of Contents

NOTE: Chapter 11 in the GFD Private Pilot textbook presents a scenario that illustrates how all of the knowledge and skills presented in Chapters 1 through 10 are applied when you plan and implement a cross-country flight. The FAA knowledge test questions that apply to the subjects included in Chapter 11's flight scenario have been placed in the chapter in which that content is introduced in the GFD Private Pilot textbook (Chapters 1 through 10). For example, although obtaining a weather briefing from Flight Service is an essential part of planning a flight and is part of the Chapter 11 scenario, the FAA knowledge questions that apply to weather briefings are included in Chapter 7, Section D — Sources of Weather Information.

Introduction

The *FAA Airman Knowledge — Private Pilot Test Guide* is designed to help you prepare for the FAA Private Pilot — Airplane Airman Knowledge Test. This guide contains examples of the types of questions that appear on the FAA knowledge test that applies to airplanes. To ensure comprehensive preparation for the knowledge test, one or more questions apply to each knowledge subject code in the Private Pilot — Airplane Airman Certification Standards. Questions about rotorcraft, gliders, balloons, powered-lift, and airships are not included.

USING THIS TEST GUIDE

The test guide is organized like the GFD *Private Pilot* textbook. Ten numbered chapters and lettered sections within each chapter. Within the chapters of this test guide, each section contains a content summary followed by sample knowledge test questions that typically appear in the same sequence as the textbook material. The question is shown in the left column with the applicable Airman Certification Standards (ACS) code(s). Answers and explanations are in the right column with references to applicable sections in the textbook and to FAA publications. The following is an example of the information that appears with each question and explanation.

[1]	[2]	[3]	[4]	[5]	[6]
4-59	**PA.I.F.K6**	**4-59.**	**Answer C.**	**GFDPP 4B**	**AIM**
(FAA Question)		*(Explanation of FAA Question)*			

[1] Jeppesen designated test guide question number. The first number is the chapter where the question is located in the test guide, which corresponds to the chapter in the GFD textbook. The second number is the question number within the chapter. In this example, the question is the 59th question in Chapter 4 of the test guide.

[2] The airman certification standards (ACS) code(s). These codes are associated with Knowledge subject areas in the FAA Private Pilot — Airplane Airman Certification Standards (ACS), which provides the standards that you must meet for your private pilot practical test (checkride). In some cases, codes that apply to Risk and Skills subject areas of the ACS are also included as applicable. As revisions to the ACS are released, the FAA might modify these codes slightly so some codes in this knowledge test guide might not match the most recent version of the ACS.

[3] The Jeppesen designated test guide number. The number is repeated in the right-hand column above the explanation.

[4] The correct answer to the question. In this example, answer C is correct.

[5] The location of content in the GFD *Private Pilot* textbook that applies to the question. In this case, the question is covered in Chapter 4, Section B of the textbook.

[6] Abbreviation for the FAA or other authoritative source document. In this example, the reference is the *Aeronautical Information Manual* (AIM).

REFERENCE ABBREVIATIONS

The following abbreviations for references are used in Jeppesen airmen knowledge test guides.

AC	—	Advisory Circular
A/FD	—	Airport/Facility Directory (section in Chart Supplement)
AFH	—	Airplane Flying Handbook, FAA-H-8083-3
AIM	—	Aeronautical Information Manual
ASI-SA##	—	Air Safety Institute (AOPA) Safety Advisor (by number ##)
AW	—	Aviation Weather, AC 00-6
AWS	—	Aviation Weather Services, AC 00-45
FAR	—	Federal Aviation Regulation (14 CFR)
GFDIC	—	Guided Flight Discovery Instrument/Commercial Textbook
GFDPP	—	Guided Flight Discovery Private Pilot Textbook
GFDPPM	—	Guided Flight Discovery Private Pilot Maneuvers
IAP	—	Instrument Approach Procedure
IFH	—	Instrument Flying Handbook, FAA-H-8083-15
IPG	—	Instrument Procedures Guide (Jeppesen)
IPH	—	Instrument Procedures Handbook, FAA·H-8083-16
NAVWEPS	—	Aerodynamics for Naval Aviators
PHB	—	Pilot's Handbook of Aeronautical Knowledge, FAA-H-8083-25
RMH	—	Risk Management Handbook, FAA-H-8083-2
WBH	—	Aircraft Weight and Balance Handbook, FAA-H-8083-1
TERPS	—	U.S. Standard for Terminal Instrument Procedures
TSA	—	Transportation Security Administration

QUESTIONS ANSWERS AND EXPLANATIONS

To the right of the question, an explanation of the correct answer and why the other answers are wrong, (if needed) is included. Wrong answers are not explained for calculated answers, unless a common error in the calculations leads to one of the wrong answers. The correct answers have been determined by Jeppesen to be the best choice of the available answers based on official reference documents. Some questions that were valid when the FAA test was developed might no longer be appropriate due to ongoing changes in regulations or official operating procedures. The knowledge test that you take can be updated at any time. Therefore, when taking the FAA test, be sure to answer the questions according to the latest regulations or official operating procedures.

APPENDICES

Two appendices are included in the back of the test guide:

- Appendix 1 FAA—Legends: the legends from the FAA *Airman Knowledge Testing Supplement for Sport Pilot, Recreational Pilot, Remote Pilot, and Private Pilot* are an important resource for answering questions about charts and information included in the Chart Supplement. For example, if you do not know the answer to a chart-related question or how to interpret an Airport/Facility Directory excerpt, refer to the legends and remember that they are available during your test.

- Appendix 2 FAA— Figures: refer to these figures from the from the FAA *Airman Knowledge Testing Supplement for Sport Pilot, Recreational Pilot, Remote Pilot, and Private Pilot* when required to answer questions about information in a figure. A copy of this supplement is available during your test.

RECREATIONAL AND SPORT PILOT COVERAGE

The recreational pilot certificate limits a person to flying a basic, single-engine airplane with no more than 180 horsepower within 50 NM of their home airport. Mainly because of the 50 NM limitation, the recreational pilot certificate has not been practical. Because the number of applicants for recreational pilot certificates is negligible, we have removed all recreational pilot questions from this test guide. If you are interested in simplified training and do not need all the private pilot privileges, consider checking out the sport pilot requirements in FAR 61, Subpart J (FAR 61.301 – FAR 61.327). The sport pilot certificate allows you to fly certain light, two-place aircraft during daylight hours and does not have the 50 NM limitation. You can use this guide, together with the sport pilot regulations, to prepare for the Sport Pilot knowledge test for airplane.

PREPARING FOR THE FAA TEST

To become a safe and competent private pilot, you need more than just the academic knowledge required to pass a test. For a comprehensive ground training program, a structured ground school with a qualified flight or ground instructor is essential. An organized course of instruction covers the content more quickly, and enables you to obtain answers to questions you think of as you learn the material. The additional instruction is beneficial in your flight training.

Use this test guide in conjunction with the *GFD Private Pilot* textbook. Follow these steps to get the most benefit from your study:

1. After reading a section of the textbook, review the summary checklist and answer the questions at the end of the section to reinforce the material.

2. Refer to the corresponding section in this study guide and test your knowledge of the subject area by answering the sample FAA questions. Cover the answers in the right-hand column, read each question, and choose what you consider the best answer. A sliding mask is provided for this purpose.

3. Move down the sliding mask and read the answer and explanation for that question.

4. Mark the questions you miss for further study and review before taking the knowledge test.

5. After you complete your study, schedule your knowledge test right away, while the information is fresh in your mind.

FAA TRACKING NUMBER

Prior to taking the FAA airman knowledge test, you must establish an FAA tracking number (FTN) within the Integrated Airman Certification and Rating Application (IACRA) system. IACRA is the web-based certification/rating application that guides the user through the FAA's airman application process. IACRA helps ensure applicants meet regulatory and policy requirements through the use of extensive data validation.

To register for an FTN in IACRA, you must visit the IACRA website and follow the instructions provided. The FTN is a number assigned to you by the FAA that stays with you throughout the course of your aviation career. If you have been issued an airman certificate in the past, then you already have an FTN. Record your FTN as it will be required later by your instructor when recommending you for the practical test and by the FAA examiner/evaluator when you take the practical test. You may also need to reference this number to inquire about your application.

SCHEDULING THE TEST

After you have your FTN, you can register to take your FAA Airman Knowledge Test by going to the registration and scheduling website operated by PSI Services LLC at: https://faa.psiexams.com/FAA/login. After the PSI system verifies your FTN, you will be able to create a user profile in the PSI system and schedule your knowledge test. When you create your account in PSI's system, choose a username and password. You will use those credentials to access your account in PSI's system and register or check the status of your knowledge test.

ELIGIBILITY

When you arrive at the testing center, you must show that you have completed the appropriate ground instruction or home study course by presenting a:

- Certificate of graduation or a statement of accomplishment certifying satisfactory completion of the ground school portion of a course from a FAA certificated pilot school. [FAR 61.71 (a)]

OR

- Written statement or logbook endorsement from an FAA authorized ground or flight instructor certifying that you are prepared to take the required knowledge test. [FAR 61.103 (d)(1 and 2)]

You must provide:

- Your federal training number (FTN).
- Identification that includes a photo, date of birth, signature, and physical, residential address. If you are a U.S. citizen or resident alien, you must provide one or more of the following:
  - Identification card issued by a U.S. state, territory, or government entity, such as a driver permit or license, government ID card, or military ID card.
  - Passport
  - Alien residency (green) card.
- Non-U.S. citizens must provide a passport **and** a U.S. driver permit or license or government ID card.

TAKING THE KNOWLEDGE TEST

The Private Pilot Airplane (PAR) test contains 60 multiple-choice questions, and you have 2 hours and 30 minutes to complete it. Each test question is independent of other questions—a correct response to one question does not depend on, or influence, the correct response to another. The minimum passing score is 70 percent.

TEST MATERIALS, REFERENCE MATERIALS, AND AIDS

You are allowed to use aids, reference materials, and test materials within specified guidelines, provided the actual test questions or answers are not revealed. The following guidelines apply:

- Aviation-oriented calculators — You may use any model of aviation-oriented calculator, including small electronic calculators that perform arithmetic functions (add, subtract, multiply and divide). Simple programmable memories, which allow addition to, subtraction from, or retrieval of one number from the memory and simple functions, such as square root and percent keys are permissible.
- Calculators — Testing centers may provide calculators to applicants and/or deny applicants' use of their personal calculators based on the following limitations:
  - Before and upon completion of the test, while in the presence of the unit member (proctor), you must actuate the "ON/OFF" switch or "RESET" button and perform any other function that ensures the erasure of any data stored in the memory circuits.
  - The use of electronic calculators incorporating permanent or continuous type memory circuits with erasure capability is prohibited. The unit member (proctor) may refuse the use of your calculator when unable to determine the calculator's erasure capability.

- If your calculator has a printer, you must surrender any printouts of data at the completion of the test.

- You may not use magnetic cards, magnetic tapes, modules, computer chips, or any other device that you can store pre-written programs or information related to the test.

- You may not use any booklet or manual containing instructions related to the use of test aids.

- Written materials — You may not take any written materials (either handwritten, printed, or electronic) other than the supplemental book provided by the unit member (proctor) into the testing area.

- Test materials — You may use scales, straightedges, protractors, plotters, navigation computers, blank log sheets, and electronic or mechanical calculators that are directly related to the test.

- Manufacturer's aids — Permanently inscribed instructions on the front and back of such aids, such as formulas, conversions, regulations, signals, weather data, holding pattern diagrams, frequencies, weight and balance formulas, and air traffic control (ATC) procedures are permissible.

- Dictionaries — You are not allowed to have a dictionary in the testing area.

- Final decision — The unit member (proctor) makes the final determination of which test materials and personal possessions you may take into the testing area.

TEST-TAKING TIPS

Before starting the actual test, the testing software gives you a few sample questions so that you can practice navigating through the test. The sample questions in your practice session have no relation to the content of the test. This "practice test" familiarizes you with the look and feel of the system screens, including how to select answers, mark questions for later review, view the remaining time in the test, and use other features of the testing software.

After you start the actual test, you answer the questions that appear on the screen. If you are prepared, you will likely have plenty of time to complete the test. After you begin the test, the screen will show you the time remaining for completion. When taking the test, keep the following points in mind:

1. Answer each question in accordance with the latest regulations and procedures. If a recent change invalidates a question, you receive credit if you answer it. However, the FAA normally deletes or updates these questions.

2. Read each question carefully before looking at the possible answers. Make sure you clearly understand the problem before attempting to solve it.

3. After formulating an answer, determine which of the alternatives most nearly corresponds with that answer. The answer chosen should completely resolve the problem.

4. A question might appear to have more than one possible answer; however, only one answer is correct and complete. The other answers are either incomplete or are derived from popular misconceptions.

5. Make sure that you select an answer for each question. Questions left unanswered are counted as incorrect.

6. If you find a certain question difficult, mark it for review and proceed to the next question. After you answer the less difficult questions, return to the questions you marked for review and answer them. The review marking procedure is explained before you start the test. Although the testing software alerts you to unanswered questions, make sure that every question has an answer recorded before submitting the test for grading.

7. After solving a calculation problem, select the answer nearest to your solution. The problem has been checked with various calculators. If you solve it correctly, your result will be closer to the correct answer than the other choices.

8. For graph type questions, you may request a printed copy of the graph on which you may draw and write to compute the answer. You must turn in all paperwork when you complete the test.

YOUR TEST RESULTS

All knowledge test data will be sent electronically to IACRA immediately after you complete the knowledge test. The testing center will also provide your airman knowledge test report (AKTR) with your score and FTN. The AKTR lists the ACS codes that apply to the subject areas of the questions that you answered incorrectly. Prior to taking the practical test, compare the codes on the AKTR to the ACS codes in the *Private Pilot—Airplane Airman Certification Standards* to determine the specific subjects that you should study and the areas in which you should obtain additional training.

You present the AKTR to the examiner before taking your practical test to prove you have completed the appropriate knowledge test within the required time frame. During the oral portion of the practical test, the examiner is required to evaluate the noted areas of deficiency based on the codes shown on the AKTR. The airman knowledge test report includes ACS codes for incorrect answers. The AKTR is valid for 24 calendar months. If the AKTR expires before you complete the practical test, you must retake the knowledge test.

RETESTING AFTER FAILURE

If you receive a score less than 70%, you may apply for retesting after an authorized instructor provides additional training and an endorsement that states you are competent to pass the test. Before retesting, you must surrender the previous test report to the unit member (proctor), who will destroy that test report after administering the retest. The results from the latest test taken are the official score.

CHAPTER 1

Discovering Aviation

NOTE: *No FAA questions apply to GFD Private Pilot textbook Chapter 1, Section D — Developing Resilience With CBTA.*

SECTION A
Pilot Training

THE TRAINING PROCESS

- The Federal Aviation Administration (FAA) oversees all regulatory aspects of flight, including the process by which you obtain your private pilot certificate.

- Your training focuses on gaining the skills and knowledge that the FAA Private Pilot Airman Certification Standards (ACS) require for issuing a private pilot certificate.

- To be eligible for a student pilot certificate, you must be at least 16 years of age and be able to read, speak, and understand the English language.

- In addition to the student pilot requirements, to be eligible for a private pilot certificate you must be at least 17 years of age, complete specific training and flight time requirements described in the FARs, pass a knowledge test, and successfully complete a practical test that consists of oral quizzing, performing pilot operations, and performing maneuvers in the airplane.

- You are required to keep an accurate record of pilot training time and aeronautical experience used to meet the requirements for a certificate, rating, or flight review. You must also log flights required for meeting recent flight experience requirements. Instructor endorsements that are required to exercise pilot privileges also must be recorded in your logbook.

- When you fly an airplane as a certificated private pilot, you must have in your possession, or readily accessible in the aircraft, the following documentation: your private pilot certificate, photo identification such as a driver's license or passport, and an appropriate medical certificate or BasicMed documents.

- You, along with each person who holds a pilot certificate or a medical certificate, shall present it for inspection upon the request of the Administrator, the National Transportation Safety Board (NTSB), or any federal, state, or local law enforcement officer.

- To continue acting as pilot in command of an aircraft, you must satisfactorily complete a flight review every 24 calendar months.

- If you change your permanent mailing address and fail to notify the FAA Airmen Certification Branch of the new address, you are entitled to exercise the privileges of the pilot certificate for a period of only 30 days after the date of the move.

- To act as pilot in command of an aircraft carrying passengers, you must be a certificated private pilot and have performed at least three takeoffs and landings in an aircraft of the same category and class (and type, if a type rating is required) within the preceding 90 days.

  ○ If the flight is to be conducted at night, the takeoffs and landings must have been made at night and to a full stop. For the purpose of recency requirements, nighttime is defined as the period beginning one hour after sunset to one hour before sunrise.

  ○ If recency of experience requirements for night flight are not met, the latest time that you may carry passengers is one hour after official sunset.

  ○ If the flight is to be conducted in a tailwheel airplane, the takeoffs and landings must have been in a tailwheel airplane and to a full stop.

- As private pilot, you may pay the pro rata share of the operating expenses of a flight with passengers.

- As private pilot, you may act as pilot in command of an aircraft used in a passenger-carrying airlift sponsored by a charitable organization, and for which the passengers make a donation to the organization.

- As a private pilot, you may act as pilot in command of an aircraft towing a glider if you have logged:

  ○ A minimum of 100 hours of pilot flight time in powered aircraft.

  ○ At least three actual or simulated glider tows while accompanied by a qualified pilot in the preceding 12 months.

MEDICAL CERTIFICATES

- The three classes of medical certificates are:
    - First-class for airline transport pilot (ATP) operations.
    - Second-class for commercial pilot operations other than airline transport.
    - Third-class for student and private pilot operations.
- Third-class medical certificate privileges expire after:
    - 24 calendar months—age 40 or over.
    - 60 calendar months—under age 40.
- Second-class medical certificate privileges expire after 12 calendar months.
- First-class medical certificate privileges expire after:
    - 6 calendar months—age 40 and over.
    - 12 calendar months—under age 40.
- To continue acting as pilot in command after your medical certificate expires, you must obtain a new exam from an aviation medical examiner or if you qualify, comply with the requirements of the BasicMed rule, which permits certain operations with a driver's license instead of a medical certificate.
- The FAA BasicMed rule allows you to operate small aircraft on certain personal flights using a driver's license instead of a medical certificate. To qualify, you must:
    - Obtain a physical exam from a state-licensed physician every 48 months and have that physician complete an FAA-provided checklist.
    - Complete an approved online medical education course every 24 months.
    - Consent to a National Driver Register check.

AIRCRAFT CATEGORIES AND CLASSES

- With respect to the certification of airmen, categories of aircraft are airplane, rotorcraft, powered lift, glider, and lighter-than-air.
- For aircraft certification, category relates to the intended use of an aircraft and sets strict limits on its operation. Examples include transport, normal, utility, acrobatic, limited, restricted, and provisional.

SECTION A ■ Pilot Training

NOTE: An asterisk appearing after an ACS code (i.e. PA.VII.B.K1*) indicates that the question subject appears more than one time in the ACS. The code shown corresponds to the first instance of the subject in the ACS.

1-1 PA.I.A.K1, K3

As part of the requirements to be able to fly solo as a student pilot, you must hold at least a

A – second-class medical certificate.

B – first-class medical certificate.

C – third-class medical certificate.

1-1. Answer C. GFDPP 1A, FAR 61.23
When exercising the privileges of a student pilot certificate, you must hold at least a third-class medical certificate. First-class medical certificates have the highest physical requirements, followed by second-class, and then third-class.

1-2   PA.I.A.K1

What is the minimum age requirement to apply for an airplane student pilot certificate?

A – 14 years of age.

B – 16 years of age.

C – 17 years of age.

1-2. Answer B. GFDPP 1A, FAR 61.83

To obtain a student pilot certificate for the airplane category, you must be at least 16 years of age. To obtain a student pilot certificate for a glider or balloon, you must be at least 14 years of age. To obtain a private pilot certificate, you must be at least 17 years of age. There is no age limitation to begin flight lessons.

1-3   PA.I.A.K1

What is the minimum age requirement to apply for an airplane private pilot certificate?

A – 16 years of age.

B – 17 years of age.

C – 18 years of age.

1-3. Answer B. GFDPP 1A, FAR 61.103

To obtain a private pilot certificate for the airplane category, you must be at least 17 years of age. To obtain a student pilot certificate in the airplane category, you must be at least 16 years of age. To obtain a private pilot certificate for a glider or balloon, you must be at least 16 years of age. There is no age limitation to begin flight lessons. To be eligible for a flight instructor certificate or rating, you must be at least 18 years of age.

1-4   PA.I.A.K1, K4

Which flight time must you record in your logbook?

A – All flight time.

B – All solo flight time.

C – Flight time needed to meet the requirements for a certificate, rating, flight review, or recency of experience.

1-4. Answer C. GFDPP 1A, FAR 61.51

You are required to keep an accurate record of pilot training time and aeronautical experience used to meet the requirements for a certificate, rating, or flight review. You must also log flights required for meeting recent flight experience requirements. Instructor endorsements that are required to exercise private pilot privileges also must be recorded in your logbook.

1-5   PA.I.A.K3

A third-class medical certificate is issued to a 36-year-old pilot on August 10, this year. To exercise the privileges of a private pilot certificate, the medical certificate will be valid until midnight on

A – August 10, three years later.

B – August 31, three years later.

C – August 31, five years later.

1-5. Answer C. GFDPP 1A, FAR 61.23

For pilots who were under 40 years of age at the time of their medical exam, a third-class medical certificate expires at the end of the 60th month after the examination.

1-6　　PA.I.A.K3

A third-class medical certificate is issued to a 51-year-old pilot on May 3, this year. To exercise the privileges of a private pilot certificate, the medical certificate will be valid until midnight on

A – May 31, two years later.

B – May 31, three years later.

C – May 31, five years later.

1-6. Answer A. GFDPP 1A, FAR 61.23
For pilots who were 40 years of age or more at the time of their medical exam, a third-class medical certificate expires at the end of the 24th month after the examination.

1-7　　PA.I.A.K3

You were 40 years old when you obtained a second-class medical certificate on March 15, 2024. For exercising private pilot privileges, when does your medical certificate expire?

A – March 15, 2026.

B – March 31, 2026.

C – March 31, 2029.

1-7. Answer B. GFDPP 1A, FAR 61.23
For pilots who were 40 years of age or more at the time of their medical exam, second class privileges expire at the end of the 12th month after the month of the examination and third-class privileges expire at the end of the 24th month after the examination.

1-8　　PA.I.A.K3

For private pilot operations, a second-class medical certificate issued to a 42-year-old pilot on July 15, 2024, will expire at midnight on

A – July 31, 2025.

B – July 15, 2026.

C – July 31, 2026.

1-8. Answer C. GFDPP 1A, FAR 61.23
For a pilot aged 40 or over to exercise the privileges of a commercial pilot, a second-class medical is valid until the end of the 12th calendar month after the date of examination. For private pilot operations, a second-class medical is valid until the end of the 24th calendar month after the date of examination.

1-9　　PA.I.A.K3

For private pilot operations, a first-class medical certificate issued to a 23-year-old pilot on October 21, this year, will expire at midnight on

A – October 31, next year.

B – October 21, two years later.

C – October 31, five years later.

1-9. Answer C. GFDPP 1A, FAR 61.23
For pilots under 40 years of age, to exercise the privileges of an ATP, a first-class medical certificate is valid until the end of the 12th calendar month after the date of examination. From the beginning of the 13th month to the end of the 60th calendar month, a first-class medical is valid only for operations requiring a third-class medical certificate.

SECTION A ■ **Pilot Training**

1-10   PA.I.A.K3

A third-class medical certificate was issued to a 19-year-old pilot on August 10, this year. To exercise the privileges of a private pilot certificate, the medical certificate will expire at midnight on

A – August 10, two years later.

B – August 31, two years later.

C – August 31, five years later.

1-10. Answer C. GFDPP 1A, FAR 61.23

If the pilot is under the age of 40 at the time of the examination, a third-class medical, which is appropriate to exercise the privileges of a private pilot, expires at the end of the 60th calendar month after the date of the examination.

1-11   PA.I.A.K5

The FAA BasicMed rule allows pilots to operate small aircraft on certain personal flights

A – without a physical examination.

B – using a driver's license instead of a medical certificate.

C – after one-time completion of an approved online medical education course.

1-11. Answer B. GFDPP 1A, FAR 61.23, 61.113

The FAA BasicMed rule allows pilots to operate small aircraft on certain personal flights using a driver's license instead of a medical certificate. To qualify, pilots must obtain a physical exam from a state-licensed physician every 48 months and have that physician complete an FAA-provided checklist; complete an approved BasicMed online medical education course every 24 months; and consent to a National Driver Register check.

1-12   PA.I.A.K5

Under the BasicMed rule, what flight operations may you conduct using a driver's license instead of an FAA medical certificate?

A – You may operate an aircraft below 18,000 feet MSL at a maximum airspeed of 250 knots.

B – You may operate an aircraft outside the United States as long as the flight is not for compensation or hire.

C – You may carry up to 6 passengers in an aircraft with a maximum certificated takeoff weight of no more than 6,000 pounds.

1-12. Answer A. GFDPP 1A, FAR 61.23, 61.113

Under the BasicMed rule, you may conduct flight operations:

- In an aircraft certificated to carry no more than 6 occupants, including the pilot, and with a maximum certificated takeoff weight of no more than 6,000 pounds.

- Below 18,000 feet MSL.

- At a maximum airspeed of 250 knots.

- That are entirely within the United States.

- That are not for compensation or hire.

Answer C is wrong because the aircraft may have a total of 6 occupants, not 6 passengers plus a pilot.

1-13 PA.I.A.K5

What requirements must you meet to utilize the BasicMed rule?

A – You must have completed an approved BasicMed exam from an aviation medical examiner within the previous 48 months.

B – You must have completed an approved medical education course within the previous 24 calendar months and a comprehensive medical exam from a physician within the previous 48 months.

C – You must have completed an approved medical education course within the previous 48 months and a comprehensive medical exam from a physician within the previous 24 calendar months.

1-13. Answer B. GFDPP 1A, FAR 61.23, 61.113

The FAA BasicMed rule allows pilots to operate small aircraft on certain personal flights using a driver's license instead of a medical certificate. To qualify, pilots must obtain a physical exam from a state-licensed physician every 48 months and have that physician complete an FAA-provided checklist; complete an approved BasicMed online medical education course every 24 months; and consent to a National Driver Register check.

1-14 PA.I.A.K1

To act as pilot in command of an aircraft carrying passengers, you must show by logbook endorsement the satisfactory completion of a flight review or completion of a pilot proficiency check within the preceding

A – 6 calendar months.

B – 12 calendar months.

C – 24 calendar months.

1-14. Answer C. GFDPP 1A, FAR 61.56

To act as pilot in command of any aircraft, whether you are carrying passengers or not, you must have, within the preceding 24 calendar months, complied with the flight review requirements.

1-15 PA.I.A.K1

If recency of experience requirements for night flight are not met and official sunset is 1830, the latest time passengers may be carried is

A – 1829.

B – 1859.

C – 1929.

1-15. Answer C. GFDPP 1A, FAR 61.57, AFH

No person may act as pilot in command of an aircraft carrying passengers during the period beginning one hour after sunset and ending one hour before sunrise unless that person meets night experience requirements.

1-16 PA.I.A.K1

Your cousin wants you to take him flying. You must have made at least three takeoffs and three landings within the preceding

A – 90 days.

B – 60 days.

C – 30 days.

1-16. Answer A. GFDPP 1A, FAR 61.57

To meet recent flight experience requirements for carrying passengers, you must have, within the preceding 90 days, made three takeoffs and landings in the same category and class of aircraft. If the aircraft to be flown is an airplane with a tailwheel, the takeoffs and landings must have been made to a full stop in an airplane with a tailwheel. If the flight is to be made at night, landings must be made to a full stop for night currency requirements.

SECTION A ■ **Pilot Training**

1-17   PA.I.A.K1

To act as pilot in command of an aircraft carrying passengers, you must have made three takeoffs and three landings within the preceding 90 days in an aircraft of the same

A – make and model.

B – category and class, but not type.

C – category, class, and type, if a type rating is required.

1-17. Answer C. GFDPP 1A, FAR 61.57

To meet the recency of experience requirements for carrying passengers, FAR 61.57(a) states that you must have made three takeoffs and landings within the preceding 90 days in an aircraft of the same category and class, and if a type rating is required, of the same type. If the aircraft to be flown is an airplane with a tailwheel, the takeoffs and landings must have been made to a full stop in an airplane with a tailwheel. If the flight is to be made at night, landings must be made to a full stop for night currency requirements.

1-18   PA.I.A.K1

The takeoffs and landings required to meet the recency of experience requirements for carrying passengers in a tailwheel airplane

A – may be touch and go or full stop.

B – must be touch and go.

C – must be to a full stop.

1-18. Answer C. GFDPP 1A, FAR 61.57, AFH

Due to their design and structure, tailwheel airplanes exhibit handling and operational characteristics that are different from those of tricycle gear aircraft. Therefore, the FAA requires full-stop landings for currency in tailwheel airplanes.

1-19   PA.I.A.K1

The three takeoffs and landings that are required to act as pilot in command carrying passengers at night must be done during the time period from

A – sunset to sunrise.

B – one hour after sunset to one hour before sunrise.

C – the end of evening civil twilight to the beginning of morning civil twilight.

1-19. Answer B. GFDPP 1A, FAR 61.57

To act as pilot in command of an aircraft carrying passengers at night (defined by FAR 61.57 as beginning one hour after sunset and ending one hour before sunrise), you must have, within the preceding 90 days, made three takeoffs and landings to a full stop during that same period—from 1 hour after sunset to one 1 hour before sunrise. The required takeoffs and landings must be in an aircraft of the same category, class, and type (if a type rating is required).

1-20   PA.I.A.K1

To meet the recency of experience requirements to act as pilot in command carrying passengers at night, you must have made at least three takeoffs and three landings to a full stop within the preceding 90 days in an aircraft of the same

A – make and model.

B – category and class, but not type.

C – category, class, and type, if a type rating is required.

1-20. Answer C. GFDPP 1A, FAR 61.57

To act as pilot in command of an aircraft carrying passengers at night (defined by FAR 61.57 as beginning one hour after sunset and ending one hour before sunrise), you must have, within the preceding 90 days, made three takeoffs and landings to a full stop during that same period—from 1 hour after sunset to one 1 hour before sunrise. The required takeoffs and landings must be in an aircraft of the same category, class, and type (if a type rating is required).

1-21 PA.I.A.K1

If a private pilot had a flight review on August 8, this year, when is the next flight review required?

A – August 31, next year.

B – August 8, two years later.

C – August 31, two years later.

1-21. Answer C. GFDPP 1A, FAR 61.56
You may not act as pilot in command of an aircraft unless you have accomplished a flight review within the preceding 24 calendar months. Calendar month means the review is good until the end of the month in which it expires.

1-22 PA.I.A.K1

Each private pilot is required to have a flight review every

A – 6 calendar months.

B – 12 calendar months.

C – 24 calendar months.

1-22. Answer C. GFDPP 1A, FAR 61.56
You may not act as pilot in command of an aircraft unless you have accomplished a flight review within the preceding 24 calendar months. Because of the two-year interval, this flight review is sometimes called a biennial flight review.

1-23 PA.I.A.K2

In regard to privileges and limitations, as a private pilot, you may

A – not be paid in any manner for the operating expenses of a flight.

B – not pay less than the pro rata share of the operating expenses of a flight with passengers, provided the expenses involve only fuel, oil, airport expenditures, or rental fees.

C – act as pilot in command of an aircraft carrying a passenger for compensation if the flight is in connection with a business or employment.

1-23. Answer B. GFDPP 1A, FAR 61.113
As a private pilot, you may not pay less than the pro rata share of the operating expenses of a flight with passengers, provided the expenses involve only fuel, oil, airport expenditures, or rental fees.

1-24 PA.I.A.K2

What exception, if any, permits a private pilot to act as pilot in command of an aircraft carrying passengers who pay for the flight?

A – If the passengers pay all the operating expenses.

B – If a donation is made to a charitable organization for the flight.

C – There is no exception.

1-24. Answer B. GFDPP 1A, FAR 61.113, FAR 91.146
As a private pilot, you may act as pilot in command of an aircraft used in a passenger-carrying airlift sponsored by a charitable organization, and for which the passengers make a donation to the organization.

SECTION A ■ **Pilot Training**

SECTION A ■ Pilot Training

1-25   PA.I.A.K1, K2

A certificated private pilot may not act as pilot in command of an aircraft towing a glider unless there is entered in the pilot's logbook a minimum of

A – 100 hours of pilot-in-command time in the aircraft category, class, and type, if required, that the pilot is using to tow a glider.

B – 200 hours of pilot-in-command time in the aircraft category, class, and type, if required, that the pilot is using to tow a glider.

C – 100 hours of pilot flight time in any aircraft, that the pilot is using to tow a glider.

1-25. Answer A. GFDPP 1A, FAR 61.69
No person may act as pilot in command for towing a glider unless that person has logged at least 100 hours of pilot-in-command time in the aircraft category, class, and type, if required, that the pilot is using to tow a glider.

1-26   PA.I.A.K1, K2

To act as pilot in command of an aircraft towing a glider, a pilot is required to have made within the preceding 24 months

A – at least three flights in a powered glider.

B – at least three flights as observer in a glider being towed by an aircraft.

C – at least three actual or simulated glider tows while accompanied by a qualified pilot.

1-26. Answer C. GFDPP 1A, FAR 61.69
The pilot in command of an aircraft towing a glider must have, within the preceding 12 months, made at least three actual, or simulated glider tows while accompanied by a qualified pilot, or made at least three flights as PIC of a glider towed by an aircraft.

1-27   PA.I.A.K4

In addition to your private pilot certificate, what other documents are you required to have in your possession during flight to exercise your private pilot privileges?

A – photo identification, medical certificate documentation.

B – medical certificate documentation, logbook.

C – photo identification, medical certificate documentation, logbook.

1-27. Answer A. GFDPP 1A, FAR 61.3
When you fly an airplane as a certificated private pilot, you must have in your possession, or readily accessible in the aircraft, the following documentation: your private pilot certificate, photo identification such as a driver's license or passport, and an appropriate medical certificate or BasicMed documents. Your logbook is only required to be in your possession when you are flying solo with a student pilot certificate. However, the FAA can request that you present your logbook with all required entries in a reasonable period of time.

1-28   PA.I.A.K2

When must a current pilot certificate be in the pilot's personal possession or readily accessible in the aircraft?

A – When acting as a crew chief during launch and recovery.

B – Only when passengers are carried.

C – Anytime when acting as pilot in command or as a required crewmember.

1-28. Answer C. GFDPP 1A, FAR 61.3
You must have a current pilot certificate in your personal possession or readily accessible in the aircraft whenever you are pilot in command or acting as a required pilot flight crewmember.

1-29   PA.I.A.K2

As a private pilot acting as pilot in command, or in any other capacity as a required pilot flight crewmember, you must have in your personal possession or readily accessible in the aircraft a current

A – endorsement on the pilot certificate to show that a flight review has been satisfactorily accomplished.

B – medical certificate if required and an appropriate pilot certificate.

C – logbook endorsement to show that a flight review has been satisfactorily accomplished.

1-29. Answer B. GFDPP 1A, FAR 61.3

To act as PIC or as a required crewmember, you must have an appropriate medical certificate, if required, and an appropriate pilot certificate. Although you must have a logbook (not pilot certificate) endorsement of completing a flight review within the preceding 24 calendar months, you are not required to carry the logbook that contains this endorsement.

1-30   PA.I.A.K4

Each person who holds a pilot certificate or a medical certificate shall present it for inspection upon the request of any

A – authorized representative of the Department of Transportation.

B – person in a position of authority.

C – local law enforcement officer.

1-30. Answer A. GFDPP 1A, FAR 61.3

By regulation you, as the pilot, are required to present your pilot certificate upon request of the Administrator, an authorized representative of the NTSB or any federal, state, or local law enforcement officer.

"Authorized representative of the Department of Transportation" does not mean the same thing as "the Administrator." Generally, FAA inspectors are the ones who represent the Administrator. Not every person "in a position of authority" is entitled to examine your certificates.

1-31   PA.I.A.K1

As a certificated pilot, if you change your permanent mailing address and fail to notify the FAA Airmen Certification Branch of the new address, you are entitled to exercise the privileges of your pilot certificate for a period of only

A – 30 days after the date of the move.

B – 60 days after the date of the move.

C – 90 days after the date of the move.

1-31. Answer A. GFDPP 1A, FAR 61.60

As a pilot, you may not exercise the privileges of your certificate after 30 days from the date of your permanent mailing address change unless you notify the FAA's Airman Certificate Branch in writing of the new address.

1-32   PA.I.A.K1

With respect to the certification of pilots, which are categories of aircraft?

A – Gyroplane, helicopter, airship, free balloon.

B – Airplane, rotorcraft, glider, lighter-than-air.

C – Single-engine land and sea, multi-engine land and sea.

1-32. Answer B. GFDPP 1A, FAR 1.1

Pilots are certificated according to five categories of aircraft: airplane, rotorcraft, glider, lighter-than-air, and powered lift. Gyroplane and helicopter are classes of aircraft within the rotorcraft category. Airship and balloon are classes of aircraft within the lighter-than-air category. Single-engine land and sea and multi-engine land and sea are the four classes within the airplane category.

SECTION A ■ **Pilot Training**

1-33   PA.I.A.K1

With respect to the certification of pilots, which are classes of aircraft?

A – Airplane, rotorcraft, glider, lighter-than-air.

B – Single-engine land and sea, multi-engine land and sea.

C – Lighter-than-air, airship, hot air balloon, gas balloon.

1-33. Answer B. GFDPP 1A, FAR 1.1
Each category of aircraft is broken down into classes. The airplane category is divided into single-engine land and sea, and multi-engine land and sea.

1-34   PA.I.A.K1

With respect to the certification of aircraft, which is a category of aircraft?

A – Normal, utility, acrobatic.

B – Airplane, rotorcraft, glider.

C – Landplane, seaplane.

1-34. Answer A. GFDPP 1A, FAR 1.1
Normal, utility, and acrobatic are three of the categories under which aircraft are certified, based on their construction and use. Airplane, rotorcraft, and glider are categories of aircraft with respect to certification of *airmen*. Landplane and seaplane are common, but incomplete, descriptions of classes of airplanes based on airmen certification.

1-35   PA.I.A.K1

With respect to the certification of aircraft, which is a class of aircraft?

A – Normal, utility, acrobatic, limited.

B – Airplane, rotorcraft, glider, balloon.

C – Transport, restricted, provisional.

1-35. Answer B. GFDPP 1A, FAR 1.1
FAR 1.1 defines "class" when used with respect to the certification of aircraft, as a broad grouping of aircraft having similar means of flight, propulsion or landing. These classes include: airplane, rotorcraft, glider, balloon, and powered-lift. In reality, however, class is not used as a designator in aircraft certification.

SECTION B

Aviation Opportunities

ADDITIONAL TRAINING, FLIGHT EXPERIENCE, CERTIFICATES AND RATINGS

- A high-performance airplane is an airplane that has an engine with more than 200 horsepower.

- Before you may act as pilot in command of a high-performance airplane, you must receive ground and flight instruction from an authorized flight instructor, who then endorses your logbook. The flight instruction must be in a high-performance airplane or in a flight simulator or flight training device that is representative of a high-performance airplane.

- A complex airplane is an airplane with retractable landing gear, flaps, and a controllable propeller or full authority digital engine control (FADEC).

- Before you may act as pilot in command of a complex airplane, you must receive ground and flight instruction from an authorized flight instructor, who then endorses your logbook. The flight instruction must be in a complex airplane or in a flight simulator or flight training device that is representative of a complex airplane.

- As pilot in command, you are required to hold a type rating for the operation of aircraft having a maximum certificated takeoff weight of more than 12,500 pounds.

NOTE: An asterisk appearing after an ACS code (i.e. PA.VII.B.K1*) indicates that the question subject appears more than one time in the ACS. The code shown corresponds to the first instance of the subject in the ACS.

1-36   PA.I.A.K1, K2

The pilot in command is required to hold a type rating in which aircraft?

A – Any aircraft involved in ferry flights, training flights, or test flights.

B – Any aircraft having a maximum certificated takeoff weight of more than 12,500 pounds.

C – Any aircraft operated under an authorization issued by the Administrator.

1-36. Answer B. GFDPP 1B, FAR 1.1, FAR 61.31
FAR 61.31 indicates that a type rating is required for a large aircraft. FAR 1.1 defines a large aircraft as having a maximum certificated takeoff weight greater than 12,500 pounds. While answers A and C could occur in an aircraft that requires a type rating, the operations listed in those choices could also occur in aircraft that do not meet the large aircraft definition.

1-37   PA.I.A.K1, K2

What is the definition of a high-performance airplane?

A – An airplane with 180 horsepower, or retractable landing gear, flaps, and a fixed-pitch propeller.

B – An airplane with a normal cruise speed of more than 200 knots.

C – An airplane with an engine of more than 200 horsepower.

1-37. Answer C. GFDPP 1B, FAR 61.31
A high-performance airplane has an engine with more than 200 horsepower. A complex airplane has retractable landing gear, flaps, and a controllable-pitch propeller or FADEC. Cruise speed is not used to determine whether an airplane is high performance or complex.

1-38   PA.I.A.K1, K2

In order to act as pilot in command of a high-performance airplane, you must have

A – passed a flight test in that airplane from an FAA inspector.

B – an endorsement in your logbook that you are competent to act as pilot in command.

C – received and logged ground and flight instruction from an authorized flight instructor who then endorses your logbook.

1-38. Answer C. GFDPP 1B, FAR 61.31
To act as pilot in command of a high-performance airplane (an airplane with an engine of more than 200 horsepower), you must receive ground and flight instruction and a logbook endorsement stating that you are proficient to pilot a high-performance airplane. Answer B is partly correct but is not the most complete answer.

1-39   PA.I.A.K1, K2, PA.I.G.K1d

What is the definition of a complex airplane?

A – An airplane with retractable landing gear, flaps, and a controllable-pitch propeller or FADEC.

B – An airplane with a normal cruise speed of more than 200 knots.

C – An airplane with an engine of more than 200 horsepower.

1-39. Answer A. GFDPP 1B, FAR 61.31
A complex airplane has retractable landing gear, flaps, and a controllable pitch propeller or FADEC. A high-performance airplane has an engine with more than 200 horsepower. Cruise speed is not used to determine whether an airplane is high performance or complex.

1-40   PA.I.A.K1, K2

In order to act as pilot in command of a complex airplane, you must have

A – received and logged ground and flight instruction from an authorized instructor in a complex airplane, who then endorsed your logbook.

B – performed and logged three solo takeoffs and landings in a complex airplane.

C – passed a flight test in a complex airplane.

1-40. Answer A. GFDPP 1B, FAR 61.31
To act as pilot in command of a complex airplane—an airplane with retractable landing gear, flaps, and a controllable pitch propeller or FADEC—you must receive ground and flight instruction from an authorized flight instructor, who then endorses your logbook. The flight instruction must be in a complex airplane or in a flight simulator or flight training device that is representative of a complex airplane.

SECTION C
Introduction to Human Factors

SINGLE-PILOT RESOURCE MANAGEMENT

Chapter 1, Section C of the *Private Pilot* textbook defines single-pilot resource management (SRM) concepts that you should be aware of as you begin flight training. Chapter 10, Section B — Single-Pilot Resource Management provides a more extensive examination of SRM as it applies to private pilot operations. The FAA questions associated with SRM concepts, such as aeronautical decision making, risk, task and automation management, situational awareness, and controlled flight into terrain awareness are presented in Chapter 10, Section B of this test guide.

AVIATION PHYSIOLOGY

Chapter 1 Section C in the *Private Pilot* textbook introduces some aviation physiology concepts that you should be aware of as you begin flight training. Chapter 10, Section A — Aviation Physiology provides a more extensive examination of the performance and limitations of the body in the flight environment. The FAA questions associated with aviation physiology are presented in Chapter 10, Section A of this test guide.

SECTION C ■ Introduction to Human Factors

SECTION C ■ **Introduction to Human Factors**

CHAPTER 2

Airplane Systems

SECTION A
Airplanes

AIRWORTHINESS REQUIREMENTS

- The airworthiness certificate of an airplane remains valid as long as the aircraft is maintained and operated as required by the Federal Aviation Regulations.
- The owner or operator of an aircraft is responsible for ensuring that an aircraft is maintained in an airworthy condition.
- An acronym you can use to remember the required certificates and documents is ARROW:
  - **A**irworthiness certificate
  - **R**egistration
  - **R**adio station license (required by the Federal Communications Commission (FCC) when transmitting to ground stations outside the United States)
  - **O**perating limitations
  - **W**eight and balance

PILOT'S OPERATING HANDBOOK

- You can find the operating limitations for an aircraft in the current, FAA-approved airplane flight manual (AFM), approved manual material, markings, and placards, or any combination of these items.
- The FAA requires all currently manufactured airplanes to be equipped with an AFM which is specifically assigned to the individual airplane and must be accessible by the pilot during flight. To satisfy the regulatory requirement, the pilot's operating handbook (POH) for most of these aircraft is also designated as the AFM.
- A special airworthiness certificate (FAA Form 8130-7) is issued for all aircraft certificated in other than the standard categories. Operating limitations, applicable to these non-standard aircraft, such as experimental, restricted, limited, provisional, and light sport aircraft (LSA), are attached to the special airworthiness certificate.

REQUIRED MAINTENANCE AND INSPECTIONS

- As the pilot in command, you are responsible for determining that an aircraft is in condition for safe flight.
- Required inspections must be conducted by an appropriately certificated aviation maintenance technician (AMT) and documented in the maintenance records for the aircraft.
- You may not fly an aircraft unless it has received an annual inspection and an ELT inspection within the previous 12 calendar months.
- When the ELT has been in use for more than one cumulative hour, or if 50 percent of the useful life of the battery has expired, the battery must be replaced or recharged.
- Transponder inspections are required within the previous 24 calendar months. If the aircraft is flown under IFR, altimeter and static system inspections are also required within the previous 24 calendar months.
- 100-hour inspections are required on aircraft that are used for flight instruction for hire and provided by the flight instructor, or that carry any person, other than a crewmember, for hire. The aircraft may be flown up to 10 additional hours enroute to a location where service can be completed. However, the next 100-hour inspection must be completed within 100 hours of the original expiration time.

- To determine the expiration date of any inspection, refer to the aircraft maintenance records.

- After preventive maintenance is performed on an aircraft, the signature, certificate number, and kind of certificate held by the person approving the work must be entered in the aircraft maintenance records. In addition, the entry must also include a description of the work and the date of completion of the work performed.

- If an alteration or repair substantially affects the operation of an aircraft in flight, that aircraft must be test flown by at least a private pilot and approved for return to service before you may fly passengers.

- When an unsafe condition might exist or develop in an aircraft or other aircraft of the same design, the FAA publishes an airworthiness directive (AD). ADs are legally enforceable rules and compliance is mandatory.

AIRCRAFT EQUIPMENT REQUIRED FOR VFR

- FAR 91.205 requires specific equipment to be installed and operational for day and night VFR flight, which includes flight instruments, engine and system monitoring indicators, and safety equipment.

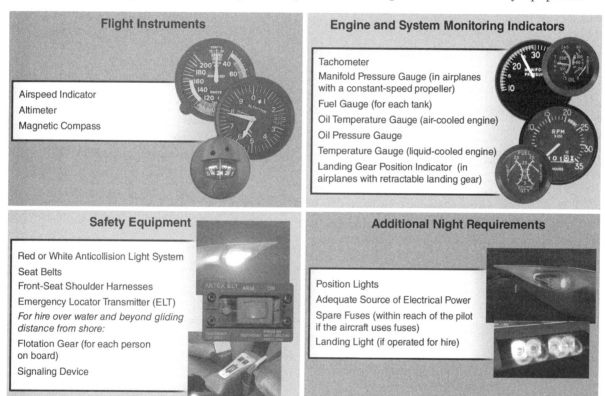

Flight Instruments

Airspeed Indicator
Altimeter
Magnetic Compass

Engine and System Monitoring Indicators

Tachometer
Manifold Pressure Gauge (in airplanes with a constant-speed propeller)
Fuel Gauge (for each tank)
Oil Temperature Gauge (air-cooled engine)
Oil Pressure Gauge
Temperature Gauge (liquid-cooled engine)
Landing Gear Position Indicator (in airplanes with retractable landing gear)

Safety Equipment

Red or White Anticollision Light System
Seat Belts
Front-Seat Shoulder Harnesses
Emergency Locator Transmitter (ELT)
For hire over water and beyond gliding distance from shore:
Flotation Gear (for each person on board)
Signaling Device

Additional Night Requirements

Position Lights
Adequate Source of Electrical Power
Spare Fuses (within reach of the pilot if the aircraft uses fuses)
Landing Light (if operated for hire)

SECTION A ■ **Airplanes**

- A minimum equipment list (MEL) takes into consideration the regulations and the specific requirements for your aircraft and flight operation and indicates the equipment that is allowed to be inoperative for a particular flight. MELs are not common for light single-engine piston-powered airplanes.

- If your airplane does not have an MEL, use the procedure described in FAR 91.213. Answer yes or no— do any of the following require the inoperative equipment?

  ○ The VFR-day type certificate requirements prescribed in the airworthiness certification regulations.

  ○ FAR 91.205 for the specific kind of flight operation (e.g. day or night VFR) or by other flight rules for the specific kind of flight to be conducted.

  ○ The aircraft's equipment list or the kinds of operations equipment list (KOEL).

  ○ An airworthiness directive (AD).

- If the answer is YES to ANY of the questions about required equipment, the airplane is not airworthy and maintenance is required. If the answer is NO to ALL of these questions, then you may fly the airplane after the inoperative equipment is:
  - Removed and the cockpit control placarded by an AMT.
  - Deactivated and placarded "inoperative" (if deactivation involves maintenance, this must be done by an AMT).
- If you determine that the airworthiness requirements are not met, the FAA may permit the airplane to be flown to a location where the needed repairs can be made by issuing a special flight permit, sometimes called a ferry permit.
- To obtain a special flight permit, an application must be submitted to the nearest FAA flight standards district office (FSDO).
- If an AD prohibits further flight until the AD is satisfied, the FSDO may not issue a special flight permit and an AMT might have be transported to the aircraft's location to resolve the AD.

NOTE: *An asterisk appearing after an ACS code (i.e. PA.VII.B.K1*) indicates that the question subject appears more than one time in the ACS. The code shown corresponds to the first instance of the subject in the ACS.*

SECTION A ■ Airplanes

2-1   PA.I.B.K1

Where can the operating limitations for an aircraft be found?

A – On the airworthiness certificate.

B – In the aircraft airframe and engine logbooks.

C – In the current, FAA-approved airplane flight manual, approved manual material, markings, and placards, or any combination thereof.

2-1. Answer C. GFDPP 2A, FAR 91.9
You can find the operating limitations in the current, FAA-approved airplane flight manual, approved manual material, markings, and placards, or any combination of these items.

2-2   PA.I.B.K1

Which of the following are always required to be carried in the aircraft during flight?

A – Pilot's information manual (PIM).

B – Radio station license.

C – FAA-approved airplane flight manual (AFM).

2-2. Answer C. GFDPP 2A, FAR 91.9
The FAA requires all currently manufactured airplanes to be equipped with an FAA approved airplane flight manual (AFM), which is specifically assigned to the individual airplane and must be accessible by the pilot during flight. The radio station license is only required when transmitting outside the United States and although the PIM is a derivative of the AFM, it does not meet the legal requirement.

2-3   PA.I.B.K3d, PA.II.F.K1c

Which is true regarding an airplane's operating limitations?

A – Operating limitations are only located in the POH.

B – Operating limitations can be in the form of placards.

C – Operating limitations must be committed to memory because they are not required to be in the airplane.

2-3. Answer B. GFDPP 2A, PHB
You can find the operating limitations for an aircraft in the current, FAA-approved airplane flight manual (AFM), approved manual material, markings, and placards, or any combination of these items.

2-4   PA.I.B.K1b, K1e

To determine the expiration date of the last annual aircraft inspection, you should refer to the

A – airworthiness certificate.

B – registration certificate.

C – aircraft maintenance records.

2-4. Answer C. GFDPP 2A, FAR 91.417

The registered owner or operator shall keep records of the maintenance, preventive maintenance, alterations, records of the annual, 100-hour, progressive, and other required or approved inspections, as appropriate, for each aircraft. This information is found in the maintenance records for the aircraft.

2-5   PA.I.B.K1a

A certificate of aircraft registration is valid for how many years?

A – 3 years from date of issuance.

B – 1 year from date of issuance.

C – 7 years from date of issuance.

2-5. Answer C. GFDPP 2A, FAR 47.40, FAR 91.203

The aircraft registration is a document that shows as proof that it has been registered with the national aviation authority of a country. A certificate of aircraft registration expires seven years after the last day of the month in which it was issued.

2-6   PA.I.B.K1, K4

How long does the airworthiness certificate of an aircraft remain valid?

A – As long as the aircraft has a current registration certificate.

B – Indefinitely unless the aircraft suffers major damage.

C – As long as the aircraft is maintained and operated as required by the Federal Aviation Regulations.

2-6. Answer C. GFDPP 2A, PHB

The airworthiness certificate remains valid only as long as the aircraft is maintained and operated in accordance with the FARs.

2-7   PA.I.A.K1, K4

In addition to a valid airworthiness certificate, what documents or records must be onboard an aircraft during flight?

A – Aircraft engine and airframe logbooks, and owner's manual.

B – Radio operator's permit, and repair and alteration forms.

C – Operating limitations and registration certificate.

2-7. Answer C. GFDPP 2A, FAR 91.9, FAR 91.203

Use the acronym, ARROW, to remember the required certificates and documents. ARROW stands for **A**irworthiness certificate; **R**egistration; **R**adio station class license (international flights only); **O**perating limitations; and **W**eight and balance.

2-8   PA.I.B.K1a

A certificate of aircraft registration is valid for how many years?

A – 3 years from date of issuance.

B – 1 year from date of issuance.

C – 7 years from date of issuance.

2-8. Answer C. GFDPP 2A, FAR 47.40, FAR 91.203

The aircraft registration is a document that shows as proof that it has been registered with the national aviation authority of a country. A certificate of aircraft registration expires seven years after the last day of the month in which it was issued.

2-9   PA.I.B.K1a, K1e, K4

Which of the following preflight actions is the pilot in command required to take in order to comply with the FARs regarding day visual flight rules (VFR)?

A – File a VFR flight plan with Flight Service.

B – Verify the airworthiness certificate is legible to passengers.

C – Verify approved position lights are not burned out.

2-9. Answer B. GFDPP 2A, AFH, FAR 91.203

According to FAR 91.203, you must ensure that the airworthiness certificate and registration certificate are in the airplane. The airworthiness certificate must be displayed at the cabin or cockpit entrance so that it is legible to passengers and crew.

Although not required for flight under VFR, filing a flight plan is strongly recommended so that you receive search and rescue protection. Position lights are not required for day VFR flight. However, if the aircraft has these lights and they are inoperative, you must follow the appropriate steps for flying with inoperative equipment or cancel the flight.

2-10   PA.I.B.K1b

An annual inspection was performed on an aircraft June 13, 2024. When is the next annual inspection due?

A – June 1, 2025.

B – June 13, 2025.

C – June 30, 2025.

2-10. Answer C. GFDPP 2A, FAR 91.409

No person may operate an aircraft unless, within the preceding 12 calendar months, it has had an annual inspection. The term "calendar month" is defined as the last day of the month.

2-11   PA.I.B.K1b

What aircraft inspections are required for rental aircraft that are also used for flight instruction?

A – Annual and 100-hour inspections.

B – Biannual and 100-hour inspections.

C – Annual and 50-hour inspections.

2-11. Answer A. GFDPP 2A, FAR 91.409

No person may operate an aircraft carrying any person (other than a crewmember) for hire, or give flight instruction for hire in an aircraft that person provides, unless within the preceding 100 hours of time in service, the aircraft has received an annual or 100-hour inspection.

2-12   PA.I.B.K1b

An aircraft had a 100-hour inspection when the tachometer read 1259.6. When is the next 100-hour inspection due?

A – 1349.6 hours.

B – 1359.6 hours.

C – 1369.6 hours.

2-12. Answer B. GFDPP 2A, FAR 91.409

No person may operate an aircraft carrying any person (other than a crewmember) for hire, or give flight instruction for hire in an aircraft that person provides, unless within the preceding 100 hours of time in service, the aircraft has received an annual or 100-hour inspection.

2-13   PA.I.B.K1b

A 100-hour inspection was due at 3303 hours. The 100-hour inspection was actually done at 3300 hours. When is the next 100-hour inspection due?

A – 3400 hours.

B – 3403 hours.

C – 3413 hours.

2-13. Answer A. GFDPP 2A, FAR 91.409

If the 100-hour inspection is delayed past the time it is due, the next inspection is due 100 hours after the *original* inspection was due. However, if the 100-hour inspection is done ahead of schedule, the next inspection is due 100 hours after the *actual* time of the inspection.

2-14   PA.I.B.K1b

A 100-hour inspection was due at 3302.5 hours. The 100-hour inspection was actually done at 3309.5 hours. When is the next 100-hour inspection due?

A – 3312.5 hours.

B – 3395.5 hours.

C – 3402.5 hours.

2-14. Answer C. GFDPP 2A, FAR 91.409

The 100-hour limitation may be exceeded by not more than 10 hours if you are enroute to a location where the 100-hour inspection is to be performed. However, the next inspection is still due 100 hours after the original inspection was due.

2-15   PA.I.B.K1b

Maintenance records show the last transponder inspection was performed on September 1, 2024. The next inspection is due no later than

A – September 30, 2025.

B – September 1, 2026.

C – September 30, 2026.

2-15. Answer C. GFDPP 2A, FAR 91.413

No person may use an ATC transponder unless, within the preceding 24 calendar months, that transponder has been tested and found to comply with the appropriate standards listed in Appendix F of FAR Part 43. The term "calendar month" refers to the end of the month when an inspection is due.

SECTION A ■ **Airplanes**

2-16   PA.I.B.K3, PA.IX.D.K1

The maximum cumulative time that an emergency locator transmitter may be operated before the rechargeable battery must be recharged is

A – 30 minutes.

B – 45 minutes.

C – 60 minutes.

2-16. Answer C. GFDPP 2A, FAR 91.207
You may not fly an aircraft unless it has received an annual inspection and an emergency locator transmitter (ELT) inspection within the previous 12 calendar months. In addition, the ELT battery must be replaced or recharged if the transmitter has been operated for a total of 1 hour or after 50 percent of useful battery life (or charge) has expired.

2-17   PA.I.B.K1b, K1e

The responsibility for ensuring that maintenance personnel make the appropriate entries in the aircraft maintenance records indicating the aircraft has been approved for return to service lies with the

A – owner or operator.

B – pilot in command.

C – mechanic who performed the work.

2-17. Answer A. GFDPP 2A, FAR 91.405
Each owner or operator of an aircraft shall ensure that maintenance personnel make appropriate entries in the aircraft maintenance record indicating the aircraft has been approved for return to service.

2-18   PA.I.B.K1b, K1e

Who is responsible for ensuring appropriate entries are made in maintenance records indicating the aircraft has been approved for return to service?

A – Repair station.

B – Certified mechanic.

C – Owner or operator.

2-18. Answer C. GFDPP 2A, FAR 91.405
Each owner or operator of an aircraft shall ensure that maintenance personnel make appropriate entries in the aircraft maintenance records indicating the aircraft has been approved for return to service.

2-19   PA.I.B.S2, PA.I.B.K1e, PA.II.A.K3

During the preflight inspection who is responsible for determining the aircraft is safe for flight?

A – The owner or operator.

B – The certificated mechanic who performed the annual inspection.

C – The pilot in command.

2-19. Answer C. GFDPP 2A, PHB
The owner or operator is responsible to make sure that the required maintenance and inspections are performed. These tasks may be assigned to a certificated maintenance technician, but the responsibility for compliance remains with the owner or operator. Preflight inspection is the responsibility of the pilot in command (PIC). Before every flight, the PIC is required to accomplish a thorough and systematic preflight to ensure that the aircraft is safe for flight. The preflight inspection should be completed according to procedures recommended by the manufacturer. Normally, this means the PIC should use a checklist for the preflight inspection.

2-20   PA.I.B.K1e, PA.II.A.K3a, K3b, K3c, PA.I.B.S2

How should an aircraft preflight inspection be accomplished for the first flight of the day?

A – Thorough and systematic means recommended by the manufacturer.

B – Quick walk around with a check of fuel and oil.

C – Any sequence as determined by the pilot-in-command.

2-20. Answer A. GFDPP 2A, PHB
The owner or operator is responsible to make sure that the required maintenance and inspections are performed. These tasks may be assigned to a certificated maintenance technician, but the responsibility for compliance remains with the owner or operator. Preflight inspection is the responsibility of the pilot in command (PIC). Before every flight, the PIC is required to accomplish a thorough and systematic preflight to ensure that the aircraft is safe for flight. The preflight inspection should be completed according to procedures recommended by the manufacturer. Normally, this means the PIC should use a checklist for the preflight inspection.

2-21   PA.I.B.K1e

Who is responsible for determining if an aircraft is in condition for safe flight?

A – A certificated aircraft mechanic.

B – The pilot in command.

C – The owner or operator.

2-21. Answer B. GFDPP 2A, FAR 91.7
The pilot in command of a civil aircraft is responsible, and has final authority, for determining whether that aircraft is in condition for safe flight.

2-22   PA.I.B.K1e, PA.II.A.K2

The responsibility for ensuring that an aircraft is maintained in an airworthy condition is primarily that of the

A – pilot in command.

B – owner or operator.

C – mechanic who performs the work.

2-22. Answer B. GFDPP 2A, FAR 91.403
The owner or operator of an aircraft is primarily responsible for maintaining that aircraft in an airworthy condition. The pilot in command is responsible for checking the condition of the aircraft before flying it. The mechanic is only responsible for the work performed.

2-23   PA.I.B.K1

If an alteration or repair substantially affects the operation of an aircraft in flight, that aircraft must be test flown by an appropriately rated pilot and approved for return to service before being operated

A – by any private pilot.

B – with passengers aboard.

C – for compensation or hire.

2-23. Answer B. GFDPP 2A, FAR 91.407
Before any person (other than a crewmember) can fly in an aircraft that has been maintained, rebuilt, or altered in a manner that may have appreciably changed its flight characteristics or substantially affected the operation in flight, an appropriately rated pilot with at least a private pilot certificate must first conduct a test flight and log the flight in aircraft records.

SECTION A ■ **Airplanes**

2-24   PA.I.B.K1

Before passengers may be carried in an aircraft that has been altered in a manner that could have appreciably changed its flight characteristics, it must be flight tested by an appropriately rated pilot who holds at least a

A – Commercial pilot certificate with an instrument rating.

B – Private pilot certificate.

C – Commercial Pilot Certificate and a mechanic's certificate.

2-24. Answer B. GFDPP 2A, FAR 91.407
An appropriately rated pilot with at least a private pilot certificate is authorized to flight test the aircraft.

2-25   PA.I.B.K1c, K1e

Which records or documents shall the owner or operator of an aircraft keep to show compliance with an applicable airworthiness directive?

A – Aircraft maintenance records.

B – Airworthiness certificate and pilot's operating handbook.

C – Airworthiness and registration certificates.

2-25. Answer A. GFDPP 2A, FAR 91.417
The owner or operator of an aircraft shall keep a record of current status of applicable airworthiness directives (ADs) in the appropriate aircraft maintenance records.

2-26   PA.I.B.K2

Which operation would be described as preventive maintenance?

A – Repair of landing gear brace struts.

B – Replenishing hydraulic fluid.

C – Repair of portions of skin sheets by making additional seams.

2-26. Answer B. GFDPP 2A, FAR 1.1, FAR 43 App A
Replenishing hydraulic fluid is listed as preventive maintenance in FAR Part 43, Appendix A. Structural repairs, such as those to landing gear brace struts, or adding seams to skin, are *not* preventive maintenance and require an appropriate aircraft mechanic certificate.

2-27   PA.I.B.K2

Preventive maintenance has been performed on an aircraft. What paperwork is required?

A – A full, detailed description of the work done must be entered in the airframe logbook.

B – The date the work was completed, and the name of the person who did the work must be entered in the airframe and engine logbook.

C – The signature, certificate number, and kind of certificate held by the person approving the work and a description of the work must be entered in the aircraft maintenance records.

2-27. Answer C. GFDPP 2A, FAR 43.9, FAR 91.417
FAR 91.417 states that records of preventive maintenance must include a description of the work performed, the date of completion, and the signature and certificate number of the person approving the aircraft for return to service. In addition, FAR 43.9 also indicates that the kind of certificate held by the person approving the work must be included in the record.

2-28   PA.I.B.K2

What regulation allows a private pilot to perform preventive maintenance?

A – 14 CFR 43.7.

B – 14 CFR 91.403.

C – 14 CFR 61.113.

2-28. Answer A. GFDPP 2A, FAR 43.7
14 CFR 43 covers preventive maintenance. If you hold at least a private pilot certificate, you may perform preventive maintenance such as replacing and servicing batteries, replacing spark plugs, servicing wheel bearings, etc.

2-29   PA.I.B.K2

Who may perform preventive maintenance on an aircraft and approve it for return to service?

A – Student or recreational pilot.

B – Private or commercial pilot.

C – None of the above.

2-29. Answer B. GFDPP 2A, FAR 43.7
14 CFR 43 covers preventive maintenance. If you hold at least a private pilot certificate, you may perform preventive maintenance such as replacing and servicing batteries, replacing spark plugs, servicing wheel bearings, etc.

2-30   PA.I.B.K1b, K1e

Completion of an annual inspection and the return of the aircraft to service should always be indicated by

A – the relicensing date on the Registration Certificate.

B – an appropriate notation in the aircraft maintenance records.

C – an inspection sticker placed on the instrument panel that lists the annual inspection completion date.

2-30. Answer B. GFDPP 2A, FAR 91.409
No person may operate an aircraft unless, within the preceding 12 calendar months, it has had an annual inspection by a person authorized to do that type of inspection and is entered as an annual inspection in the required maintenance records.

2-31   PA.I.B.K1c, K1e

What should an owner or operator know about airworthiness directives (ADs)?

A – They are for informational purposes only.

B – They are mandatory.

C – They are voluntary.

2-31. Answer B. GFDPP 2A, FAR 39
ADs are published as part of the FARs. You may not operate an aircraft to which an airworthiness directive applies, except in accordance with that airworthiness directive.

2-32   PA.I.B.K1c, K1e

Who is responsible for ensuring that airworthiness directives (ADs) are complied with?

A – Mechanic with inspection authorization (IA).

B – Owner or operator.

C – Repair station.

2-32. Answer B. GFDPP 2A, FAR 91.403
The owner or operator of an aircraft is primarily responsible for maintaining that aircraft in an airworthy condition, including compliance with 14 CFR 39 (ADs).

SECTION A ■ **Airplanes**

2-33   PA.I.B.K1a, K4

What is required for an airworthiness certificate to remain valid?

A – The airworthiness certificate remains valid as long as required maintenance and inspections are performed on the aircraft within the specified time periods.

B – The airworthiness certificate remains valid as long as it is in the aircraft.

C – The airworthiness certificate is valid from the day it is issued by the FAA, and it remains valid for a specified period that is determined by the FAA.

2-33. Answer A. GFDPP 2A, FAR 21.181, FAR 91.409
This certificate remains valid as long as the required maintenance and inspections are performed on the aircraft within the specified time periods.

2-34   PA.I.B.K3b

A minimum equipment list (MEL) is

A – based on the regulations and the specific requirements for your aircraft and flight operation and indicates the equipment that is allowed to be inoperative for a particular flight.

B – is a required document for all aircraft.

C – lists only the minimum equipment required by 91.205 for a specific flight operation.

2-34. Answer A. GFDPP 2A, FAR 91.213
If a minimum equipment list (MEL) exists for your aircraft, you must use it. The MEL takes into consideration the regulations and the specific requirements for your aircraft and flight operation and indicates the equipment that is allowed to be inoperative for a particular flight. MELs are not common for light single-engine piston-powered airplanes.

2-35   PA.I.B.K3b

What is true regarding the use of a minimum equipment list (MEL)?

A – If your airplane is equipped with an MEL, you must use it, as it takes into account all regulations and requirements for your aircraft.

B – The kinds of operation equipment list (KOEL) can be used in lieu of an MEL if your airplane has one.

C – The MEL is advisory only and should not be used to determine the requirements of your airplane.

2-35. Answer A. GFDPP 2A, FAR 91.213
If a minimum equipment list (MEL) exists for your aircraft, you must use it. The MEL takes into consideration the regulations and specific requirements for your aircraft and flight operation, and indicates the equipment that is allowed to be inoperative for a particular flight.

2-36   PA.I.B.K3a, K3c; PA.II.A.K3b, K3c, Kd

During the preflight inspection, you discover inoperative equipment on an airplane that you are planning to fly. Under what conditions can you complete the flight under VFR?

A – The equipment is not required by FAR 91.205.

B – As the pilot in command with at least a private pilot certificate, you determine the airplane is in a safe condition for flight.

C – The equipment is not required by FAR 91.205, an equipment list or KOEL, the VFR-day type certificate requirements, or an AD.

2-36. Answer C. GFDPP 2A, FAR 91. 213

You must determine that the inoperative equipment is not required by:

- The VFR-day type certificate requirements prescribed in the airworthiness certification regulations.

- FAR 91.205 for the specific kind of flight operation to be conducted.

- The equipment list or the kinds of operations equipment list (KOEL) for the aircraft.

- An airworthiness directive (AD).

The inoperative equipment must be deactivated and placarded as inoperative.

2-37   PA.I.B.K3a, K3c; PA.II.A.K3b, K3c, Kd

You discover inoperative equipment on an airplane that you are planning to fly. The airplane does not have an MEL. You have determined that—according to the FARs—you may legally fly the airplane. What additional actions must be taken?

A – No further action is required.

B – You must obtain a special flight permit.

C – The equipment must be removed or deactivated and placarded "inoperative."

2-37. Answer C. GFDPP 2A, 91.213

If your airplane does not have an MEL, use the procedure described in FAR 91.213. Answer yes or no—do any of the following require the inoperative equipment?

- The VFR-day type certificate requirements; FAR prescribed in the airworthiness certification regulations.

- FAR 91.205 for the specific kind of flight operation (e.g. day or night VFR) or by other flight rules for the specific kind of flight to be conducted.

- The aircraft's equipment list or the kinds of operations equipment list (KOEL).

- An airworthiness directive (AD).

If the answer is YES to ANY of these questions, the airplane is not airworthy and maintenance is required. If the answer is NO to ALL of these questions, then you may fly the airplane after the inoperative equipment is:

- Removed and the cockpit control placarded by an AMT.

- Deactivated and placarded "inoperative" (if deactivation involves maintenance, this must be done by an AMT).

SECTION A ■ **Airplanes**

2-38   PA.I.B.K1d, K4

If you determine that the airworthiness requirements are not met, a special flight permit

A – allows the aircraft to be flown to a location where the needed repairs can be made.

B – allows the aircraft to be flown with inoperative equipment until the next annual inspection.

C – can be issued by an AMT and allows the aircraft to be flown until scheduled repairs can be made.

2-38. Answer A. GFDPP 2A, FAR 21.197

If you determine that the airworthiness requirements are not met, the FAA may permit the airplane to be flown to a location where the needed repairs can be made by issuing a special flight permit, sometimes called a ferry permit.

To obtain this permit, an application must be submitted to the nearest FAA flight standards district office (FSDO). However, if an airworthiness directive prohibits further flight until the AD is satisfied, the FSDO may not issue a ferry permit. In this case, an AMT might have be transported to the aircraft's location to resolve the AD.

2-39   PA.I.B.K3d, K4

If a special flight permit is needed, you can

A – submit an application to the nearest FAA FSDO.

B – obtain the permit from an AMT.

C – notify your local FSDO, and no application is required.

2-39. Answer A. GFDPP 2A, PHB

If you determine that the airworthiness requirements are not met, the FAA may permit the airplane to be flown to a location where the needed repairs can be made by issuing a special flight permit, sometimes called a ferry permit.

To obtain this permit, an application must be submitted to the nearest FAA flight standards district office (FSDO). However, if an airworthiness directive prohibits further flight until the AD is satisfied, the FSDO may not issue a ferry permit. In this case, an aviation maintenance technician (AMT) might have be transported to the aircraft's location to resolve the AD.

SECTION B
Powerplant and Related Systems

INDUCTION SYSTEMS

- Float-type carburetors operate based on the difference in air pressure at the venturi throat and the air inlet. The decreased pressure caused by air flowing rapidly through the venturi tube draws fuel from the float chamber.

- Carburetors are more susceptible to icing than fuel-injected engines due to the sharp temperature drop caused by fuel vaporization and decreasing air pressure in the venturi of the carburetor.

- Carburetor icing is most likely to occur when temperatures are below 21°C (70°F) and the relative humidity is above 80%. Although uncommon, carburetor icing is possible at temperatures as high as 38°C (100°F) and humidity as low as 50%.

- On an airplane with a fixed-pitch propeller, the first indication of carburetor ice would most likely be a loss of engine RPM.

- Carbureted engines are equipped with carburetor heat to prevent ice from forming in the venturi. Turn on the heat when flying in conditions conducive to carburetor icing to heat the air entering the engine. The heated air is less dense, which enrichens the mixture and reduces engine performance and RPM.

- If carburetor ice is present, applying carburetor heat will cause a temporary decrease in RPM, followed by a gradual increase as the ice melts.

- Fuel injection, compared to a carburetor, increases engine efficiency, offering lower fuel consumption, increased horsepower, lower operating temperatures, and longer engine life. The most significant safety advantage is the reduced risk of induction icing.

THE IGNITION SYSTEM

- The main purpose of a dual-ignition system on an aircraft is to provide system redundancy.

- Another advantage of dual-ignition systems is improved engine performance.

ABNORMAL COMBUSTION

- Detonation occurs in a reciprocating aircraft engine when the unburned charge in the cylinders explodes instead of burning normally.

- Detonation is most likely to occur at high power settings and when the engine is running hot.

- Using a grade of fuel that is lower than specified for the engine can cause detonation.

- If you suspect that the engine is detonating during climb-out, enrichen the mixture and lower the nose slightly to increase airspeed.

- Pre-ignition occurs when the fuel/air mixture ignites too soon, in advance of normal spark ignition. Pre-ignition can be caused by hot spots in the cylinder, or by using too low a grade of fuel.

FUEL SYSTEMS

- Because engine-driven fuel pumps operate only when the engine is running, an electric fuel pump, which is controlled by a switch in the cockpit, provides fuel under pressure for engine starting and as a backup, should the engine-driven fuel pump malfunction.

- On aircraft equipped with fuel pumps, running a fuel tank dry before switching tanks is not recommended because the engine-driven or electric fuel pump can draw air into the fuel system and cause vapor lock.

SECTION B ■ Powerplant and Related Systems

- Using fuel of a lower-than-specified grade can cause cylinder head and engine oil temperature gauges to exceed their normal operating ranges. Fuel of the next higher octane may be substituted if the recommended octane is not available.

- Filling the fuel tanks after the last flight of the day will prevent moisture condensation by eliminating air in the tanks.

- A fuel-injected engine can be difficult to restart a few minutes after shutdown—the residual heat in the engine can boil the fuel in the injection system's lines and components, and the resulting bubbles can cause vapor lock, which interferes with fuel metering and pumping.

EXHAUST SYSTEMS

- In many light airplanes, the exhaust system is used to provide heat for the cabin—hot exhaust gases heat the muffler and metal shrouds around the muffler capture the heat and duct it to the cabin.

- If the muffler is cracked, exhaust gases can enter the cabin. If you smell exhaust gases, you can assume that carbon monoxide is present. Turn off the heater, open the fresh air vents, and use supplemental oxygen, if it is available.

OIL SYSTEMS

- For internal cooling, reciprocating aircraft engines rely on the circulation of lubricating oil.

- The oil pressure gauge provides a direct indication of the oil system operation. An abnormally high oil temperature indication can be caused by the oil level being too low, accompanied by a low oil pressure reading.

COOLING SYSTEMS

- One way to cool an engine that is overheating is to enrichen the fuel mixture.

- Excessively high engine temperatures can cause a loss of power, excessive oil consumption, and possible permanent internal engine damage.

- To aid engine cooling during climb, lower the nose, reduce the rate of climb and increase airspeed.

PROPELLERS

- Two basic types of propellers exist on single-engine airplanes—fixed-pitch and constant-speed.

- Fixed-pitch propellers are normally optimized either for climb or for cruise.

- A constant-speed propeller enables selecting an RPM that results in a blade pitch angle for most efficient performance.

- To operate the engine on an airplane with a constant-speed propeller, use the throttle to control power output, as shown on the manifold pressure gauge, and the propeller control to regulate engine RPM.

- When operating an engine equipped with a constant-speed propeller, avoid a high manifold pressure setting with low RPM, as specified in your POH.

- Full authority digital engine control (FADEC) is a computer with associated systems that manages the engine and propeller of an aircraft, simplifying the engine controls to a single power lever. Because the engine will stop running if the FADEC engine control unit (ECU) fails, a back-up ECU is required.

ELECTRICAL SYSTEMS AND MALFUNCTIONS

- Consider these actions to manage an alternator failure:
    - Shed electrical load to preserve battery power for essential equipment.
    - Notify ATC of the situation and use ATC services.
    - Be prepared to control the airplane and land without digital instruments, radio communication, lights, or flaps, and to perform a manual landing gear extension, if applicable.
- Because many light aircraft have electrically actuated flaps, in the event of an alternator failure, you may also lose the ability to extend your flaps. Consider these factors as you prepare to perform a no-flap approach and landing:
    - The glide path is not as steep as with flaps extended, so the higher nose attitude on final can cause errors in your judgment of height and distance.
    - Landing distance is substantially increased.
    - Floating during the flare is likely.
- If you experience an electric trim malfunction, immediately grasp the control wheel to control the airplane. If you cannot correct the problem and are experiencing runaway trim, pull the applicable circuit breakers as recommended by the AFM/POH and land as soon as practical.
- Managing smoke and fire in the cabin typically involves turning off the master, using a fire extinguisher to extinguish any flames, opening the air vents after the fire is out, and landing as soon as possible.

EMERGENCY DESCENT

- An emergency descent is a maneuver for descending to a lower altitude as rapidly as possible within the structural limitations of the airplane. You might need to perform an emergency descent due to an uncontrollable fire, smoke in the cockpit, a sudden loss of cabin pressurization, or any other situation that demands an immediate rapid descent.

NOTE: An asterisk appearing after an ACS code (i.e. PA.VII.B.K1*) indicates that the question subject appears more than one time in the ACS. The code shown corresponds to the first instance of the subject in the ACS.

SECTION B ■ Powerplant and Related Systems

2-40   PA.I.G.K1c, PA.IX.C.K1
Excessively high engine temperatures will

A – cause damage to heat-conducting hoses and warping of the cylinder cooling fins.

B – cause loss of power, excessive oil consumption, and possible permanent internal engine damage.

C – not appreciably affect an aircraft engine.

2-40. Answer B. GFDPP 2B, PHB
High temperature can cause detonation and a resulting loss of power, excessive oil consumption, and engine damage, including scoring of the cylinders and damage to pistons, rings, and valves.

2-41   PA.I.G.K1c, PA.I.G.K2
If the engine oil temperature and cylinder head temperature gauges have exceeded their normal operating range, the pilot may have been operating with

A – the mixture set too rich.

B – higher-than-normal oil pressure.

C – too much power and with the mixture set too lean.

2-41. Answer C. GFDPP 2B, PHB
High power settings with the mixture set too lean can cause overheating, which is indicated by high engine oil and cylinder head temperatures.

2-42   PA.I.G.K1e

Float-type carburetors operate based on

A – automatic metering of air at the venturi as the aircraft gains altitude.

B – the difference in air pressure at the venturi throat and the air inlet.

C – increase in air velocity in the throat of a venturi causing an increase in air pressure.

2-42. Answer B. GFDPP 2B, PHB

The decreased pressure caused by air flowing rapidly through the venturi tube draws fuel from the float chamber.

2-43   PA.I.G.K1e

The basic purpose of adjusting the fuel/air mixture at altitude is to

A – decrease the amount of fuel in the mixture in order to compensate for increased air density.

B – decrease the fuel flow in order to compensate for decreased air density.

C – increase the amount of fuel in the mixture to compensate for the decrease in pressure and density of the air.

2-43. Answer B. GFDPP 2B, PHB

If fuel flow is not decreased with altitude, the mixture becomes too rich with fuel. Therefore, the fuel mixture must be leaned to maintain the proper fuel/air ratio.

2-44   PA.I.G.K1e, PA.II.F.K1b, K1c

During the runup at a high elevation airport, you notice a slight engine roughness that is not affected by the magneto check but grows worse during the carburetor heat check. Under these circumstances, what would be the most logical initial action?

A – Check the results obtained with a leaner setting of the mixture.

B – Taxi back to the flight line for a maintenance check.

C – Reduce manifold pressure to control detonation.

2-44. Answer A. GFDPP 2B, PHB

In this case, engine roughness is probably caused by the mixture set too rich for the high altitude. When the carburetor heat is turned on, the warmer air entering the carburetor is less dense, and the mixture is further enriched. As a result, the engine roughness increases. The problem can usually be corrected by leaning the mixture.

2-45   PA.I.G.K1e

While cruising at 9,500 feet MSL, the fuel/air mixture is properly adjusted. What will occur if a descent to 4,500 feet MSL is made without readjusting the mixture?

A – The fuel/air mixture can become excessively lean.

B – There can be more fuel in the cylinders than is needed for normal combustion, and the excess fuel will absorb heat and cool the engine.

C – The excessively rich mixture creates higher cylinder head temperatures and can cause detonation.

2-45. Answer A. GFDPP 2B, PHB

With a decrease in altitude, air density increases. This means you will have to enrich the mixture as you descend, otherwise the fuel/air mixture can become excessively lean.

2-46   PA.I.G.K1c, K1e, PA.IX.C.K1

Which condition is most favorable to the development of carburetor icing?

A – Any temperature below freezing and a relative humidity of less than 50 percent.

B – Temperature between 32°F and 50°F and low humidity.

C – Temperature between 20°F and 70°F and high humidity.

2-46. Answer C. GFDPP 2B, PHB

Carburetor icing is most probable between 20°F and 70°F with high humidity or visible moisture. Although uncommon, carburetor icing is possible at temperatures as high as 38°C (100°F) and humidity as low as 50%. Carburetor icing is not a high risk at low temperatures, as the air cannot hold much moisture.

2-47   PA.I.G.K1c, K1e, PA.IX.C.K1

The risk of carburetor ice is

A – high when the ambient temperature is between 20°F and 70°F and the relative humidity is high.

B – nonexistent at 95°F even when there is visible moisture.

C – high at 0°F when the relative humidity is high.

2-47. Answer A. GFDPP 2B, PHB

Carburetor icing is most probable between 20°F and 70°F with high humidity or visible moisture. Although uncommon, carburetor icing is possible at temperatures as high as 38°C (100°F) and humidity as low as 50%. Carburetor icing is not a high risk at low temperatures, as the air cannot hold much moisture.

SECTION B ■ **Powerplant and Related Systems**

2-48   PA.I.G.K1c, K1e, PA.IX.C.K1

If an aircraft is equipped with a fixed-pitch propeller and a float-type carburetor, the first indication of carburetor ice would most likely be

A – a drop in oil temperature and cylinder head temperature.

B – engine roughness.

C – loss of RPM.

2-48. Answer C. GFDPP 2B, PHB
The restricted airflow through the carburetor causes an enriched mixture and a loss of RPM.

2-49   PA.I.G.K1c, K1e, PA.IX.C.K1

Applying carburetor heat will

A – result in more air going through the carburetor.

B – enrich the fuel/air mixture.

C – not affect the fuel/air mixture.

2-49. Answer B. GFDPP 2B, PHB
When the carburetor heat is turned on, the warmer air entering the carburetor is less dense, and with the same amount of fuel entering the carburetor, the mixture is thereby enriched.

2-50   PA.I.G.K1c, K1e, PA.IX.C.K1

What change occurs in the fuel/air mixture when carburetor heat is applied?

A – A decrease in RPM results from the lean mixture.

B – The fuel/air mixture becomes richer.

C – The fuel/air mixture becomes leaner.

2-50. Answer B. GFDPP 2B, PHB
When the carburetor heat is turned on, the warmer air entering the carburetor is less dense, and with the same amount of fuel entering the carburetor, the mixture is thereby enriched.

2-51   PA.I.G.K1c, K1e, PA.IX.C.K1

The use of carburetor heat tends to

A – decrease engine performance.

B – increase engine performance.

C – have no effect on engine performance.

2-51. Answer A. GFDPP 2B, PHB
Because the warmer air entering the carburetor is less dense, the fuel/air mixture is enriched and power decreases.

2-52   PA.I.G.K1c, K1e, PA.IX.C.K1

The presence of carburetor ice in an aircraft equipped with a fixed-pitch propeller can be verified by applying carburetor heat and noting

A – an increase in RPM and then a gradual decrease in RPM.

B – a decrease in RPM and then a constant RPM indication.

C – a decrease in RPM and then a gradual increase in RPM.

2-52. Answer C. GFDPP 2B, PHB
When carburetor heat is first applied, the mixture is enriched, and RPM decreases. Then, as the ice melts, airflow into the carburetor increases, as does the engine RPM.

2-53   PA.I.G.K1c, K1e, PA.IX.C.K1

With regard to carburetor ice, float-type carburetor systems in comparison to fuel injection systems are generally considered to be

A – more susceptible to icing.

B – equally susceptible to icing.

C – less susceptible to icing.

2-53. Answer A. GFDPP 2B, PHB

Because the fuel is introduced directly into the hot cylinders instead of vaporizing in a cold venturi throat, fuel injection systems are not as susceptible to icing as float-type carburetors.

2-54   PA.I.G.K1c

One purpose of the dual-ignition system on an aircraft engine is to provide for

A – improved engine performance.

B – uniform heat distribution.

C – balanced cylinder head pressure.

2-54. Answer A. GFDPP 2B, PHB

Dual-ignition systems fire two spark plugs on each cylinder. In addition to system redundancy, this improves combustion of the fuel/air mixture and results in slightly more power.

2-55   PA.I.G.K1c, PA.II.C.K1, K3

What should be the first action after starting an aircraft engine?

A – Adjust for proper RPM and check for desired indications on the engine gauges.

B – Place the magneto or ignition switch momentarily in the OFF position to check for proper grounding.

C – Test each brake and the parking brake.

2-55. Answer A. GFDPP 2B, PHB

Immediately after starting an engine, set the proper RPM and check engine gauges for proper indications. Turning the magnetos off could cause a backfire and testing the brakes occurs after the airplane starts moving.

2-56   PA.I.G.K1c, K1e, K2

If the grade of fuel used in an aircraft engine is lower than specified for the engine, it will most likely cause

A – detonation.

B – a mixture of fuel and air that is not uniform in all cylinders.

C – lower cylinder head temperatures.

2-56. Answer A. GFDPP 2B, PHB

The higher the grade of fuel, the more pressure it can withstand without detonating. Conversely, lower fuel grades are more prone to detonation with accompanying higher engine temperatures.

SECTION B ■ Powerplant and Related Systems

2-57   PA.I.G.K1c, K1e

Detonation occurs in a reciprocating aircraft engine when

A – the spark plugs are fouled or shorted out or the wiring is defective.

B – hot spots in the combustion chamber ignite the fuel/air mixture in advance of normal ignition.

C – the unburned charge in the cylinders explodes instead of burning normally.

2-57. Answer C. GFDPP 2B, PHB

Detonation occurs when the fuel/air mixture suddenly explodes in the cylinders instead of burning smoothly. Detonation is caused by excessively lean mixtures, while hot spots in the cylinder cause preignition.

2-58   PA.I.G.K1c, PA.II.C.K1

If it becomes necessary to hand-prop an airplane engine, it is extremely important that a competent pilot

A – call "contact" before touching the propeller.

B – be at the controls in the cockpit.

C – be in the cockpit and call out all commands.

2-58. Answer B. GFDPP 2B, PHB

When hand-propping an airplane, a qualified pilot must be at the controls to prevent the airplane from moving and to set the engine controls properly. The person who is turning the propeller calls out the commands.

2-59   PA.I.G.K1c

Detonation may occur at high-power settings as

A – the fuel mixture ignites instantaneously instead of burning progressively and evenly.

B – an excessively rich fuel mixture causes an explosive gain in power.

C – the fuel mixture is ignited too early by hot carbon deposits in the cylinder.

2-59. Answer A. GFDPP 2B, PHB

Detonation occurs when the fuel/air mixture suddenly explodes in the cylinders instead of burning smoothly. Detonation is caused by excessively lean mixtures, while hot spots in the cylinder cause preignition.

2-60   PA.I.G.K1c

You suspect that your engine (with a fixed-pitch propeller) is detonating during climbout after takeoff. Your initial corrective action would be to

A – lean the mixture.

B – lower the nose slightly to increase airspeed.

C – apply carburetor heat.

2-60. Answer B. GFDPP 2B, PHB

Detonation can occur when the engine overheats. During a climb, one action that you can take to help cool the engine is to increase airspeed by lowering the nose. This action will reduce the rate of climb and increase the cooling airflow around the engine.

2-61　PA.I.G.K1c, K2

The uncontrolled firing of the fuel/air charge in advance of normal spark ignition is known as

A – combustion.

B – preignition.

C – detonation.

2-61. Answer B. GFDPP 2B, PHB

Preignition occurs when the fuel/air mixture is ignited in advance of the normal timed ignition.

2-62　PA.I.G.K1c

Which would most likely cause the cylinder head temperature and engine oil temperature gauges to exceed their normal operating ranges?

A – Using fuel that has a lower-than-specified fuel rating.

B – Using fuel that has a higher-than-specified fuel rating.

C – Operating with higher-than-normal oil pressure.

2-62. Answer A. GFDPP 2B, PHB

Lower grade fuels detonate under less pressure. Using fuel with a lower fuel rating than specified can cause detonation, which in turn leads to excessive engine temperatures.

2-63　PA.II.C.K1, K2, K3, K4

Which is true regarding starting a fuel-injected engine?

A – Vapor lock typically occurs in cold weather conditions if the engine has not been started for a significant period of time.

B – Using the electric/auxiliary fuel pump to prime the engine can help purge vapors from the injection system.

C – A typical procedure for starting a flooded engine involves initially setting the throttle to idle and setting the mixture control to the idle-cutoff position.

2-63. Answer B. GFDPP 2B, PHB

Priming the engine provides a combustible mixture for starting and purges any vapors, or air pockets, from the injection system. A fuel-injected engine can be difficult to restart a few minutes after shutdown—the residual heat in the engine can boil the fuel in the injection system's lines and components, and the resulting bubbles can cause vapor lock, which interferes with fuel metering and pumping.

The objective of the flooded engine starting procedure is to purge the excess fuel from the cylinders until you obtain a combustible fuel/air ratio. This procedure typically involves opening the throttle fully to provide maximum airflow through the cylinders and setting the mixture control to the idle-cutoff position before cranking the engine.

SECTION B ■ **Powerplant and Related Systems**

2-64   PA.I.G.K1c, K1e

What type fuel can be substituted for an aircraft if the recommended octane is not available?

A – The next higher octane aviation gas.

B – The next lower octane aviation gas.

C – Unleaded automotive gas of the same octane rating.

2-64. Answer A. GFDPP 2B, PHB

Typically, if you follow the manufacturer's recommendations, you may use the next higher grade of fuel. You may not use automotive gas unless the aircraft has a supplemental type certificate (STC) permitting its use.

2-65   PA.I.G.K1e

Filling the fuel tanks after the last flight of the day is considered a good operating procedure because this will

A – force any existing water to the top of the tank away from the fuel lines to the engine.

B – prevent expansion of the fuel by eliminating airspace in the tanks.

C – prevent moisture condensation by eliminating airspace in the tanks.

2-65. Answer C. GFDPP 2B, PHB

As the airplane cools overnight, water condenses in the tanks from vapor in the air and enters the fuel. Filling the fuel tanks eliminates the airspace and prevents this condensation buildup.

2-66   PA.I.G.K1e

To properly purge water from the fuel system of an aircraft equipped with fuel tank sumps and a fuel strainer quick drain, it is necessary to drain fuel from the

A – fuel strainer drain.

B – lowest point in the fuel system.

C – fuel strainer drain and the fuel tank sumps.

2-66. Answer C. GFDPP 2B, PHB

Fuel needs to be drained from both the fuel tank sumps and the fuel strainer drain. The fuel strainer drain might be the lowest point in the fuel system, but draining it without draining the fuel tank sumps is insufficient.

2-67   PA.I.G.K1e, K2

An abnormally high engine oil temperature indication may be caused by

A – the oil level being too low.

B – operating with a too high viscosity oil.

C – operating with an excessively rich mixture.

2-67. Answer A. GFDPP 2B, PHB

Engine oil lubricates moving parts, reduces friction, and carries heat from interior portions of the engine to the oil cooler. A low oil level is less effective at cooling the engine indicated by high engine oil temperature.

2-68   PA.I.G.K1c

What action can you take to aid in cooling an engine that is overheating during a climb?

A – Reduce rate of climb and increase airspeed.

B – Reduce climb speed and increase RPM.

C – Increase climb speed and increase RPM.

2-68. Answer A. GFDPP 2B, PHB
During a climb, one action that you can take to help cool the engine is to increase airspeed by lowering the nose. This action will reduce the rate of climb and increase the cooling airflow around the engine.

2-69   PA.I.G.K1c, K2

Excessively high engine temperatures, either in the air or on the ground, will

A – increase fuel consumption and may increase power due to the increased heat.

B – result in damage to heat-conducting hoses and warping of cylinder cooling fans.

C – cause loss of power, excessive oil consumption, and possible permanent internal engine damage.

2-69. Answer C. GFDPP 2B, PHB
High temperatures can cause detonation and a resulting loss of power, excessive oil consumption, and engine damage, including scoring of the cylinders and damage to piston, rings, and valves.

2-70   PA.I.G.K1c

For internal cooling, reciprocating aircraft engines are especially dependent on

A – a properly functioning thermostat.

B – air flowing over the exhaust manifold.

C – the circulation of lubricating oil.

2-70. Answer C. GFDPP 2B, PHB
Engine oil lubricates moving parts, reduces friction, and carries heat from interior portions of the engine to the oil cooler.

2-71   PA.I.G.K1c, K2

What is one procedure to aid in cooling an engine that is overheating?

A – Enrichen the fuel mixture.

B – Increase the RPM.

C – Reduce the airspeed.

2-71. Answer A. GFDPP 2B, PHB
A richer fuel mixture burns at a lower temperature and reduces heat in the engine. Increasing the RPM does not significantly affect the airflow through the engine and on a fixed-pitch propeller, increasing power worsens the overheating problem. Reducing airspeed reduces the airflow through the engine and worsens the problem.

SECTION B ■ **Powerplant and Related Systems**

2-72   PA.I.G.K1i, K2

If you smell exhaust gases in flight, assume that

A – Carbon monoxide is present—close the fresh air vents and turn the heater on so filtered air enters the cabin.

B – Carbon monoxide is present—turn off the heater, open the fresh air vents, and use supplemental oxygen, if it is available.

C – The cabin heat is working properly because heat is transferred to incoming air from the hot engine exhaust gases.

2-72. Answer B. GFDPP 2B, PHB

In many light airplanes, the exhaust system is used to provide heat for the cabin—hot exhaust gases heat the muffler and metal shrouds around the muffler capture the heat and duct it to the cabin. If the muffler is cracked, exhaust gases can enter the cabin. If you smell exhaust gases, you can assume that carbon monoxide is present. Turn off the heater, open the fresh air vents, and use supplemental oxygen, if it is available.

2-73   PA.I.G.K1c

How is engine operation controlled on an engine equipped with a constant-speed propeller?

A – The throttle controls power output as registered on the manifold pressure gauge and the propeller control regulates engine RPM.

B – The throttle controls power output as registered on the manifold pressure gauge and the propeller control regulates a constant blade angle.

C – The throttle controls engine RPM as registered on the tachometer and the mixture control regulates the power output.

2-73. Answer A. GFDPP 2B, PHB

The throttle controls the power output of the engine, which is indicated on the manifold pressure gauge. The propeller control changes engine RPM, which is indicated on the tachometer, by adjusting the pitch of the propeller.

2-74   PA.I.G.K1c

What is an advantage of a constant-speed propeller?

A – Permits the pilot to select and maintain a desired cruising speed.

B – Permits the pilot to select the blade angle for the most efficient performance.

C – Provides a smoother operation with stable RPM and eliminates vibrations.

2-74. Answer B. GFDPP 2B, PHB

By setting propeller RPM, you indirectly set blade angle. Selecting the optimal blade angle enables a high percentage of engine power to be converted into thrust over a wide range of RPM and airspeed combinations, which enables the best performance to be gained from the engine.

2-75   PA.I.G.K1c

A precaution for the operation of an engine equipped with a constant-speed propeller is to

A – avoid high RPM settings with high manifold pressure.

B – avoid high manifold pressure settings with low RPM.

C – always use a rich mixture with high RPM settings.

2-75. Answer B. GFDPP 2B, PHB

For a given RPM setting, there is a maximum allowable manifold pressure. Generally, you should avoid high manifold pressures with low RPM to prevent internal stress within the engine.

2-76   PA.I.G.K1c

Which preflight checks should you make on an airplane with full-authority digital engine control (FADEC)?

A – Verify proper operation of both ECUs as well as their back-up power sources.

B – Ensure that engine RPM drops sufficiently when the propeller control is pulled back.

C – Ensure that the mixture control is adjusted for highest RPM with smooth engine operation.

2-76. Answer A. GFDPP 2B, PHB

On an airplane equipped with FADEC, the engine control units (ECUs) completely manage the engine and propeller operation, eliminating the need for a propeller or mixture control. Because the engine stops running if an ECU fails, redundant systems are required with back-up power. These systems should be checked for proper operation before flight.

2-77   PA.II.C.K2

Which procedures are best practices that can aid in starting your airplane's engine in cold temperatures?

A – Preheating; hand propping.

B – Priming; following vapor lock procedure.

C – Preheating; using external power.

2-77. Answer C. GFDPP 2B, PHB

In cold weather—then air temperatures are below 20°F (–6°C)—the AFM/POH typically recommend using an external preheater and/or an external power source to start the engine. Hand-propping is dangerous and is generally discouraged. Vapor lock is associated with high engine temperatures when bubbles of fuel vapor form in the fuel lines or fuel pump.

SECTION B ■ **Powerplant and Related Systems**

NOTE: The following questions refer to Emergency Operations. Although Chapter 2 Section B, of the GFD Private Pilot textbook addresses content regarding systems on the airplane, Private Pilot Maneuvers, Chapter 4 — Emergency Operations covers equipment malfunctions and maneuvers such as an emergency descent.

2-78   PA.I.G.K1f, PA.IX.C.K2a

Which is an action to manage an alternator failure?

A – Shed electrical load to preserve battery power for essential equipment.

B – Prepare to perform an emergency approach and landing.

C – Turn off the master switch to avoid a fire and prepare to land without radio communication, lights, or flaps.

2-78. Answer A. GFDPPM, PHB

Consider these actions to manage an alternator failure:

- Shed electrical load to preserve battery power for essential equipment.

- Notify ATC of the situation and use ATC services.

- Be prepared to control the airplane and land without digital instruments, radio communication, lights, or flaps, and to perform a manual landing gear extension, if applicable.

2-79   PA.IX.A.K1

Which situations would require an emergency descent?

A – Loss of cabin pressurization; engine failure after takeoff; smoke on the flight deck.

B – Smoke on the flight deck; alternator failure; engine fire.

C – Loss of cabin pressurization; engine fire; smoke on the flight deck.

2-79. Answer C. GFDPPM, AFH

An emergency descent is a maneuver for descending to a lower altitude as rapidly as possible within the structural limitations of the airplane. You might need to perform an emergency descent due to an uncontrollable engine or wing fire, smoke in the cockpit, a sudden loss of cabin pressurization, or any other situation that demands an immediate rapid descent.

2-80   PA.IX.A.K2, K3, K4

Steps for performing an emergency descent typically include to establish

A – best glide speed.

B – a standard-rate turn.

C – a descending turn at 30° to 45°.

2-80. Answer C. GFDPPM, AFH

Steps to perform an emergency descent typically include:

1. Configure the airplane to descend.

2. Establish a descending turn at 30° to 45°.

3. Maintain the bank angle and maximum allowable airspeed (V_{NE}, V_A, V_{FE} or V_{LE}).

4. Return to straight-and-level flight.

5. Return to cruise flight or prepare for landing.

2-81   PA.IX.C.K3

Turning off the master switch is an appropriate immediate action to take when

A – smoke or fire exists in the cabin.

B – the engine is running roughly.

C – you want to power only essential equipment after an alternator failure.

2-81. Answer A. GFDPPM, AFH

Managing smoke and fire in the cabin typically involves these actions:

- Turn off the master switch to remove the possible source of the fire.

- If flames exist, use the fire extinguisher to put out the fire.

- After extinguishing the fire, open the air vents to clear the cabin of smoke and fumes.

- Land as soon as possible.

2-82   PA.IX.C.K2f

Immediately grasping the control wheel and pulling the applicable circuit breakers as recommended by the AFM/POH is an action to manage

A – runaway trim.

B – loss of elevator control due to linkage failure.

C – a flap malfunction.

2-82. Answer A. GFDPPM, AFH

An aural and/or visual warning normally signals you to an electric trim malfunction associated with either manual or autopilot trim. If you experience a trim malfunction, immediately grasp the control wheel to control the airplane. If you cannot correct the problem and are experiencing runaway trim, pull the applicable circuit breakers as recommended by the AFM/POH and land as soon as practical.

If the linkage between the cabin and the elevator fails in flight—leaving the elevator free to weathervane in the wind—you typically can use trim to raise or lower the elevator within limits.

2-83   PA.IX.C.K2e

If the flaps do not extend as you prepare for landing

A – the glide path is not as steep and floating during the flare is likely.

B – the landing distance will be decreased.

C – the glide path is steeper so be alert to errors in your judgment of height and distance.

2-83. Answer A. GFDPPM, AFH

Consider these factors as you prepare to perform a no-flap approach and landing:

- The glide path is not as steep as with flaps extended, so the higher nose attitude on final can cause errors in your judgment of height and distance.

- Landing distance is substantially increased.

- Floating during the flare is likely.

SECTION B ■ Powerplant and Related Systems

SECTION C
Flight Instruments

PITOT-STATIC INSTRUMENTS

- Pitot-static instruments measure air pressure, and are also affected by temperature.

- The standard temperature and pressure at sea level are 59°F (15°C) and 29.92 inches Hg (1013.2 hPa).

- The pitot-static system supplies ambient air pressure to operate the altimeter and vertical speed indicator (VSI), and both ambient and ram air (impact pressure) to operate the airspeed indicator.

- The static system is connected to three instruments—the altimeter, VSI, and airspeed indicator.

- If the pitot tube becomes clogged, only the airspeed indicator is affected; if the static vents are clogged, the altimeter, airspeed indicator, and vertical speed indicator are all affected.

AIRSPEED INDICATOR

- The red line on an airspeed indicator represents never-exceed speed—the maximum speed at which the airplane may be operated under any conditions.

- The yellow arc is the caution range—operating at these speeds is permissible only in smooth air.

- The green arc is the normal operating range, with the bottom of the arc representing the power-off stalling speed in a specified configuration, and the upper limit representing the maximum structural cruising speed.

- The white arc identifies the normal flap operating range. The bottom of the white arc is the power-off stall speed in the landing configuration.

- The indicated airspeed at which a given airplane stalls in a particular configuration is not affected by altitude.

- The upper and lower limits of the colored arcs correspond to some airspeed limitations, called V-speeds.

   - V_{FE}—maximum flap operating speed.
   - V_{LE}—maximum landing gear extended speed.
   - V_{NO}—maximum structural cruising speed.
   - V_{S0}—stalling speed or minimum steady flight speed in the landing configuration.

- V_A is defined as maneuvering speed—an airspeed limitation that is not marked on the airspeed indicator. Avoid abrupt control movements above V_A, and also limit your airplane to V_A to protect the airframe during significant turbulence.

TYPES OF ALTITUDE

- Types of altitude include:

   - True altitude—vertical distance of the aircraft above sea level.
   - Absolute altitude—vertical distance of the aircraft above the surface.
   - Pressure altitude—altitude indicated when the barometric pressure scale is set to 29.92.
   - Density altitude—pressure altitude corrected for nonstandard temperature.
   - Indicated altitude—same as true altitude when at sea level under standard conditions.
   - Pressure altitude—true altitude under standard conditions.

- The altimeter setting is the value to which the barometric pressure scale of the altimeter is set so that the altimeter indicates true altitude at field elevation.

- One inch of change of Hg in the altimeter causes 1,000 feet of altitude change in the same direction.

- Variations in temperature affect the altimeter—on warm days, pressure levels are raised and the indicated altitude is lower than the true altitude.

- The true altitude of an aircraft is lower than the indicated altitude when you fly into an area of colder-than-standard air temperature, or an area of lower pressure.

- An increase in ambient temperature increases the density altitude at a given airport.

GYROSCOPIC INSTRUMENTS

- Gyroscopic instruments include the turn coordinator, attitude indicator and heading indicator, which operate off of a gyro's tendency to remain rigid in space.

- A turn coordinator provides an indication of the rate of movement of an aircraft about the yaw and roll axes.

- To properly adjust the attitude indicator during level flight, align the miniature airplane to the horizon bar.

- You determine the direction of bank from the attitude indicator by the relationship of the miniature airplane to the deflected horizon bar.

- You must periodically realign a nonslaved heading indicator with the magnetic compass as the gyro precesses.

- In an airplane with analog instruments, a vacuum/pressure system malfunction affects the attitude indicator and heading indicator.

MAGNETIC COMPASS

- The magnetic compass contains a bar magnet that swings freely to align with the Earth's magnetic field.

- The angular difference between the true and magnetic poles at a given point is referred to as magnetic variation.

- Deviation in a magnetic compass is caused by the magnetic fields in the aircraft distorting the lines of magnetic force.

- In the Northern Hemisphere, a magnetic compass shows a turn toward the west if rolling into a right turn from a north heading, and a turn toward the east if rolling into a left turn.

- While on an east or west heading, the compass indicates a turn toward the north if the aircraft is accelerating, and a turn toward the south the aircraft is decelerating.

- The compass indicates correctly if the aircraft is accelerating or decelerating on a north or south heading.

- Generally, in-flight indications of a magnetic compass are correct only when the aircraft is in straight-and-level, unaccelerated flight.

ELECTRONIC FLIGHT DISPLAYS

- The electronic flight instrument systems used in light general aviation airplanes typically have two screens: a primary flight display (PFD) and a multifunction display (MFD).

- The PFD contains digital versions of traditional flight instruments, including the attitude indicator, airspeed indicator, altimeter, vertical speed indicator, and HSI.

- The attitude and heading reference system (AHRS) uses inertial sensors such as electronic gyroscopes, accelerometers, and a magnetometer to determine the attitude of an aircraft relative to the horizon and heading.

- A magnetometer senses the earth's magnetic field to function as a magnetic compass, but without some of the errors associated with a conventional compass.

- The AHRS instruments are the attitude indicator, HSI and turn indicator.

- A slip/skid indicator below the roll pointer of the attitude indicator helps you maintain coordinated flight.

SECTION C ■ **Flight Instruments**

- In addition to a compass card, the digital HSI displays the current airplane heading in a window, a course indicator arrow and CDI, and a rate-of-turn indicator.

- When the AHRS system detects a problem, it places a red X over the display of the affected instruments to alert you that the indications are unreliable.

- The pitot tube, static source, and outside air temperature probe provide information to the air data computer (ADC), which drives the airspeed indicator, altimeter, and VSI.

- On the digital airspeed indicator, a window and a pointer on the airspeed tape show the indicated airspeed.

- A trend vector on the digital airspeed indicator shows the future airspeed (typically reached in six seconds) if the acceleration or deceleration continues at the same rate.

- The trend vector on the digital altimeter shows the future altitude (typically reached in six seconds) if it continues to climb or descend at the same rate.

- The trend vector on the heading indicator (HSI) shows the future heading (typically reached in six seconds) if the aircraft continues turning at the same rate. This trend vector acts as a rate-of-turn indicator with calibration marks for a standard-rate turn.

- On the digital VSI, you read vertical speed in feet per minute in a window that also serves as a pointer.

- If one or more sensors stops providing input, or if the ADC determines that its own internal operations are not correct, it places a red X over the display of the affected instrument.

- On an electronic flight display, if the PFD fails, you can typically display PFD instruments on the MFD screen.

- If you lose all electric power to a digital flight display, you must use your back-up instruments. The back-up attitude indicator is vacuum- or electrically-powered. If it is electrically powered, that instrument must have a separate power source that is isolated from the main electrical system.

NOTE: *An asterisk appearing after an ACS code (i.e. PA.VII.B.K1*) indicates that the question subject appears more than one time in the ACS. The code shown corresponds to the first instance of the subject in the ACS.*

2-84   PA.I.G.K1h

If an altimeter setting is not available before flight, to which altitude should you adjust the altimeter?

A – The elevation of the nearest airport corrected to mean sea level.

B – The elevation of the departure area.

C – Pressure altitude corrected for nonstandard temperature.

2-84. Answer B. GFDPP 2C, FAR 91.121, PHB
If unable to obtain a local altimeter setting, you should set the altimeter to the departure field elevation as shown in flight information sources. Field elevation is already referenced to mean sea level (true altitude) and does not need to be corrected. Use the elevation of your airport, not a nearby airport. Pressure altitude corrected for nonstandard temperature is density altitude, which should not be displayed on an altimeter.

2-85   PA.I.G.K1h

Prior to takeoff, the altimeter should be set to

A – the current local altimeter setting, if available, or the departure airport elevation.

B – the corrected density altitude of the departure airport.

C – the corrected pressure altitude for the departure airport.

2-85. Answer A. GFDPP 2C, FAR 91.121, PHB
If unable to obtain a local altimeter setting, you should set the altimeter to the departure field elevation as shown in flight information sources. Pressure altitude corrected for nonstandard temperature is density altitude, which should not be displayed on an altimeter. Corrected density altitude and corrected pressure altitudes are not valid terms.

2-86   PA.I.G.K1h, PA.IX.C.K2c, PA.VIII.A.K1a*, K1c*

If the pitot tube and outside static vents become clogged, which instruments would be affected?

A – The altimeter, airspeed indicator, and turn-and-slip indicator.

B – The altimeter, airspeed indicator, and vertical speed indicator.

C – The altimeter, attitude indicator, and turn-and-slip indicator.

2-86. Answer B. GFDPP 2C, PHB
The altimeter, the airspeed indicator, and the vertical speed indicator all use static air and would therefore be affected. The turn-and-slip indicator and attitude indicator are gyroscopic instruments and are not affected by any clogs in the pitot-static system.

2-87   PA.I.G.K1h, PA.IX.C.K2c, PA.VIII.A.K1a*, K1c*

Which instrument becomes inoperative if the pitot tube becomes clogged?

A – Altimeter.

B – Vertical speed.

C – Airspeed.

2-87. Answer C. GFDPP 2C, PHB
The airspeed indicator is the only pitot-static instrument that is connected to the pitot tube. It operates by sensing impact pressure (ram air) in the pitot tube and comparing it to static pressure.

2-88   PA.I.G.K1h, PA.VIII.A.K1c*

The pitot system provides impact pressure for which instrument?

A – Altimeter.

B – Vertical-speed indicator.

C – Airspeed indicator.

2-88. Answer C. GFDPP 2C, PHB
The airspeed indicator is the only pitot-static instrument that is connected to the pitot tube. It operates by sensing impact pressure (ram air) in the pitot tube and comparing it to static pressure.

2-89   PA.I.G.K1h, PA.VIII.A.K1a*, K1c*

Which instrument(s) become inoperative if the static vents become clogged?

A – Airspeed only.

B – Altimeter only.

C – Airspeed, altimeter, and vertical speed.

2-89. Answer C. GFDPP 2C, PHB
The altimeter, the airspeed indicator, and the vertical speed indicator all use static air and would therefore be affected.

2-90   PA.I.G.K1h, PA.V.A.K2c, PA.IX.A.K4

Which V-speed represents maneuvering speed?

A – V_A.

B – V_{LO}.

C – V_{NE}.

2-90. Answer A. GFDPP 2C, FAR 1.2, PHB
V_A is defined as the design maneuvering speed. V_{LO} is maximum gear operating speed and V_{NE} is never-exceed speed.

SECTION C ■ **Flight Instruments**

2-91   PA.I.G.K1h, PA.I.G.K1d, PA.IX.C.K2e

Which V-speed represents maximum landing gear extended speed?

A – V_{LE}.

B – V_{LO}.

C – V_{FE}.

2-91. Answer A. GFDPP 2C, FAR 1.2, PHB

V_{LE} is defined as maximum landing gear extended speed. V_{LO} is maximum gear operating speed and V_{FE} is maximum flap extended speed.

2-92   PA.I.G.K1h, PA.IX.A.K3

V_{NO} is defined as the

A – normal operating range.

B – never-exceed speed.

C – maximum structural cruising speed.

2-92. Answer C. GFDPP 2C, FAR 1.2, PHB

V_{NO} is defined as maximum structural cruising speed. It is at the upper end of the green arc— the normal operating range—on the airspeed indicator.

2-93   PA.I.G.K1h

Which V-speed represents maximum flap extended speed?

A – V_{FE}.

B – V_{LOF}.

C – V_{FC}.

2-93. Answer A. GFDPP 2C, FAR 1.2, PHB

V_{FE} is defined as maximum flap extended speed.

2-94   PA.I.G.K1h, PA.VII.B.K1*

V_{S0} is defined as the

A – stalling speed or minimum steady flight speed in the landing configuration.

B – stalling speed or minimum steady flight speed in a specified configuration.

C – stalling speed or minimum takeoff safety speed.

2-94. Answer A. GFDPP 2C, FAR 1.2, PHB

V_{S0} is stalling speed or minimum steady flight speed in a landing configuration. V_{S1} is the stalling speed in a specified configuration. Takeoff safety speed (V_1) normally does not apply to light general aviation airplanes.

2-95   PA.I.G.K1b, K1h

(Refer to Figure 4.) What is the full flap operating range for the airplane?

A – 55 to 100 knots.

B – 55 to 208 knots.

C – 55 to 165 knots.

2-95. Answer A. GFDPP 2C, PHB

The white arc indicates the flap operating range. On this instrument, it is 55 to 100 knots.

2-96   PA.I.G.K1h, PA.VII.B.K1*

As altitude increases, the indicated airspeed at which a given airplane stalls in a particular configuration will

A – decrease as the true airspeed decreases.

B – decrease as the true airspeed increases.

C – remain the same regardless of altitude.

2-96. Answer C. GFDPP 2C, PHB
Because airspeed indicators are calibrated to read true airspeed only under standard sea level conditions, the indicated airspeed does not reflect lower air density at higher altitudes. As a result, the indicated airspeed of a stall remains the same.

2-97   PA.I.G.K1h

What does the red line on an airspeed indicator represent?

A – Maneuvering speed.

B – Turbulence or rough-air speed.

C – Never-exceed speed.

2-97. Answer C. GFDPP 2C, PHB
The red line on an airspeed indicator is the never-exceed speed.

2-98   PA.I.G.K1h

(Refer to Figure 4.) Which marking identifies the never-exceed speed?

A – Upper limit of the green arc.

B – Upper limit of the white arc.

C – The red radial line.

2-98. Answer C. GFDPP 2C, PHB
The red line is the never-exceed speed (V_{NE}), the yellow arc is the caution range, the green arc is the normal operating range, and the white arc is the flap operating range.

2-99   PA.I.G.K1h

(Refer to Figure 4.) Which color identifies the power-off stalling speed in a specified configuration?

A – Upper limit of the green arc.

B – Upper limit of the white arc.

C – Lower limit of the green arc.

2-99. Answer C. GFDPP 2C, PHB
The lower limit of the green arc represents the power-off stall speed in a specified configuration (V_{S1}), which is usually flaps up and gear retracted.

2-100   PA.I.G.K1h

(Refer to Figure 4.) The maximum speed at which the airplane can be operated in smooth air is

A – 100 knots.

B – 165 knots.

C – 208 knots.

2-100. Answer C. GFDPP 2C, PHB
In smooth air, an airplane can be operated in the yellow arc up to the red line, in this case, 208 knots.

SECTION C ■ **Flight Instruments**

2-101   PA.I.G.K1b, K1h

(Refer to Figure 4.) What is the maximum flaps-extended speed?

A – 58 knots.

B – 100 knots.

C – 165 knots.

2-101. Answer B. GFDPP 2C, PHB

The maximum flaps-extended speed (V_{FE}) is represented by the upper limit of the white arc, which in this case is 100 knots.

2-102   PA.I.G.K1h

(Refer to Figure 4.) What is the caution range of the airplane?

A – 0 to 55 knots.

B – 100 to 165 knots.

C – 165 to 208 knots.

2-102. Answer C. GFDPP 2C, PHB

The yellow arc indicates the caution range. On this instrument, it is 165 to 208 knots.

2-103   PA.I.G.K1h, PA.V.A.K2c

What is an important airspeed limitation that is not color coded on airspeed indicators?

A – Never-exceed speed.

B – Maneuvering speed.

C – Maximum structural cruising speed.

2-103. Answer B. GFDPP 2C, PHB

Maneuvering speed (V_A) is not shown on the airspeed indicator as it varies with weight. You can find V_A in the AFM/POH or on placards in the airplane.

2-104   PA.I.G.K1h

(Refer to Figure 4.) Which color identifies the normal flap operating range?

A – The lower limit of the white arc to the upper limit of the green arc.

B – The green arc.

C – The white arc.

2-104. Answer C. GFDPP 2C, PHB

The white arc indicates the normal flap operating range.

2-105   PA.I.G.K1h

(Refer to Figure 4.) What is the maximum structural cruising speed?

A – 100 knots.

B – 165 knots.

C – 208 knots.

2-105. Answer B. GFDPP 2C, PHB

Maximum structural cruising speed (V_{NO}) is indicated by the upper limit of the green arc, which in this case is 165 knots.

2-106   PA.I.G.K1h

(Refer to Figure 4.) Which color identifies the power-off stalling speed with wing flaps and landing gear in the landing configuration?

A – Upper limit of the green arc.

B – Upper limit of the white arc.

C – Lower limit of the white arc.

2-106. Answer C. GFDPP 2C, PHB

Power-off stalling speed in the landing configuration (V_{S0}) is represented by the lower limit of the white arc.

2-107   PA.I.G.K1h, PA.VIII.A.K1d

(Refer to Figure 3.) Altimeter 1 indicates

A – 500 feet.

B – 1,500 feet.

C – 10,500 feet.

2-107. Answer C. GFDPP 2C, PHB

The small 10,000 ft pointer is just above the 1, indicating that the altitude is above 10,000 feet. The wide 1,000 ft pointer is between 0 and 1, which indicates less than 1,000 feet. Finally, the 100 ft pointer is on 5. Therefore, the altimeter reading is 10,500 feet.

2-108   PA.I.G.K1h, PA.VIII.A.K1d

(Refer to Figure 3.) Altimeter 2 indicates

A – 1,500 feet.

B – 4,500 feet.

C – 14,500 feet.

2-108. Answer C. GFDPP 2C, PHB

The 10,000 ft pointer is above 1, the 1,000 ft pointer is above 4, and the 100 ft pointer is on 5. This indicates an altitude of 14,500 feet.

2-109   PA.I.G.K1h, PA.VIII.A.K1d

(Refer to Figure 3.) Altimeter 3 indicates

A – 9,500 feet.

B – 10,950 feet.

C – 15,940 feet.

2-109. Answer A. GFDPP 2C, PHB

The 10,000 ft pointer is near 1, the 1,000 ft pointer is above 9, and the 100 ft pointer is on 5. This indicates the altitude is 9,500 feet.

2-110   PA.I.G.K1h, PA.VIII.A.K1d

(Refer to Figure 3.) Which altimeter(s) indicate(s) more than 10,000 feet?

A – 1, 2, and 3.

B – 1 and 2 only.

C – 1 only.

2-110. Answer B. GFDPP 2C, PHB

- Altimeter 1—10,500 feet. The small 10,000 ft pointer is just above the 1, which indicates that the altitude is above 10,000 feet. The wide 1,000 ft pointer is between 0 and 1, which indicates less than 1,000 feet. The 100 ft pointer is on 5.

- Altimeter 2—14,500 feet. The 10,000 ft pointer is above 1, the 1,000' pointer is above 4, and the 100' pointer is on 5.

- Altimeter 3— 9,500 feet. The 10,000 ft pointer is near 1, the 1,000' pointer is above 9, and the 100 ft pointer is on 5. This indicates the altitude is.

2-111   PA.I.G.K1h

Altimeter setting is the value to which the barometric pressure scale of the altimeter is set so the altimeter indicates

A – calibrated altitude at field elevation.

B – absolute altitude at field elevation.

C – true altitude at field elevation.

2-111. Answer C. GFDPP 2C, AW, PHB

When the current altimeter setting is set on the ground, the altimeter reads true altitude of the field, which is the actual height above mean sea level. Calibrated altitude is not a valid term and absolute altitude is height above ground level (AGL). If the airplane is sitting on the ground at the field, the absolute altitude would be zero.

2-112   PA.I.F.K2a

If the outside air temperature (OAT) at a given altitude is warmer than standard, the density altitude is

A – equal to pressure altitude.

B – lower than pressure altitude.

C – higher than pressure altitude.

2-112. Answer C. GFDPP 2C, PHB

At standard temperature, density altitude is equal to pressure altitude. However, when the ambient air temperature is above standard, the air density is reduced, density altitude is higher than pressure altitude, and aircraft performance degrades.

2-113   PA.I.G.K1h

How do variations in temperature affect the altimeter?

A – Pressure levels are raised on warm days and the indicated altitude is lower than true altitude.

B – Higher temperatures expand the pressure levels and the indicated altitude is higher than true altitude.

C – Lower temperatures lower the pressure levels and the indicated altitude is lower than true altitude.

2-113. Answer A. GFDPP 2C, PHB

Because atmospheric pressure levels are raised on warm days, the aircraft is at a higher altitude than indicated. In other words, the indicated altitude is lower than true altitude. When lower temperatures lower the pressure levels, the indicated altitude is higher, not lower, than true altitude.

2-114   PA.I.G.K1h

What is true altitude?

A – The vertical distance of the aircraft above sea level.

B – The vertical distance of the aircraft above the surface.

C – The height above the standard datum plane.

2-114. Answer A. GFDPP 2C, PHB

True altitude is the actual height (vertical distance) above mean sea level. The vertical distance above the surface is absolute altitude and the height above the standard datum plane is pressure altitude.

2-115   PA.I.G.K1h

What is absolute altitude?

A – The altitude read directly from the altimeter.

B – The vertical distance of the aircraft above the surface.

C – The height above the standard datum plane.

2-115. Answer B. GFDPP 2C, PHB

Absolute altitude, often referred to as altitude above ground level (AGL), is the height (vertical distance) above the surface. The altitude read directly from the altimeter is indicated altitude, and the height above the standard datum plane is pressure altitude.

2-116   PA.I.G.K1h

What is density altitude?

A – The height above the standard datum plane.

B – The pressure altitude corrected for nonstandard temperature.

C – The altitude read directly from the altimeter.

2-116. Answer B. GFDPP 2C, PHB

You can find density altitude by applying a correction for nonstandard temperature to the pressure altitude. Pressure altitude is the height above the standard datum plane, which is indicated when you set 29.92 on the scale in the altimeter setting window. Indicated altitude is the altitude read directly from the altimeter.

2-117   PA.I.G.K1h

What is pressure altitude?

A – The indicated altitude corrected for position and installation error.

B – The altitude indicated when the barometric pressure scale is set to 29.92.

C – The indicated altitude corrected for nonstandard temperature and pressure.

2-117. Answer B. GFDPP 2C, PHB

Pressure altitude is the height above the standard datum plane, which is indicated when you set 29.92 on the scale in the altimeter setting window.

2-118   PA.I.G.K1h

Under what condition is pressure altitude and density altitude the same value?

A – At sea level, when the temperature is 0°F.

B – When the altimeter has no installation error.

C – At standard temperature.

2-118. Answer C. GFDPP 2C, PHB

Because density altitude is pressure altitude corrected for nonstandard temperature, density altitude and pressure altitude are equal only at standard temperature.

2-119   PA.I.G.K1h

Which condition would cause the altimeter to indicate a lower altitude than true altitude?

A – Air temperature lower than standard.

B – Atmospheric pressure lower than standard.

C – Air temperature warmer than standard.

2-119. Answer C. GFDPP 2C, PHB

When the air temperature is warmer than standard, indicated altitude is lower than actual (true) altitude.

SECTION C ■ **Flight Instruments**

2-120   PA.I.G.K1h

Under what condition will true altitude be lower than indicated altitude?

A – In colder than standard air temperature.

B – In warmer than standard air temperature.

C – When density altitude is higher than indicated altitude.

2-120. Answer A. GFDPP 2C, PHB

When the air is colder than standard, the actual (true) altitude of an aircraft is lower than indicated.

2-121   PA.I.G.K1h

If a flight is made from an area of high pressure into an area of low pressure without the altimeter setting being adjusted, the altimeter will indicate

A – lower than the actual altitude above sea level.

B – higher than the actual altitude above sea level.

C – the actual altitude above sea level.

2-121. Answer B. GFDPP 2C, PHB

The aircraft is at a lower true (actual) altitude above sea level than is indicated because the altimeter indicates higher than actual altitude. Remember, "from high to low, look out below."

2-122   PA.I.G.K1h

If a flight is made from an area of low pressure into an area of high pressure without the altimeter setting being adjusted, the altimeter will indicate

A – the actual altitude above sea level.

B – higher than the actual altitude above sea level.

C – lower than the actual altitude above sea level.

2-122. Answer C. GFDPP 2C, PHB

The aircraft is at a higher true (actual) altitude above sea level than is indicated because the altimeter indicates lower than the actual altitude.

2-123   PA.I.G.K1h

Under which condition will pressure altitude be equal to true altitude?

A – When the atmospheric pressure is 29.92 inches Hg.

B – When standard atmospheric conditions exist.

C – When indicated altitude is equal to the pressure altitude.

2-123. Answer B. GFDPP 2C, PHB

Pressure altitude equals true altitude only when standard atmospheric conditions of pressure and temperature exist.

2-124   PA.I.G.K1h

Under what condition is indicated altitude the same as true altitude?

A – If the altimeter has no mechanical error.

B – When at sea level under standard conditions.

C – When at 18,000 feet MSL with the altimeter set at 29.92.

2-124. Answer B. GFDPP 2C, PHB

When properly set, the altimeter indicates true altitude as closely as possible. However, some amount of error always occurs due to pressure and temperature variations. If standard conditions (29.92" Hg and 15°C) exist at sea level, indicated altitude is the same as true altitude, pressure altitude, and density altitude. When the altimeter is set to 29.92 at any altitude, it indicates pressure altitude.

2-125   PA.I.G.K1h, PA.I.F.K2a

What are the standard temperature and pressure values for sea level?

A – 15°C and 29.92 inches Hg.

B – 59°C and 1013.2 millibars.

C – 59°F and 29.92 millibars.

2-125. Answer A. GFDPP 2C, PHB

The standard atmosphere at sea level is a temperature of 15°C (59°F) and 29.92" Hg (1013.2 millibars).

2-126   PA.I.G.K1h

If it is necessary to set the altimeter from 29.15 to 29.85, what change occurs?

A – 70-foot decrease in indicated altitude.

B – 700-foot decrease in indicated altitude.

C – 700-foot increase in indicated altitude.

2-126. Answer C. GFDPP 2C, PHB

The indicated altitude changes by about 1,000 feet for each one-inch change in the altimeter's setting and in the same direction. Increasing the altimeter setting by 0.10 inch increases the indicated altitude by about 100 feet. In this case, you increase the altimeter setting by 0.70 inches (29.85 – 29.15 = 0.70), therefore, the indicated altitude increases by 700 feet.

2-127   PA.I.G.K1h

If you change the altimeter setting from 30.11 to 29.96, what is the approximate change in indication?

A – Altimeter will indicate .15 inches Hg higher.

B – Altimeter will indicate 150 feet higher.

C – Altimeter will indicate 150 feet lower.

2-127. Answer C. GFDPP 2C, PHB

The indicated altitude changes by about 1,000 feet for each one-inch change in the altimeter's setting and in the same direction. Decreasing the altimeter setting by 0.100 inch decreases the indicated altitude by about 100 feet. In this case, you decrease the altimeter setting by 0.15 inches (30.11 – 29.96 = 0.15), therefore, the indicated altitude decreases by 150 feet.

2-128   PA.I.G.K1h, PA.VI.A.K5d

Which factor would tend to increase the density altitude at a given airport?

A – An increase in barometric pressure.

B – An increase in ambient temperature.

C – A decrease in relative humidity.

2-128. Answer B. GFDPP 2C, PHB

The factors that decrease air density, increasing density altitude, are increased ambient temperature, decreased barometric pressure, and increased relative humidity.

SECTION C ■ **Flight Instruments**

2-129   PA.I.G.K1f, PA.VIII.A.K1b*, K1d*

(Refer to Figure 5.) A turn coordinator provides an indication of the

A – angle of bank up to but not exceeding 30°.

B – movement of the aircraft about the yaw and roll axes.

C – attitude of the aircraft with reference to the longitudinal axis.

2-129. Answer B. GFDPP 2C, PHB
The turn coordinator senses movement about both the vertical axis (yaw) and the longitudinal axis (roll).

2-130   PA.VIII.A.K1a, PA.IX.C.K2b

What is an advantage of an electric turn coordinator if the airplane has a vacuum system for other gyroscopic instruments?

A – It is a backup in case of vacuum system failure.

B – It is more reliable than the vacuum-driven indicators.

C – It will not tumble, unlike vacuum-driven turn indicators.

2-130. Answer A. GFDPP 2C, PHB
On many light airplanes, the vacuum system supplies power to the attitude and heading indicators, while the electrical system powers the turn coordinator. This configuration provides a backup in case one system fails.

2-131   PA.I.G.K1h, PA.VIII.A.K1b*

(Refer to Figure 7.) The proper adjustment to make on the attitude indicator during level flight is to align the

A – horizon bar to the level-flight indication.

B – horizon bar to the miniature airplane.

C – miniature airplane to the horizon bar.

2-131. Answer C. GFDPP 2C, PHB
Set the adjustable miniature airplane to match the level flight indication of the horizon bar.

2-132   PA.I.G.K1h, PA.VIII.A.K1b*

(Refer to Figure 7.) How should a pilot determine the direction of bank from an attitude indicator such as the one illustrated?

A – By the direction of deflection of the banking scale (A).

B – By the direction of deflection of the horizon bar (B).

C – By the relationship of the miniature airplane (C) to the deflected horizon bar (B).

2-132. Answer C. GFDPP 2C, PHB
As the airplane banks, the relationship between the miniature airplane and the horizon bar depict the direction of turn.

SECTION C ■ **Flight Instruments**

2-133   PA.VI.A.K2

The angular difference between true north and magnetic north is

A – magnetic deviation.

B – magnetic variation.

C – compass acceleration error.

2-133. Answer B. GFDPP 2C, 9A, PHB
Magnetic variation occurs because the earth's magnetic poles do not coincide with its geographic poles, and a magnetic compass aligns with the magnetic poles. You can determine local magnetic variation by referencing the isogonic lines on aeronautical charts, which are represented by dashed magenta lines.

2-134   PA.VI.A.K2, PA.VIII.A.K1a*

Deviation in a magnetic compass is caused by the

A – presence of flaws in the permanent magnets of the compass.

B – difference in the location between true north and magnetic north.

C – magnetic fields within the aircraft distorting the lines of magnetic force.

2-134. Answer C. GFDPP 2C, PHB
Metal and electronic components in the aircraft create magnetic fields that distort the lines of magnetic force. These distortions cause deviation errors in the compass readings.

2-135   PA.VI.A.K2, PA.VIII.A.K1a*

In the Northern Hemisphere, a magnetic compass will normally indicate initially a turn toward the west if

A – a left turn is entered from a north heading.

B – a right turn is entered from a north heading.

C – an aircraft is accelerated while on a north heading.

2-135. Answer B. GFDPP 2C, PHB
When turning from a northerly heading, the compass initially indicates a turn in the opposite direction. When starting a right turn toward the east, the compass begins to show a turn to the west.

2-136   PA.VI.A.K2, PA.VIII.A.K1a*

In the Northern Hemisphere, a magnetic compass will normally indicate initially a turn toward the east if

A – an aircraft is decelerated while on a south heading.

B – an aircraft is accelerated while on a north heading.

C – a left turn is entered from a north heading.

2-136. Answer C. GFDPP 2C, PHB
When turning from a northerly heading, the compass initially indicates a turn in the opposite direction. When starting a left turn toward the west, the compass begins to show a turn to the east.

SECTION C ■ **Flight Instruments**

2-137   PA.VI.A.K2, PA.VIII.A.K1a*

In the Northern Hemisphere, a magnetic compass will normally indicate a turn toward the north if

A – a right turn is entered from an east heading.

B – an aircraft is decelerated while on an east or west heading.

C – an aircraft is accelerated while on an east or west heading.

2-137. Answer C. GFDPP 2C, PHB

Acceleration error is most pronounced on east/west headings. In the northern hemisphere, acceleration will show a turn to the north, and deceleration will show a turn to the south. The memory aid, ANDS (accelerate north, decelerate south), can help you recall how acceleration error affects the compass.

2-138   PA.VI.A.K2, PA.VIII.A.K1a*

In the Northern Hemisphere, the magnetic compass will normally indicate a turn toward the south when

A – a left turn is entered from an east heading.

B – a right turn is entered from a west heading.

C – the aircraft is decelerated while on a west heading.

2-138. Answer C. GFDPP 2C, PHB

Acceleration error is most pronounced on east/west headings. In the northern hemisphere, acceleration will show a turn to the north, and deceleration will show a turn to the south. The memory aid, ANDS (accelerate north, decelerate south), can help you recall how acceleration error affects the compass.

2-139   PA.VI.A.K2, PA.VIII.A.K1a*

In the Northern Hemisphere, if an aircraft is accelerated or decelerated, the magnetic compass will normally indicate

A – a turn momentarily.

B – correctly when on a north or south heading.

C – a turn toward the south.

2-139. Answer B. GFDPP 2C, PHB

Because acceleration and deceleration errors are most pronounced on east/west headings, accelerating or decelerating on a north or south heading will not show much of an error on the magnetic compass.

2-140   PA.VI.A.K2, PA.VIII.D.K1a*, K1b*

During flight, when are the indications of a magnetic compass accurate?

A – Only in straight-and-level unaccelerated flight.

B – As long as the airspeed is constant.

C – During turns if the bank does not exceed 18°.

2-140. Answer A. GFDPP 2C, PHB

Magnetic dip causes turning and acceleration/deceleration errors. For this reason, magnetic compass indications are accurate only in straight-and-level unaccelerated flight.

2-141   PA.VI.A.K2, PA.VIII.A.K1a*

What should be the indication on the magnetic compass as you roll into a standard rate turn to the right from a south heading in the Northern Hemisphere?

A – The compass will initially indicate a turn to the left.

B – The compass will indicate a turn to the right, but at a faster rate than is actually occurring.

C – The compass will remain on south for a short time, then gradually catch up to the magnetic heading of the airplane.

2-141. Answer B. GFDPP 2C, PHB
When turning from a southerly heading, the compass moves the correct direction but at a faster rate than the actual change in heading.

2-142   PA.VI.A.K2, PA.VIII.A.K1a*

Deviation error of the magnetic compass is caused by

A – a northerly turning error.

B – certain metals and electrical systems within the aircraft.

C – the difference in location of true north and magnetic north.

2-142. Answer B. GFDPP 2C, PHB
Metal and electronic components in the aircraft create magnetic fields that cause deviation errors in the compass readings. The difference between true north and magnetic north is *variation*, not deviation.

2-143   PA.VI.A.K2, PA.VIII.A.K1a*

In the Northern Hemisphere, a magnetic compass will normally indicate a turn toward the north if

A – a left turn is entered from a west heading.

B – an aircraft is decelerated while on an east or west heading.

C – an aircraft is accelerated while on an east or west heading.

2-143. Answer C. GFDPP 2C, PHB
Acceleration error is most pronounced on east/west headings. In the northern hemisphere, acceleration will show a turn to the north, and deceleration will show a turn to the south. The memory aid, ANDS (accelerate north, decelerate south), can help you recall how acceleration error affects the compass.

SECTION C ■ **Flight Instruments**

2-144   PA.I.G.K1g, PA.IX.C.K2d, PA.VIII.A.K1a*

If the AHRS detects a problem with the integrity of the sensor information for the instruments on a PFD, what occurs?

A – The system reverts to reversionary mode and PFD information is displayed on the MFD.

B – A red X is placed over the display of the affected instrument (attitude indicator or HSI).

C – After an alert message appears, you must determine the affected instrument by comparing the indications of all instruments.

2-144. Answer B. GFDPP 2C, PHB
The AHRS monitors itself constantly, comparing the data from different inputs and checking the integrity of its information. When the system detects a problem, it places a red X over the display of the affected instrument to alert you that the indications are unreliable. When this happens, use the back-up instruments.

2-145   PA.IX.C.K2d, PA.I.G.K1f, K1g, PA.VIII.A.K1a*

You are flying an airplane with an electronic flight display and the PFD screen fails. What is a back-up options for the PFD instruments?

A – Display the PFD instruments on the MFD screen.

B – Monitor traditional back-up instruments that are powered from the main electrical bus.

C – Reset the AHRS and the ADC and observe whether the PFD turns back on.

2-145. Answer A. GFDPP 2C, PHB
The electronic flight display is configured so that the functions of the PFD can be transferred to the MFD screen in reversionary mode. In the event that the PFD instruments do not automatically appear on the MFD, most systems enable you to manually switch to reversionary mode. The back-up airspeed indicator and altimeter are pitot-static instruments that do not require electrical power. The back-up attitude indicator can be vacuum powered or electronically powered but must have a power source that is isolated from the electrical system of the aircraft.

2-146   PA.I.G.K1h, PA.VIII.D.K1b

How do you perform a standard-rate turn in an airplane with instruments on a PFD?

A – Monitor the digital turn coordinator and adjust the angle of bank as needed.

B – Monitor the trend vector on the HSI and adjust the angle of bank as needed.

C – Monitor the trend indicator on the digital attitude indicator and adjust the bank as needed.

2-146. Answer B. GFDPP 2C, PHB
On a PFD, use the turn rate indicator on the HSI to perform a standard-rate turn. The turn rate indicator consists of a trend vector that shows what the airplane's heading will be in six seconds and index marks on either side of the heading lubber line to provide a reference for making standard rate turns.

2-147   PA.I.G.K1g, PA.IX.C.K2d

Which digital flight instruments are related to the air data computer (ADC)?

A – Attitude indicator, heading indicator, and turn indicator.

B – Compass, magnetometer, and ground track indicator.

C – Airspeed indicator, altimeter, and VSI.

2-147. Answer C. GFDPP 2C, PHB
In an electronic flight display system, the pitot tube, static source, and outside air temperature probe provide information to the ADC. The ADC uses these pressure and temperate inputs to determine the appropriate readings for the airspeed indicator, altimeter, and VSI.

CHAPTER 3

Aerodynamic Principles

SECTION A
Four Forces of Flight

FORCES ACTING ON AN AIRPLANE

- The four forces that act on an airplane in flight are lift, weight, thrust, and drag.
- The forces are in equilibrium when the airplane is in unaccelerated flight—neither accelerating nor decelerating.
- During straight-and-level flight, lift equals weight, and thrust equals drag. This relationship is also true during climbs and descents and any time the airplane is in unaccelerated flight.

LIFT

- The airplane wing's shape is designed to take advantage of Newton's laws and Bernoulli's principle.
  - Bernoulli's principle states that as the velocity of a fluid (including air) increases, its pressure decreases. This principle provides an explanation of lift produced by an airfoil that is curved more on the top than on the bottom. Air travels faster over the curved upper surface, causing lower pressure on the top surface.
  - The airflow pattern around the wing causes a downward flow of air behind the wing called downwash. The reaction to this downwash results in an upward force on the wing, demonstrating Newton's third law of motion.
- The angle of attack is the angle between the wing chord line and the relative wind.
- Total lift depends on the combined effects of airspeed and angle of attack. When speed decreases, you must increase the angle of attack to maintain the same amount of lift.
- Angle of incidence refers to the angle between the wing chord line and a line parallel to the longitudinal axis of the airplane.
- A stall is caused by the separation of airflow from the wing's upper surface. For a given airplane, a stall always occurs at the critical angle of attack, regardless of airspeed, flight attitude, or weight.
- Flaps enable steeper approaches to a landing without increasing airspeed.

GROUND EFFECT

- Ground effect is the result of the interference of the surface of the Earth with the airflow patterns about an airplane.
- Ground effect reduces induced drag on the airplane, increasing performance.
- An adverse consequence of ground effect is that an airplane can become airborne before reaching the recommended takeoff speed, but then be unable to climb out of ground effect.
- Another undesirable consequence of ground effect is that any excess speed at the point of flare can cause considerable floating.

NOTE: *An asterisk appearing after an ACS code (i.e. PA.VII.B.K1*) indicates that the question subject appears more than one time in the ACS. The code shown corresponds to the first instance of the subject in the ACS.*

3-1   PA.I.F.K3

The four forces acting on an airplane in flight are

A – lift, weight, thrust, and drag.

B – lift, weight, gravity, and thrust.

C – lift, gravity, power, and friction.

3-1. Answer A. GFDPP 3A, PHB

Lift is the upward force created by airflow over and under the wings. Weight, caused by the downward pull of gravity, opposes lift. Thrust is the forward force that propels the airplane forward, and drag is the retarding force that opposes thrust.

3-2   PA.I.F.K3

When are the four forces that act on an airplane in equilibrium?

A – During unaccelerated flight.

B – When the aircraft is accelerating.

C – When the aircraft is at rest on the ground.

3-2. Answer A. GFDPP 3A, PHB

In straight-and-level, unaccelerated flight, the four forces are in equilibrium. Lift equals weight, and thrust equals drag.

3-3   PA.I.F.K3

What is the relationship of lift, drag, thrust, and weight when the airplane is in straight-and-level flight?

A – Lift, drag, and weight equal thrust.

B – Lift and weight equal thrust and drag.

C – Lift equals weight and thrust equals drag.

3-3. Answer C. GFDPP 3A, PHB

Assuming the airplane is in unaccelerated flight, thrust equals drag, and lift equals weight.

3-4   PA.I.F.K3

Which statement relates to Bernoulli's principle?

A – For every action there is an equal and opposite reaction.

B – An additional upward force is generated as the lower surface of the wing deflects air downward.

C – Air traveling faster over the curved upper surface of an airfoil causes lower pressure on the top surface.

3-4. Answer C. GFDPP 3A, PHB

Bernoulli's principle states that as the velocity of a fluid (including air), increases, its pressure decreases. The explanations of lift that relate to Bernoulli focus on the higher velocity of air over the upper surface of an airfoil. The other two answers are also valid explanations for lift, but are related to Newton's Third Law of Motion rather than Bernoulli's principle.

3-5   PA.I.F.K3

(Refer to Figure 1.) The acute angle A is the angle of

A – incidence.

B – attack.

C – dihedral.

3-5. Answer B. GFDPP 3A, PHB

The angle between the chord line and the relative wind is the angle of attack.

SECTION A ■ **Four Forces of Flight**

3-6   PA.I.F.K3

The term "angle of attack" is defined as the angle between the

A – chord line of the wing and the relative wind.

B – airplane's longitudinal axis and that of the air striking the airfoil.

C – airplane's center line and the relative wind.

3-6. Answer A. GFDPP 3A, PHB
The angle of attack is the angle between the chord line of the wing and the relative wind. The other answers are wrong because the angle of attack is *not* measured from parts of the airplane other than the wing.

3-7   PA.I.F.K3

The angle between the chord line of an airfoil and the relative wind is known as the angle of

A – lift.

B – attack.

C – incidence.

3-7. Answer B. GFDPP 3A, PHB
The angle of attack is the angle between the wing chord line and the relative wind. Angle of incidence refers to the angle between the wing chord line and a line parallel to the longitudinal axis of the airplane. The wing is usually attached to the fuselage with the chord line inclined upward at a slight angle.

3-8   PA.I.G.K1b

One of the main functions of flaps during approach and landing is to

A – decrease the angle of descent without increasing the airspeed.

B – enable a touchdown at a higher indicated airspeed.

C – increase the angle of descent without increasing the airspeed.

3-8. Answer C. GFDPP 3A, PHB
Flaps increase both lift and drag, enabling a steeper angle of descent without increasing airspeed. The increased lift also enables touchdown at a lower indicated airspeed.

3-9   PA.I.G.K1b

What is one purpose of wing flaps?

A – To decrease wing area to vary the lift.

B – To relieve the pilot of maintaining continuous pressure on the controls.

C – To enable you to make steeper approaches to a landing without increasing the airspeed.

3-9. Answer C. GFDPP 3A, PHB
Flaps increase both lift and drag, enabling a steeper approach without increasing airspeed. Trim controls, not flaps, relieve the pilot of maintaining continuous pressure on the controls. Some types of flaps increase wing area when deployed to increase lift.

3-10   PA.VII.B.K1*

The angle of attack at which an airplane wing stalls

A – increases if the CG is moved forward.

B – changes with an increase in gross weight.

C – remains the same regardless of gross weight.

3-10. Answer C. GFDPP 3A, PHB

The critical angle of attack (angle of attack at which an airplane wing stalls) is determined by the lift coefficient of a particular wing configuration. An airplane stalls when the critical angle of attack is exceeded, regardless of weight or airspeed.

3-11   PA.IV.C.K4, K5

What is ground effect?

A – The result of the interference of the surface of the Earth with the airflow patterns about an airplane.

B – The result of an alteration in airflow patterns increasing induced drag about the wings of an airplane.

C – The result of the disruption of the airflow patterns about the wings of an airplane to the point where the wings can no longer support the airplane in flight.

3-11. Answer A. GFDPP 3A, PHB

When flying close to the ground, the airflow around an airplane is altered by interference with the surface of the earth. The resulting ground effect reduces the induced drag on the airplane.

3-12   PA.IV.C.K4

Floating caused by the phenomenon of ground effect is most realized during an approach to land when at

A – a higher-than-normal angle of attack.

B – less than the length of the wingspan above the surface.

C – twice the length of the wingspan above the surface.

3-12. Answer B. GFDPP 3A, PHB

Ground effect becomes noticeable when the height of the airplane above the ground is less than the length of the wingspan.

SECTION A ■ **Four Forces of Flight**

3-13   PA.IV.C.K4

What must you be aware of as a result of ground effect?

A – Wingtip vortices increase, creating wake turbulence problems for arriving and departing aircraft.

B – Induced drag decreases; therefore, any excess speed at the point of flare may cause considerable floating.

C – A full stall landing requires less up elevator deflection than a full stall that is done free of ground effect.

3-13. Answer B. GFDPP 3A, PHB
Because ground effect decreases induced drag, the airplane tends to float during the flare. This especially affects low-wing airplanes.

3-14   PA.IV.C.K4

Ground effect is most likely to result in which problem?

A – Settling to the surface abruptly during landing.

B – Becoming airborne before reaching recommended takeoff speed.

C – Inability to get airborne even though airspeed is sufficient for normal takeoff needs.

3-14. Answer B. GFDPP 3A, PHB
The decreased induced drag while in ground effect enables the airplane to become airborne at a lower airspeed than the recommended takeoff speed. However if you attempt to climb out of ground effect before reaching the speed for normal climb, the airplane might sink back to the surface. On landing, ground effect can lead to floating, especially on low-wing airplanes.

SECTION B
Stability

STABILITY CHARACTERISTICS

- A completely stable airplane would be impossible to maneuver. However, some stability is needed for desirable handling characteristics.
- An aircraft that is inherently stable requires less effort to control.

THREE AXES OF FLIGHT

- Longitudinal stability involves the pitching motion or tendency of the airplane to move about its lateral axis.
- The center of pressure is a point along the wing chord line where lift is considered to be concentrated—sometimes referred to as the center of lift (CL).

LONGITUDINAL STABILITY

- The location of the center of gravity (CG), with respect to the center of lift (CL), determines the longitudinal stability of the airplane. The airplane is more stable when the CG is forward of the CL.
- Loading an aircraft to the most aft CG causes the airplane to be less stable at all speeds.
- To recover from a stall is difficult in an airplane loaded with the CG aft of the approved CG range.
- When you reduce power without adjusting the controls, an airplane pitches nose down because the downwash on the elevator from the propeller slipstream is reduced, which reduces elevator effectiveness.

STALLS

- The inherent stability of an airplane is important because it relates to the aircraft's ability to avoid stalls and spins.
- Shifting the center of gravity towards the rear of the airplane can make the airplane more unstable, which increases the tendency of the airplane to pitch up, potentially leading to a stall.
- As an airplane approaches a stall, one of the key indications is that the control surfaces (ailerons, elevator, and rudder) become less responsive. This effect is due to the reduced airflow over the wings and control surfaces as the airplane's speed decreases.
- In addition to less responsive controls, signs of an impending stall can include a high nose attitude, a decrease in sound from the airflow and engine, and the stall warning indicator.
- The first fundamental step in stall recovery is to lower the nose to decrease the angle of attack. This action helps to restore airflow over the wings, which is crucial for regaining lift.

SECTION B ■ **Stability**

SPINS

- You must stall an airplane before it can spin. During a spin, both wings are stalled, but one wing is more stalled than the other.

- In light training airplanes, a complete spin maneuver consists of three phases—incipient, fully developed, and recovery.

- The incipient spin is that portion of a spin from the time the airplane stalls and rotation starts until the spin is fully developed.

- The incipient spin usually occurs rapidly in light airplanes (about 4 to 6 seconds) and consists of approximately the first two turns.

- The fully developed spin begins somewhere in the second turn as the airplane continues to rotate and the angular rotation rates, airspeed, and vertical speed become stabilized with a nearly vertical flight path.

UNUSUAL ATTITUDES

- When controlling the airplane solely by instrument reference, recover from a nose-high unusual attitude by performing these actions almost simultaneously but in the following sequence:
    1. Add power.
    2. Lower nose to place the aircraft symbol on the horizon bar of the attitude indicator.
    3. Level the wings using the attitude indicator as a reference.

- When controlling the airplane solely by instrument reference, recover from a nose-low unusual attitude by performing these actions almost simultaneously but in the following sequence:
    1. Reduce power.
    2. Level the wings using the attitude indicator as a reference.
    3. Raise the nose to place the aircraft symbol on the horizon bar of the attitude indicator.

NOTE: An asterisk appearing after an ACS code (i.e. PA.VII.B.K1*) indicates that the question subject appears more than one time in the ACS. The code shown corresponds to the first instance of the subject in the ACS.

3-15   PA.I.F.K3
An airplane said to be inherently stable

A – will not spin.

B – is difficult to stall.

C – requires less effort to control.

3-15. Answer C. GFDPP 3B, PHB
An airplane that is inherently stable tends to return to its original attitude after it has been displaced, which makes it easier to control.

3-16   PA.I.F.K3
Changes in the center of pressure of a wing affect the aircraft's

A – lift/drag ratio.

B – lifting capacity.

C – aerodynamic balance and controllability.

3-16. Answer C. GFDPP 3B, PHB
A wing's center of pressure moves forward and back with changing angles of attack (forward for high angles and back for lower). This movement changes the position of the air loads on the wing, which results in changes to an airplane's aerodynamic balance and controllability.

3-17   PA.I.F.K2e, K3

What determines the longitudinal stability of an airplane?

A – The location of the CG with respect to the center of lift.

B – The effectiveness of the horizontal stabilizer, rudder, and rudder trim tab.

C – The relationship of thrust and lift to weight and drag.

3-17. Answer A. GFDPP 3B, PHB

The longitudinal stability of an airplane is determined primarily by the location of the center of gravity (CG) in relation to the center of lift (CL). The airplane is more stable when the CG is forward of the CL.

3-18   PA.I.F.K2e

An airplane has been loaded in such a manner that the CG is located aft of the aft CG limit. One undesirable flight characteristic a pilot might experience with this airplane would be

A – a longer takeoff run.

B – difficulty in recovering from a stalled condition.

C – stalling at higher-than-normal airspeed.

3-18. Answer B. GFDPP 3B, PHB

With a CG aft of the rear CG limit, the airplane becomes tail heavy and unstable in pitch because the horizontal stabilizer is less effective. This condition makes it difficult, if not impossible, to recover from a stall or spin. The other choices are incorrect because an aft CG, although dangerous, results in reduced drag, which could result in a shorter takeoff run and stalling at a lower than normal airspeed.

3-19   PA.I.F.K2e

Loading an airplane to the most aft CG causes the airplane to be

A – less stable at all speeds.

B – less stable at slow speeds, but more stable at high speeds.

C – less stable at high speeds, but more stable at low speeds.

3-19. Answer A. GFDPP 3B, PHB

In an airplane loaded to the aft CG limit, the horizontal stabilizer is less effective, causing the airplane to be less stable at all speeds.

3-20   PA.I.F.K3, PA.I.G.K1a

What causes an airplane (except a T-tail) to pitch nose down when power is reduced and controls are not adjusted?

A – The CG shifts forward when thrust and drag are reduced.

B – The downwash on the elevator from the propeller slipstream is reduced and elevator effectiveness is reduced.

C – When thrust is reduced to less than weight, lift is also reduced and the wings can no longer support the weight.

3-20. Answer B. GFDPP 3B, PHB

At higher power settings, in airplanes other than T-tail designs, the propeller slipstream causes a greater downward force on the horizontal stabilizer. When power is reduced, this downward force on the tail is also reduced, and the nose pitches down.

SECTION B ■ **Stability**

3-21   PA.VII.A.K1

Which of the following changes would increase the airplane's tendency to pitch up, potentially leading to a stall when flying at slow airspeeds?

A – Decreasing the angle of attack while maintaining the same power setting.

B – Shifting the center of gravity towards the rear of the airplane.

C – Decreasing the power setting while maintaining the same airspeed.

3-21. Answer B. GFDPP 3B, PHB

Shifting the center of gravity towards the rear of the airplane can make the airplane more unstable, which increases the tendency of the airplane to pitch up, potentially leading to a stall. The other options listed would not have the same effect. Decreasing the angle of attack or the power setting, would not increase the airplane's tendency to pitch up.

3-22   PA.VII.B.K2*, PA.I.G.K1a

Which of the following is a common indication of an impending stall?

A – The engine noise becomes significantly louder.

B – The airplane's speed increases dramatically.

C – The flight controls become less responsive.

3-22. Answer C. GFDPP 3B, PHB

As an airplane approaches a stall, one of the key indications is that the control surfaces (ailerons, elevator, and rudder) become less responsive. This effect is due to the reduced airflow over the wings and control surfaces as the airplane's speed decreases. Other signs of an impending stall can include a high nose attitude, a decrease in sound from the airflow and engine, and the stall warning indicator.

3-23   PA.VII.B.K3

You are practicing emergency landings with the engine power at idle. Which of the following situations could lead to a power-off stall?

A – You are too high and push the nose down.

B – You are too low and increase your angle of attack to try to stretch your glide.

C – You are flying with a strong tailwind.

3-23. Answer B. GFDPP 3B, PHB

When practicing emergency landings with the engine power at idle, an increase in your angle of attack will decrease your airspeed and potentially cause the airplane to stall. Pushing the nose down would decrease your angle of attack and increase your airspeed. A tailwind would not cause an increase in your angle of attack.

3-24   PA.VII.B.K4, PA.VII.C.K4,

You are practicing flight maneuvers and you have just recognized signs of an impending stall. What is the first fundamental step in stall recovery?

A – Reduce the power.

B – Lower the nose to decrease the angle of attack.

C – Level the wings using the ailerons.

3-24. Answer B. GFDPP 3B, PHB

The first fundamental step in stall recovery is to lower the nose to decrease the angle of attack. This action helps to restore airflow over the wings, which is crucial for regaining lift. Increasing engine power can also aid in recovery, but it is not the first step. Leveling the wings is important for maintaining control during recovery, but it does not directly address the stall condition.

SECTION B ■ Stability

3-25   PA.VII.D.K3

Which is true regarding spin recovery procedures?

A – The same spin recovery procedure applies to every airplane.

B – Recovery from a flat spin might be impossible due to ineffective ailerons.

C – Control inputs involve stopping the rotation and reducing the angle of attack to recover from the stall.

3-25. Answer C. GFDPP 3B, PHB

Anti-spin control inputs result in a slowing and eventual cessation of rotation coupled with a decrease in angle of attack below CL_{max}. Spin recovery techniques vary for different aircraft so you must follow the recovery procedures outlined in the POH for your airplane. A flat spin is characterized by a near level pitch and roll attitude with the spin axis near the CG of the airplane. The upward flow over the tail could render the elevators and rudder ineffective, making recovery impossible.

3-26   PA.VII.D.K3

In small airplanes, normal recovery from spins might become difficult if the

A – CG is too far rearward, and rotation is around the longitudinal axis.

B – CG is too far rearward, and rotation is around the CG.

C – spin is entered before the stall is fully developed.

3-26. Answer B. GFDPP 3B, PHB

Because moving the CG aft decreases longitudinal stability and reduces pitch control forces in most airplanes, an aft CG tends to make the airplane easier to stall. After a spin is entered, an airplane with its CG farther aft tends to have a flatter spin attitude. A flat spin is characterized by a near level pitch and roll attitude with the spin axis near the CG of the airplane. The upward flow over the tail could render the elevators and rudder ineffective, making recovery impossible.

3-27   PA.VII.D.K2

In what flight condition must an aircraft be placed in order to spin?

A – Partially stalled with one wing low.

B – In a steep diving spiral.

C – Stalled.

3-27. Answer C. GFDPP 3B, PHB

An airplane must be stalled before it can spin. The spin progresses when one wing becomes more stalled than the other, which leads to rotation. A steep diving spiral is different than a spin because the airplane is not stalled, even though it is descending and turning.

3-28   PA.VII.D.K1

During a spin to the left, which wing is stalled?

A – Both wings are stalled.

B – Neither wing is stalled.

C – Only the left wing is stalled.

3-28. Answer A. GFDPP 3B, PHB

In any spin, both wings are stalled but in a spin to the left, the left wing is more stalled.

SECTION B ■ **Stability**

3-29   PA.VII.D.K2

Which is a characteristic of the incipient phase of a spin in light airplanes?

A – Consists of at least 4 turns.

B – Usually occurs within about 4 to 6 seconds.

C – Angular rotation rate, airspeed, and vertical speed are stabilized with a nearly vertical flight path.

3-29. Answer B. GFDPP 3B, PHB

In light, training airplanes, a complete spin maneuver consists of three phases—incipient, fully developed, and recovery. The incipient spin is that portion of a spin from the time the airplane stalls and rotation starts until the spin is fully developed. The incipient spin usually occurs rapidly in light airplanes (about 4 to 6 seconds) and consists of approximately the first two turns.

The fully developed spin begins somewhere in the second turn as the airplane continues to rotate and the angular rotation rates, airspeed, and vertical speed become stabilized with a nearly vertical flight path.

NOTE: *The following questions refer to the procedures to recover from unusual attitudes. Chapter 3, Section B of the Private Pilot textbook explains how to recover from stalls and spins. However, nose-high and nose-low unusual attitudes are covered in Private Pilot Maneuvers, Chapter 8 — Special Flight Operations, Maneuver 32 Basic Instrument Maneuvers.*

3-30   PA.VIII.E.K2, K4

To recover from a nose-high unusual attitude, perform these actions almost simultaneously but in the following sequence:

A – Add power; lower the nose; level the wings.

B – Reduce power; level the wings; raise the nose.

C – Add power; raise the nose; level the wings

3-30. Answer A. GFDPP 3B, PHB

Because a rapid decrease in airspeed can quickly result in a stall, prompt recognition and recovery from a nose-high unusual attitude is essential. When controlling the airplane solely by instrument reference, recover from a nose-high unusual attitude by performing these actions almost simultaneously but in the following sequence:

1. Add power.

2. Lower nose to place the aircraft symbol on the horizon bar of the attitude indicator.

3. Level the wings using the attitude indicator as a reference.

3-31   PA.VIII.E.K2, K4

To recover from a nose-low unusual attitude, perform these actions almost simultaneously but in the following sequence:

A – Add power; lower the nose; level the wings.

B – Reduce power; level the wings; raise the nose.

C – Add power; raise the nose; level the wings

3-31. Answer A. GFDPP 3B, PHB

When controlling the airplane solely by instrument reference, recover from a nose-low unusual attitude by performing these actions almost simultaneously but in the following sequence:

1. Reduce power.

2. Level the wings using the attitude indicator as a reference.

3. Raise the nose to place the aircraft symbol on the horizon bar of the attitude indicator.

If you attempt to raise the nose before you roll wings level, the increased load factor can result in an accelerated stall, a spin, or a force exceeding the airplane's design limits.

SECTION B ■ **Stability**

SECTION C
Aerodynamics of Maneuvering Flight

LEFT-TURNING TENDENCIES

In addition to the basic aerodynamic forces present in a climb, a combination of physical and aerodynamic forces can contribute to a left-turning tendency in propeller-driven airplanes. These forces are:

- Torque—the clockwise action of a spinning propeller causes a torque reaction, which tends to rotate the airplane counterclockwise about its longitudinal axis. Torque is:
    - Explained by Newton's third law of motion: "For every action there is an equal and opposite reaction."
    - Greatest at low airspeeds, high power settings, and high angles of attack.

- Spiraling slipstream—propeller rotation produces a backward flow of air, or slipstream, wrapping around the airplane and striking the left side of the vertical fin.

- Gyroscopic precession—the propeller exhibits characteristics similar to a gyroscope—rigidity in space and precession. The reaction to a force applied to a gyro acts in the direction of rotation and approximately 90° ahead of the point where the force is applied.

- Asymmetric thrust or P-factor—the propeller blade descending on the right produces more thrust than the ascending blade on the left. P-factor is:
    - The most prominent of the left-turning tendencies.
    - Most pronounced at low airspeeds, high power settings, and high angles of attack.

DESCENDING FLIGHT

- The angle of attack resulting in the least drag on your airplane will give the maximum lift-to-drag ratio (L/D_{max}), the best glide angle, and the maximum gliding distance.

- At a given weight, L/D_{max} will correspond to a certain airspeed called the best glide speed.

- Best glide speed typically provides the maximum gliding distance. In the event of an engine failure, maintaining any speed other than the best glide speed creates more drag and affects gliding distance.

TURNS AND LOAD FACTOR

- The horizontal component of lift is what makes an airplane turn.

- Load factor is the ratio of the load supported by the airplane's wings to the actual weight of the aircraft and its contents.

- Turns increase the load factor on an airplane, as compared to straight-and-level flight.

- The amount of excess load that can be imposed on an airplane depends on its speed.

- At 60 degrees of bank, 2 Gs are required to maintain level flight. To determine how much weight the airplane's wing structure must support, multiply the airplane's weight by the number of Gs.

- An increased load factor, whether it is experienced in a turn or by a pull-up out of a descent, causes an airplane to stall at a higher airspeed.

ENERGY MANAGEMENT

- When energy is exchanged, altitude and airspeed change in opposite directions (absent any other energy or control inputs). Energy management is the process of managing this relationship.

- The total mechanical energy of an airplane in flight is the sum of its potential energy from altitude, and kinetic energy from airspeed.

- A total energy error occurs when your airplane has too much or too little energy—altitude and airspeed deviate in the same direction (low and slow, or high and fast). To correct total energy errors, increase or decrease energy by adding or reducing power.

- An energy distribution error occurs when your airplane has the right amount of total energy, but the distribution over altitude and airspeed is incorrect—altitude and airspeed deviate in opposite directions (high and slow, or low and fast). Correct an energy distribution error by adjusting pitch to exchange energy between altitude and airspeed.

- You can easily visualize energy management concepts using a graph showing an airplane's power required curve (commonly called the power curve).

- On the back side of the power curve, flight at low airspeeds requires a high angle of attack and a great amount of power to overcome the resulting induced drag. Any reduction in airspeed requires an increase in power to maintain level flight.

- Flying on the back side of the curve is discouraged, because reducing speed can demand more power than the engine can supply or an unplanned reduction of power could result in an involuntary descent.

- Understanding energy management concepts helps you adjust pitch and power to safely fly a stabilized approach to land.

- Descending below the proper glide path and allowing the airspeed to decay below the approach speed is especially hazardous. Increase the power anytime the airplane is low and slow.

NOTE: An asterisk appearing after an ACS code (i.e. PA.VII.B.K1*) indicates that the question subject appears more than one time in the ACS. The code shown corresponds to the first instance of the subject in the ACS.

3-32   PA.I.F.K3

What are the four flight fundamentals that are involved in maneuvering an aircraft?

A – Aircraft power, pitch, bank, and trim.

B – Starting, taxiing, takeoff, and landing.

C – Straight-and-level flight, turns, climbs, and descents.

3-32. Answer C. GFDPP 3C, PHB
All controlled flight consists of one of the four fundamental maneuvers—straight-and-level flight, turns, climbs, and descents—or some combination of them.

3-33   PA.I.F.K3, PA.IV.C.K6

In what flight conditions are torque effects more pronounced in a single-engine airplane?

A – Low airspeed, high power, high angle of attack.

B – Low airspeed, low power, low angle of attack.

C – High airspeed, high power, high angle of attack.

3-33. Answer A. GFDPP 3C, PHB
Torque effect is greatest at low airspeeds, high power settings, and high angles of attack.

3-34   PA.I.F.K3, PA.IV.C.K6

The left-turning tendency of an airplane caused by P-factor is the result of the

A – clockwise rotation of the engine and the propeller turning the airplane counter-clockwise.

B – propeller blade descending on the right, producing more thrust than the ascending blade on the left.

C – gyroscopic forces applied to the rotating propeller blades acting 90° in advance of the point the force was applied.

3-34. Answer B. GFDPP 3C, PHB

P-factor, or asymmetric propeller loading, normally occurs at a high angle of attack. The descending propeller blade on the right side takes a larger "bite" of the air, and produces more thrust than the ascending blade on the left. The result is a left-turning tendency of the airplane.

3-35   PA.I.F.K3, PA.IV.C.K6

When does P-factor cause the airplane to yaw to the left?

A – When at low angles of attack.

B – When at high angles of attack.

C – When at high airspeeds.

3-35. Answer B. GFDPP 3C, PHB

P-factor, or asymmetric propeller loading, normally occurs at a high angle of attack. The descending propeller blade on the right side takes a larger "bite" of the air, and produces more thrust than the ascending blade on the left. The result is a left-turning tendency of the airplane.

3-36   PA.IX.B.K2a, K2b, K2c

Which is true about your approach after an engine failure?

A – Flying upwind increases glide distance and flying downwind decreases glide distance.

B – A stabilized approach at best glide speed provides the least drag and maximum gliding distance.

C – Flying at minimum sink speed provides the same gliding distance as flying at best glide speed.

3-36. Answer B. GFDPP 3C, AFH

Attaining best glide speed is an immediate action item after an engine failure. Best glide speed provides the maximum gliding distance. Any speed other than the best glide speed creates more drag so it is critical that you fly a stabilized approach to the landing site and avoid making large pitch changes. Turning upwind reduces your groundspeed and glide distance. Conversely, turning downwind increases your groundspeed and glide distance.

At minimum sink speed, the airplane loses altitude at the lowest rate. However, flying at this speed results in less distance traveled than when at best glide speed. Minimum sink speed is useful when time in flight is more important than distance flown, such as ditching an airplane at sea. This speed is not often published but generally is a few knots less than best glide speed.

SECTION C ■ Aerodynamics of Maneuvering Flight

SECTION C ■ Aerodynamics of Maneuvering Flight

3-37   PA.IX.B.K2a, K4

When executing an emergency approach to land in a single-engine airplane, it is important to maintain a constant glide speed because variations in glide speed will

A – increase the chances of shock cooling the engine.

B – assure the proper descent angle is maintained until entering the flare.

C – nullify all attempts at accuracy in judgment of gliding distance and landing spot.

3-37. Answer C. GFDPP 3C, AFH

Maintain a constant gliding speed because variations of gliding speed nullify all attempts at accuracy in judgment of gliding distance and the landing spot. Factors such as altitude, obstructions, wind direction, landing direction, landing surface and gradient, and landing distance of the airplane determines the pattern and approach procedures to use.

3-38   PA.I.F.K1

(Refer to Figure 2.) If an airplane weighs 2,300 pounds, what approximate weight would the airplane structure be required to support during a 60° banked turn while maintaining altitude?

A – 2,300 pounds.

B – 3,400 pounds.

C – 4,600 pounds.

3-38. Answer C. GFDPP 3C, PHB

At 60° of bank, the load factor is 2. The wing loading is 2,300 lb × 2 = 4,600 lb.

3-39   PA.I.F.K1

(Refer to Figure 2.) If an airplane weighs 3,300 pounds, what approximate weight would the airplane structure be required to support during a 30° banked turn while maintaining altitude?

A – 1,200 pounds.

B – 3,100 pounds.

C – 3,800 pounds.

3-39. Answer C. GFDPP 3C, PHB

At 30° of bank, the load factor is 1.154. The wing loading is 3,300 lb × 1.154 = 3,810 lb.

3-40   PA.I.F.K1

(Refer to Figure 2.) If an airplane weighs 4,500 pounds, what approximate weight would the airplane structure be required to support during a 45° banked turn while maintaining altitude?

A – 4,500 pounds.

B – 6,400 pounds.

C – 7,200 pounds.

3-40. Answer B. GFDPP 3C, PHB

At 45° of bank, the load factor is 1.414. The wing loading is 4,500 lb × 1.414 = 6,365 lb.

3-41   PA.I.F.K3

The amount of excess load that can be imposed on the wing of an airplane depends upon the

A – position of the CG.

B – speed of the airplane.

C – abruptness at which the load is applied.

3-41. Answer B. GFDPP 3C, PHB

The amount of excess load that can be imposed on an airplane depends on its speed. If abrupt control movements or strong gusts are applied at low airspeeds, the airplane stalls before the load becomes excessive. At higher airspeeds, the increased airflow causes a greater lifting capacity. A sudden control input or gust at a high airspeed can result in an excessive load factor beyond safe limits.

3-42   PA.I.F.K3, PA.V.A.K2d

Which basic flight maneuver increases the load factor on an airplane as compared to straight-and-level flight?

A – Climbs.

B – Turns.

C – Stalls.

3-42. Answer B. GFDPP 3C, PHB

In a level turn, lift must be increased to compensate for the loss of vertical lift and overcome centrifugal force. Because the wings must support not only the airplane's weight, but also the load imposed by centrifugal force, the load factor is greater than one G.

3-43   PA.I.F.K3

What force makes an airplane turn?

A – The horizontal component of lift.

B – The vertical component of lift.

C – Centrifugal force.

3-43. Answer A. GFDPP 3C, PHB

In a turn, lift has both a vertical and a horizontal component. The horizontal component of lift, which is also referred to as centripetal force, opposes centrifugal force and causes the airplane to turn. The vertical component of lift opposes gravity and enables the airplane to maintain altitude.

3-44   PA.V.A.K1

When practicing a steep turn during private pilot training, you should

A – Maintain a 30 degree bank angle.

B – Increase the angle of attack to maintain altitude.

C – Decrease the power to maintain altitude.

3-44. Answer B. GFDPP 3C, PHB

A steep turn is a high-performance maneuver normally executed at a bank angle of 45 degrees for private pilot training and 50 degrees for commercial training. During a turn, the force of lift is divided into two components: one acting horizontally and the other vertically to oppose weight. During the turn, you must compensate for the decreased vertical component of lift by increasing the angle of attack to maintain altitude.

SECTION C ■ **Aerodynamics of Maneuvering Flight**

3-45   PA.I.G.K1a, PA.V.A.K2a, K2b

What is the cause of overbanking tendency during a turn?

A – Additional lift on the outside, or raised, wing.

B – Uncoordinated flight.

C – Spiraling slipstream.

3-46   PA.V.A.K2e

Which is correct with respect to rate and radius of turn for an airplane flown in a coordinated turn at a constant altitude?

A – For a specific angle of bank and airspeed, the rate and radius of turn will not vary.

B – To maintain a steady rate of turn, the angle of bank must be increased as the airspeed is decreased.

C – The faster the true airspeed, the faster the rate and larger the radius of turn regardless of the angle of bank the effects of an increased load factor.

3-47   PA.V.A.K1. K2e

To increase the rate of turn and at the same time decrease the radius, you should?

A – increase the bank and decrease airspeed.

B – increase the bank and increase airspeed.

C – maintain the bank and decrease airspeed.

3-45. Answer A. GFDPP 3C, PHB

As you enter a turn and increase the angle of bank, you may notice the tendency of the airplane to continue rolling into a steeper bank, even though you neutralize the ailerons.

This overbanking tendency is caused by additional lift on the outside, or raised, wing. Because the outside wing is traveling faster than the inside wing, it produces more lift and the airplane tends to roll beyond the desired bank angle. To correct for overbanking tendency, you can use a small amount of opposite aileron, away from the turn, to maintain your desired angle of bank.

3-46. Answer A. GFDPP 3C, PHB

Airspeed and bank angle are the two variables in determining rate and radius of turn. If these variables remain constant, rate and radius of turn will remain constant. The rate of turn varies directly with bank angle and is inversely proportional to airspeed.

3-47. Answer C. GFDPP 3C, PHB

Rate and radius of turn vary with both airspeed and bank angle. For example:

When the angle of bank is held constant, decreasing the airspeed:

• decreases the turn radius.

• increases the turn rate.

When the airspeed is held constant, increasing the bank angle:

• decreases the turn radius.

• increases the turn rate.

3-48　PA.V.A.K1, K2e

If the airspeed is decreased from 98 knots to 85 knots during a coordinated level 45° banked turn, the load factor will

A – Remain the same, but the radius of turn will decrease.

B – Decrease, and the rate of turn will decrease.

C – Remain the same, but the radius of turn will increase.

3-48. Answer A. GFDPP 3C, PHB

Rate and radius of turn vary with both airspeed and bank angle but load factor remains constant if no change in bank angle occurs. For example:

When the angle of bank is held constant, decreasing the airspeed:

- decreases the turn radius.

- increases the turn rate.

- has no effect on the load factor.

3-49　PA.V.A.K2d

During an approach to a stall, an increased load factor causes the airplane to

A – stall at a higher airspeed.

B – have a tendency to spin.

C – be more difficult to control.

3-49. Answer A. GFDPP 3C, PHB

Stall speed increases in proportion to the square root of the load factor. Added G-forces cause an airplane to stall at an airspeed higher than the normal 1 G airspeed. Stalls that occur at higher airspeeds than the normal one-G speed are accelerated stalls.

3-50　PA.V.A.K2d, PA.VII.C.K3

Which is a situation that could lead to an inadvertent accelerated stall?

A – Improperly performing a steep turn.

B – Leveling off too slowly from a steep descent.

C – Flying at too slow of an airspeed on final approach.

3-50. Answer A. GFDPP 3C, AFH

An accelerated stall occurs when an airplane stalls at a higher indicated airspeed when excessive maneuvering loads are imposed by steep turns, pull-ups, or other abrupt changes in its attitude. You increase the risk of experiencing an inadvertent accelerated stall during improperly performed turns, stall and spin recoveries, pullouts from steep dives, or when overshooting a base to final turn.

3-51　PA.IV.B.K1*

At any given time, the energy state of the airplane is determined by the amount and distribution of energy stored as

A – thrust and drag.

B – fuel and combustion.

C – altitude and airspeed.

3-51. Answer C. GFDPP 3C, AFH

An airplane gains energy from engine thrust, and loses energy to aerodynamic drag. The difference between the two, is the net change, which determines whether total mechanical energy, stored as altitude and airspeed, increases, decreases, or remains the same.

3-52　PA.IV.B.K1*

How does an airplane gain and lose energy?

A – The airplane gains energy from heat and loses energy to ambient temperature.

B – The airplane gains energy from engine thrust and loses energy to aerodynamic drag.

C – The airplane gains energy from lift and loses energy from aerodynamic drag.

3-52. Answer B. GFDPP 3C, AFH

An airplane gains energy from engine thrust and loses energy to aerodynamic drag. The difference between the two, is the net change, which determines whether total mechanical energy, stored as altitude and airspeed, increases, decreases, or remains the same.

SECTION C ■ **Aerodynamics of Maneuvering Flight**

3-53   PA.IV.B.K1*

Which is true regarding energy management?

A – The total mechanical energy of an airplane in flight is the sum of its potential energy from airspeed, and kinetic energy from altitude.

B – An airplane that is low and slow on final approach is an example of a total energy error.

C – An airplane that is high and slow on final approach is an example of a energy distribution error.

3-53. Answer B. GFDPP 3C, AFH

The total mechanical energy of an airplane in flight is the sum of its potential energy from altitude, and kinetic energy from airspeed.

A total energy error occurs when your airplane has too much or too little energy—altitude and airspeed deviate in the same direction (low and slow, or high and fast).

An energy distribution error occurs when your airplane has the right amount of total energy, but the distribution over altitude and airspeed is incorrect—altitude and airspeed deviate in opposite directions (high and slow, or low and fast).

3-54   PA.IV.B.K1*

If you are flying "on the back side of the power curve"

A – the airplane is at a low angle of attack and a low airspeed.

B – a great amount of power is required to overcome parasite drag.

C – any reduction in airspeed requires an increase in power to maintain level flight.

3-54. Answer C. GFDPP 3C, AFH

You can easily visualize energy management concepts using a graph showing an airplane's power required curve (commonly called the power curve).

On the back side of the power curve, flight at low airspeeds requires a high angle of attack and a great amount of power to overcome the resulting induced drag. Any reduction in airspeed requires an increase in power to maintain level flight.

Flying on the back side of the curve is discouraged, because reducing speed can demand more power than the engine can supply or an unplanned reduction of power could result in an involuntary descent.

CHAPTER 4

The Flight Environment

SECTION A
Safety of Flight

PILOT IN COMMAND

- The pilot in command is the final authority as to the operation of an aircraft.

- The pilot in command is directly responsible for the pre-takeoff briefing of the passengers for a flight.

- In addition to other preflight actions, the pilot in command is required to determine runway lengths at airports of intended use and the takeoff and landing distance data for that aircraft.

- The preflight action for flights away from the vicinity of an airport include checking weather reports and forecasts, fuel requirements, alternatives available if the flight cannot be completed as planned, and any known traffic delays.

- When an ATC clearance has been obtained, you may not deviate from that clearance, unless you obtain an amended clearance. The exception to this regulation is in an emergency.

- If an in-flight emergency requires immediate action, you may deviate from the FARs to the extent required to meet that emergency. A written report is not required unless requested by the FAA.

- If you deviate from a clearance, and are given priority by ATC because of that emergency, you must submit a detailed report of that emergency within 48 hours to the manager of that facility, if requested by ATC.

COLLISION AVOIDANCE

The majority of midair collisions occur during daylight hours, in VFR conditions, and within five miles of an airport.

VISUAL SCANNING

- You might not notice objects in your peripheral vision unless there is some relative motion.

- If there is no apparent relative motion between another aircraft and yours, you are probably on a collision course.

- Empty field myopia occurs when you are looking at a featureless sky that is devoid of objects, contrasting colors, or patterns and your eyes tend to focus at only 10 to 0 feet.

- In daylight conditions, the most effective way to scan is through a series of short, regularly-spaced eye movements in 10° sectors.

- Haze reduces visibility, and makes objects appear to be farther away than they really are.

- When climbing or descending under VFR along an airway, perform gentle banks left and right to enable continuous scanning of the area in front of the airplane.

- Blind spots make it difficult to see conflicting traffic. In both high-wing and low-wing designs, portions of your view are blocked by the fuselage and wings.

- Clearing turns allow you to see areas blocked by blind spots and make it easier to maintain visual contact with other aircraft in the practice area.

AIRPORT OPERATIONS

- Operation Lights On encourages you to use your landing lights during departures and approaches, both day and night, especially when operating within 10 miles of an airport, or in conditions of reduced visibility.

- During sunset to sunrise, except in Alaska, lighted position lights must be displayed on your aircraft.

FLIGHT DECK TRAFFIC DISPLAYS

- A cockpit display of traffic information (CDTI) is a dedicated screen, MFD, or GPS moving map that depicts traffic threats.

- General aviation traffic systems are advisory only, to help you locate traffic. Continuously scan for traffic by looking outside and cross check the CDTI to learn what areas need increased attention.

FORMATION FLIGHT

No person may operate an aircraft in formation flight except by prior arrangement with the pilot in command of each aircraft.

RIGHT-OF-WAY RULES

- An aircraft in distress has the right-of-way over all other aircraft.

- Primarily, there are three situations where right-of-way rules apply: converging with another aircraft, approaching another aircraft head-on, or overtaking another aircraft.

- When aircraft of different categories are converging: a glider has the right-of-way over an airship, powered parachute, weight-shift-control aircraft, airplane, or rotorcraft; and an airship has the right-of-way over a powered parachute, weight-shift-control aircraft, airplane, or rotorcraft.

- An aircraft that is towing or refueling another has the right-of-way over other engine-driven aircraft.

- When aircraft are approaching head-on, each shall give way to the right.

- When two or more aircraft are approaching the airport with the intention of landing, the one at the lower altitude has the right-of-way.

AEROBATIC FLIGHT

- To perform aerobatics, you must be at least 1,500 feet above ground level (AGL) with at least 3 statute miles of visibility.

- You may not operate an aircraft in aerobatic flight when over any congested area of a city, town, or settlement.

- You may not perform aerobatics within the lateral boundaries of any controlled airspace (Class B, C, D, and E) extending to the surface around an airport or in the airspace designated for Federal airways.

- In addition to aerobatic flight, operating an experimental or restricted aircraft is prohibited over densely populated areas and on congested Federal airways.

- A parachute with natural canopy, shroud, and harness components must have been packed by a certificated and appropriately rated parachute rigger within the preceding 60 days.

- A parachute with synthetic canopy, shroud, and harness components must have been packed by a certificated and appropriately rated parachute rigger within the preceding 180 days.

- No pilot of a civil aircraft carrying any person (other than a crewmember) may carry out any intentional maneuver that exceeds a nose-up or nose-down attitude of 30 degrees relative to the horizon unless each occupant is wearing an approved parachute.

MINIMUM SAFE ALTITUDES

- You must maintain FAA-designated minimum safe altitudes at all times except during takeoffs and landings. Complying with safe altitude rules also minimizes your risk of a wire strike accident.

- Over a congested area, such as a city, you must maintain an altitude of 1,000 feet above the highest obstacle within a horizontal distance of 2,000 feet.

- Over other-than congested areas, you must maintain an altitude of at least 500 feet above the surface, except over sparsely populated areas or open water, where you must stay at least 500 feet away from any person, vessel, vehicle, or structure.

- If you are unable to obtain a local altimeter setting before departing, set the altimeter to the local field elevation.

SECTION A ■ Safety of Flight

FLIGHT OVER HAZARDOUS TERRAIN

Mountain flying and flight over open water require specialized training from experienced instructors who are familiar with the area over which the flights will be conducted.

DROPPING OBJECTS

Objects may be dropped from an aircraft if precautions are taken to avoid injury or damage to persons or property on the surface.

TAXIING IN WIND

- While taxiing in wind, proper use of the aileron and elevator controls will help you maintain control of the airplane.

- When taxiing with a quartering headwind, hold the aileron up on the side from which the wind is blowing, and hold the elevator neutral.

- A quartering tailwind is the most critical wind condition to a tricycle-gear, high-wing airplane.

- When taxiing with a quartering tailwind, hold the aileron down on the side from which the wind is blowing, and hold the elevator down.

- When taxiing a tailwheel airplane, position the ailerons the same as you do for a tricycle-gear airplane. However, to help keep the tailwheel on the ground, hold the elevator control aft (elevator up) in a headwind, and in a tailwind, hold the elevator control forward (elevator down).

SAFETY BELTS

- As a flight crewmember, you are required to keep your safety belt and shoulder harness fastened during takeoffs and landings. Safety belts must stay fastened while enroute.

- As pilot in command, you must brief your passengers on the use of safety belts and notify them to fasten their safety belts during taxi, takeoff, and landing. Passengers must have their safety belts fastened during taxi, takeoffs, and landings.

POSITIVE EXCHANGE OF FLIGHT CONTROLS

To ensure that it is clear as to who has control of the aircraft, the FAA strongly recommends the use of a three-step process when exchanging the flight controls.

1. The pilot passing control says: *"You have the flight controls."*

2. The pilot taking control says: *"I have the flight controls."*

3. The pilot passing control says: *"You have the flight controls."*

NTSB 830

- If an aircraft is involved in an accident or serious injury, the nearest NTSB field office must be notified immediately.

- An aircraft accident is an occurrence that takes place between the time any person boards the aircraft with the intention of flight and all persons have disembarked in which any person suffers death or serious injury, or in which the aircraft receives substantial damage.

- Examples of serious incidents that require immediate notification to the nearest NTSB field office are a flight control system malfunction or failure and an in-flight fire.

- An event such as a cabin door opening in flight would not require a report. However, this occurrence has led to accidents so you must maintain control of the airplane and close the door safely on the ground.

- Aircraft wreckage may be moved prior to the time the NTSB takes custody only to protect the wreckage from further damage.

- The owner of an aircraft that has been involved in an accident is required to file an accident report within ten days.

- An overdue aircraft that is believed to be involved in an accident must be immediately reported to the nearest NTSB field office.

- The operator of an aircraft that has been involved in an incident is required to submit a report to the nearest NTSB field office when requested.

EMERGENCY EQUIPMENT AND SURVIVAL GEAR

- Designed to activate automatically if armed and subjected to crash-generated forces, emergency locator transmitters (ELTs) emit a distinctive audio tone on designated emergency frequencies.

- If necessary after a crash landing, you can manually activate the ELT—most aircraft have a remote ELT switch in the cockpit or are designed to provide pilot access to the ELT. The transmitters should operate continuously for at least 48 hours.

- Although not required by regulation, to enhance safety, you should:

    ◦ Pack your own gear or buy a commercial survival kit to keep in the airplane.

    ◦ Ensure the survival gear is appropriate for the terrain and weather along your route.

    ◦ Properly maintain and stow the equipment/gear.

    ◦ Brief your passengers on the location and use of all emergency equipment and survival gear.

- The only survival gear required by the FARs are flotation gear and a signaling device if the aircraft is operated for hire over water and beyond power-off gliding distance from shore. However, any complete survival kit should contain a basic core of survival supplies around which you can assemble the additional items that are appropriate to the terrain and weather you would encounter in an emergency, such as:

    ◦ Cold environments, including mountainous terrain—protective, warm clothing; a collapsible lightweight shelter.

    ◦ Overwater operations—a personal flotation device for you and each passenger.

    ◦ Hot desert conditions—sun block; clothing and a shelter for protection from the sand and sun.

NOTE: An asterisk appearing after an ACS code (i.e. PA.VII.B.K1*) indicates that the question subject appears more than one time in the ACS. The code shown corresponds to the first instance of the subject in the ACS.

4-1   PA.I.D.R3

In addition to other preflight actions for a VFR flight away from the vicinity of the departure airport, regulations specifically require the pilot in command to

A – review traffic control light signal procedures.

B – check the accuracy of the navigation equipment and the emergency locator transmitter (ELT).

C – determine runway lengths at airports of intended use and the takeoff and landing distance data for the aircraft.

4-1. Answer C. GFDPP 4A, FAR 91.103
As pilot in command, for any flight, you must determine runway lengths at the airports of intended use and the takeoff and landing distance data for your airplane.

SECTION A ■ Safety of Flight

4-2   PA.I.D.S1, PA.VI.C.K1

Preflight action, as required for all flights away from the vicinity of an airport, shall include

A – the designation of an alternate airport.

B – a study of arrival procedures at airports/heliports of intended use.

C – an alternate course of action if the flight cannot be completed as planned.

4-2. Answer C. GFDPP 4A, FAR 91.103

The preflight action for flights away from the vicinity of an airport include checking weather reports and forecasts, fuel requirements, alternatives available if the flight cannot be completed as planned, and any known traffic delays.

4-3   PA.I.A.K2

The final authority as to the operation of an aircraft is the

A – Federal Aviation Administration.

B – pilot in command.

C – aircraft manufacturer

4-3. Answer B. GFDPP 4A, FAR 91.3

As indicated in the regulation, the pilot in command of an aircraft is directly responsible for, and is the final authority as to, the operation of that aircraft.

4-4   PA.VI.C.K2

When an ATC clearance has been obtained, no pilot in command may deviate from that clearance, unless that pilot obtains an amended clearance. The one exception to this regulation is

A – an emergency.

B – when the clearance states "at pilot's discretion."

C – if the clearance contains a restriction.

4-4. Answer A. GFDPP 4A, FAR 91.3, 91.123

When you receive an ATC clearance, you may not deviate from that clearance, except in an emergency, unless you obtain an amended clearance. A "pilot's discretion" clearance still must be followed, even though you have more latitude in complying with that clearance.

4-5   PA.VI.C.K2

As pilot in command of an aircraft, under which situation may you deviate from an ATC clearance?

A – In an emergency.

B – When operating in Class A airspace at night.

C – If an ATC clearance is not understood and in VFR conditions.

4-5. Answer A. GFDPP 4A, FAR 91.3

As pilot in command, you are expected to do what is necessary in an emergency. For example, in rare instances you might have to deviate from an ATC clearance to avoid a collision with terrain or with another aircraft. Any pilot deviating from an ATC clearance must report the deviation to ATC as soon as possible and obtain a new clearance.

4-6   PA.VI.C.K2

What action, if any, is appropriate if you deviate from an ATC instruction during an emergency and are given priority?

A – Take no special action because you are pilot in command.

B – File a detailed report within 48 hours to the chief of the appropriate ATC facility, if requested.

C – File a report to the FAA Administrator, as soon as possible

4-6. Answer B. GFDPP 4A, FAR 91.3
A pilot in command who is given priority by ATC because of an emergency shall submit a detailed report of that emergency within 48 hours to the manager of that ATC facility, if requested by ATC.

4-7   PA.VI.C.K2

If an in-flight emergency requires immediate action, as pilot in command, you may

A – deviate from the FARs to the extent required to meet the emergency, but must submit a written report to the Administrator within 24 hours.

B – deviate from the FARs to the extent required to meet that emergency.

C – not deviate from the FARs unless prior approval for the deviation is granted by the Administrator.

4-7. Answer B. GFDPP 4A, FAR 91.3. 91.123
In an in-flight emergency requiring immediate action, you may deviate from any rule to the extent required to meet that emergency. A written report is not required unless requested by the FAA.

4-8   PA.VI.C.K2

When must a pilot who deviates from a regulation during an emergency send a written report of that deviation to the Administrator?

A – Within 7 days.

B – Within 10 days.

C – Upon request.

4-8. Answer C. GFDPP 4A, FAR 91.3, 91.123
The regulations states that a written report is not required unless a report is requested from the FAA.

4-9   PA.VI.C.K2

When would a you be required to submit a detailed report of an emergency that caused you to deviate from an ATC clearance?

A – When requested by ATC.

B – Immediately.

C – Within 7 days.

4-9. Answer A. GFDPP 4A, FAR 91.3, 91.123
A pilot in command who deviates from a clearance, and is then given priority by ATC because of that emergency, shall submit a detailed report of that emergency within 48 hours to the manager of that ATC facility, if requested by ATC.

SECTION A ■ Safety of Flight

4-10   PA.V.A.S1*

Before starting each maneuver, you should

A – check altitude, airspeed, and heading indications.

B – visually scan the entire area for collision avoidance.

C – announce your intentions on the nearest CTAF.

4-10. Answer B. GFDPP 4A, AIM
To ensure you can see other aircraft that might be blocked by blind spots, make clearing turns and scan the area before and during your practice of maneuvers.

4-11   PA.III.B.R1*

What collision avoidance procedure is recommended when climbing or descending VFR on an airway?

A – Execute gentle banks, left and right for continuous visual scanning of the airspace.

B – Advise the nearest FSS of the altitude changes.

C – Fly away from the centerline of the airway before changing altitude.

4-11. Answer A. GFDPP 4A, AIM
Because of potential traffic on airways, it is important to scan. Making shallow turns enables you to compensate for blind spots.

4-12   PA.III.B.R1*

Eye movements during daytime collision avoidance scanning should

A – not exceed 10 degrees and view each sector at least 1 second.

B – be 30 degrees and view each sector at least 3 seconds.

C – use peripheral vision by scanning small sectors and utilizing off-center viewing.

4-12. Answer A. GFDPP 4A, AIM
The eyes are able to focus clearly only on a small area, approximately 10°, so a series of short eye movements is most effective. Peripheral vision and off-center viewing are more effective at night time.

4-13   PA.III.B.R1*

Which technique should you use to scan for traffic to the right and left during straight-and-level flight?

A – Systematically focus on different segments of the sky for short intervals.

B – Concentrate on relative movement detected in the peripheral vision area.

C – Continuous sweeping of the windshield from right to left.

4-13. Answer A. GFDPP 4A, AIM
The eyes are able to focus clearly only on a small area, approximately 10°, so a series of short eye movements is most effective.

4-14   PA.III.B.R1*

What is an effective way to scan for traffic in front of your airplane during climb?

A – Perform 90 degree turns every 1–2 miles to check for traffic along your flight path.

B – Use your cockpit display of traffic information (CDTI) to avoid aircraft that cannot be seen in a nose-high attitude.

C – Lower the nose periodically and clear the area, and then transition to cruise climb when altitude permits.

4-14. Answer C. GFDPP 4A, AIM
Your forward visibility is reduced in the climb. To clear the area, lower the nose periodically for just a moment, and then return to the climb attitude. Another method is to perform small S-turns (not 90-degree turns) to check for traffic along your flight path. Transition to a cruise climb when altitude permits in order to improve forward visibility and engine cooling, and to increase your groundspeed. A CDTI should be used as a back-up, not a replacement, for your scanning technique

4-15   PA.III.B.R1*

How can you determine if another aircraft is on a collision course with your aircraft?

A – The other aircraft always appears to get larger and closer at a rapid rate.

B – The nose of each aircraft is pointed at the same point in space.

C – No relative motion is apparent between your aircraft and the other aircraft.

4-15. Answer C. GFDPP 4A, AIM
A lack of relative movement can indicate that the two aircraft are moving toward one another on a collision course.

4-16   PA.III.B.R1*, PA.I.H.R3

Most midair collision accidents occur during

A – foggy days.

B – clear days.

C – cloudy nights.

4-16. Answer B. GFDPP 4A, AFH
The majority of midair collisions occur during daylight hours, in VFR conditions, and within 5 miles of an airport.

4-17   PA.III.B.R1*

The Aeronautical Information Manual (AIM) specifically encourages pilots to turn on their landing lights when operating below 10,000 feet, day or night, and especially when operating

A – in Class B airspace.

B – in conditions of reduced visibility.

C – within 15 miles of a towered airport.

4-17. Answer B. GFDPP 4A, AIM
The FAA's voluntary pilot safety program, *Operation Lights On*, encourages you to turn on your landing lights when operating below 10,000 feet, day or night, especially within 10 miles of any airport; also in conditions of reduced visibility and in areas where flocks of birds can be expected, such as coastal areas, lake areas, and around trash dumps.

SECTION A ■ **Safety of Flight**

4-18   PA.III.B.R1*

What is the appropriate way to use a cockpit display of traffic information (CDTI) to avoid a collision?

A – Monitor the CDTI and if you receive a traffic alert, turn away from target shown on the display.

B – Continuously scan for traffic by looking outside and cross check the CDTI to learn what areas need increased attention.

C – To avoid complacency, use the CDTI only after you are instrument rated and flying IFR.

4-18. Answer B. GFDPP 4A, AIM
General aviation traffic systems are advisory only, to help you locate traffic. You may not fly any avoidance maneuvers without first seeing the traffic out the window.

4-19   PA.XI.A.K3

Except in Alaska, during what time period should lighted position lights be displayed on an aircraft?

A – End of evening civil twilight to the beginning of morning civil twilight.

B – 1 hour after sunset to 1 hour before sunrise.

C – Sunset to sunrise.

4-19. Answer C. GFDPP 4A, FAR 91.209
No person may, during the period from sunset to sunrise, operate an aircraft unless it has lighted position lights.

4-20   PA.I.A.K2

No person may operate an aircraft in formation flight

A – over a densely populated area.

B – in Class D Airspace under special VFR.

C – except by prior arrangement with the pilot in command of each aircraft.

4-20. Answer C. GFDPP 4A, FAR 91.111
No person may operate an aircraft in formation flight except by arrangement with the pilot in command of each aircraft in the formation.

4-21   PA.III.B.K3

Which aircraft has the right-of-way over all other air traffic?

A – A balloon.

B – An aircraft in distress.

C – An aircraft on final approach to land.

4-21. Answer B. GFDPP 4A, FAR 91.113
An aircraft in distress has the right-of-way over all other aircraft.

4-22   PA.III.B.K3

What action is required when two aircraft of the same category converge, but not head-on?

A – The faster aircraft shall give way.

B – The aircraft on the left shall give way.

C – Each aircraft shall give way to the right.

4-22. Answer B. GFDPP 4A, FAR 91.113
The aircraft on the right has the right-of-way and the aircraft on the left shall give way.

4-23   PA.III.B.K3

Which aircraft has the right-of-way over the other aircraft listed?

A – Glider.

B – Airship.

C – Aircraft refueling other aircraft.

4-23. Answer A. GFDPP 4A, FAR 91.113

In general, the least maneuverable or non-powered aircraft have the right-of-way. A glider has the right-of-way over an airship, airplane, or rotorcraft. An aircraft that is towing or refueling another aircraft has the right-of-way over all other engine-driven aircraft (but not a glider).

4-24   PA.III.B.K3

An airplane and an airship are converging. If the airship is left of the airplane's position, which aircraft has the right-of-way?

A – The airship.

B – The airplane.

C – Each pilot should alter course to the right.

4-24. Answer A. GFDPP 4A, FAR 91.113

In general, the least maneuverable aircraft normally has the right-of-way. A glider has the right-of-way over an airship, airplane, or rotorcraft. An aircraft that is towing or refueling another aircraft has the right-of-way over all other engine-driven aircraft (but not a glider). Because an airship is less maneuverable than an airplane, the airship has the right-of-way.

4-25   PA.III.B.K3

Which aircraft has the right-of-way over the other aircraft listed?

A – Airship.

B – Aircraft towing other aircraft.

C – Gyroplane.

4-25. Answer B. GFDPP 4A, FAR 91.113

An aircraft towing or refueling another aircraft has the right-of-way over all other engine-driven aircraft.

4-26   PA.III.B.K3

What action should the pilots of a glider and an airplane take if on a head-on collision course?

A – The airplane pilot should give way to the left.

B – The glider pilot should give way to the right.

C – Both pilots should give way to the right.

4-26. Answer C. GFDPP 4A, FAR 91.113

When any aircraft are approaching each other head-on, both pilots should alter their course to the right. For aircraft approaching head-on, the FARs do not differentiate between aircraft categories.

4-27   PA.III.B.K3

When two or more aircraft are approaching an airport for landing, the right-of-way belongs to the aircraft

A – that has the other to its right.

B – that is the least maneuverable.

C – at the lower altitude, but it shall not take advantage of this rule to cut in front of or to overtake another.

4-27. Answer C. GFDPP 4A, FAR 91.113

When two or more aircraft are approaching an airport for landing, the one at the lower altitude has the right-of-way, but shall not use this rule to cut in front of another aircraft.

SECTION A ■ Safety of Flight

4-28   PA.I.E.K1

No person may operate an aircraft in aerobatic flight when

A – flight visibility is less than 5 miles.

B – over any congested area of a city, town, or settlement.

C – less than 2,500 feet AGL.

4-28. Answer B. GFDPP 4A, FAR 91.303

No person may operate an aircraft in aerobatic flight over any congested area of a city, town, or settlement.

4-29   PA.I.E.K1

In which class of airspace is aerobatic flight prohibited?

A – Class E airspace not designated for Federal Airways above 1,500 feet AGL.

B – Class E airspace below 1,500 feet AGL.

C – Class G airspace above 1,500 feet AGL.

4-29. Answer B. GFDPP 4A, FAR 91.303

No person may operate an aircraft in aerobatic flight within class B, C, or D airspace, class E airspace designated for an airport, or within

4 NM of the centerline of any Federal Airway. Aerobatic flight is also prohibited below 1,500 feet AGL and when the flight visibility is less than 3 statute miles.

4-30   PA.I.E.K1

What is the lowest altitude permitted for aerobatic flight?

A – 1,000 feet AGL.

B – 1,500 feet AGL.

C – 2,000 feet AGL.

4-30. Answer B. GFDPP 4A, FAR 91.303

No person may operate an aircraft in aerobatic flight below an altitude of 1,500 feet above the surface.

4-31   PA.I.E.K1

No person may operate an aircraft in aerobatic flight when the flight visibility is less than

A – 3 miles.

B – 5 miles.

C – 7 miles.

4-31. Answer A. GFDPP 4A, FAR 91.303

No person may operate an aircraft in aerobatic flight when flight visibility is less than three statute miles.

4-32   PA.I.B.K3

An approved parachute constructed of natural materials must have been packed by a certificated and appropriately rated parachute rigger within the preceding

A – 60 days.

B – 90 days.

C – 120 days.

4-32. Answer A. GFDPP 4A, FAR 91.307

Parachutes must be repacked periodically to satisfy the FARs. The materials used determine the interval. No pilot of a civil aircraft may allow an emergency parachute to be carried in that aircraft unless it is an approved type and has been packed by a certificated and appropriately rated parachute rigger within the preceding 60 days if its canopy, shrouds, and harness are composed exclusively of natural fiber or materials.

4-33   PA.I.B.K3

An approved synthetic parachute may be carried in an aircraft for emergency use if it has been packed by an appropriately rated parachute rigger within the preceding

A – 120 days.

B – 180 days.

C – 365 days.

4-33. Answer B. GFDPP 4A, FAR 91.307

Parachutes must be repacked periodically to satisfy the FARs. The materials used determine the interval. No pilot of a civil aircraft may allow an emergency parachute to be carried in that aircraft unless it is an approved type and has been packed by a certificated and appropriately rated parachute rigger within the preceding 180 days if its canopy, shrouds, and harness are composed exclusively of nylon, rayon, or other similar synthetic fiber or materials.

4-34   PA.I.B.K3

With certain exceptions, when must each occupant of an aircraft wear an approved parachute?

A – When a door is removed from the aircraft to facilitate parachute jumpers.

B – When intentionally pitching the nose of the aircraft up or down 30° or more.

C – When intentionally banking in excess of 30°.

4-34. Answer B. GFDPP 4A, FAR 91.307

Unless each occupant of the aircraft is wearing an approved parachute, no pilot of a civil aircraft, carrying any person (other than a crewmember) may carry out any intentional maneuver that exceeds a nose-up or nose-down attitude of 30 degrees relative to the horizon.

4-35   PA.I.B.K1c

Which is normally prohibited when operating a restricted category civil aircraft?

A – Flight under instrument flight rules.

B – Flight over a densely populated area.

C – Flight within Class D airspace.

4-35. Answer B. GFDPP 4A, FAR 91.313

No person may operate a restricted category civil aircraft within the United States over a densely populated area.

4-36   PA.I.B.K1c

Unless otherwise specifically authorized, no person may operate an aircraft that has an experimental certificate

A – beneath the floor of Class B airspace.

B – over a densely populated area or in a congested airway.

C – from the primary airport with Class D airspace.

4-36. Answer B. GFDPP 4A, FAR 91.319

Unless otherwise authorized by the Administrator in special operating limitations, no person may operate an aircraft that has an experimental certificate over a densely populated area or in a congested airway.

4-37   PA.I.D.K2

Except when necessary for takeoff or landing, what is the minimum safe altitude for a pilot to operate an aircraft anywhere?

A – An altitude allowing, if a power unit fails, an emergency landing without undue hazard to persons or property on the surface.

B – An altitude of 500 feet above the surface and no closer than 500 feet to any person, vessel, vehicle, or structure.

C – An altitude of 500 feet above the highest obstacle within a horizontal radius of 1,000 feet.

4-37. Answer A. GFDPP 4A, FAR 91.119
Maintain enough altitude to allow an emergency landing in the event of an engine failure without undue hazard to people or property on the surface.

4-38   PA.I.D.K2

Except when necessary for takeoff or landing, what is the minimum safe altitude required for a pilot to operate an aircraft over congested areas?

A – An altitude of 1,000 feet above any person, vessel, vehicle, or structure.

B – An altitude of 500 feet above the highest obstacle within a horizontal radius of 1,000 feet.

C – An altitude of 1,000 feet above the highest obstacle within a horizontal radius of 2,000 feet.

4-38. Answer C. GFDPP 4A, FAR 91.119
The minimum safe altitude required over a congested area is 1,000 feet above any obstacle within a horizontal radius of 2,000 feet of the aircraft.

4-39   PA.I.D.K2

Except when necessary for takeoff or landing, what is the minimum safe altitude for a pilot to operate an aircraft over other than a congested area?

A – An altitude of 1,000 feet above the highest obstacle within a horizontal radius of 2,000 feet.

B – An altitude of 500 feet AGL, except over open water or a sparsely populated area, which requires 500 feet from any person, vessel, vehicle, or structure.

C – An altitude of 500 feet above the highest obstacle within a horizontal radius of 1,000 feet.

4-39. Answer B. GFDPP 4A, FAR 91.119
The minimum safe altitude over a uncongested area is 500 feet AGL, except over open water or a sparsely populated area, which requires 500 feet (horizontally or vertically) from any person, vessel, vehicle, or structure.

4-40   PA.I.D.K2
Except when necessary for takeoff or landing, an aircraft may not be operated closer than what distance from any person, vessel, vehicle, or structure?

A – 500 feet.

B – 700 feet.

C – 1,000 feet.

4-41   PA.I.A.K2
Under what conditions may objects be dropped from an aircraft?

A – Only in an emergency.

B – If precautions are taken to avoid injury or damage to persons or property on the surface.

C – If prior permission is received from the Federal Aviation Administration.

4-42   PA.II.D.S3
When taxiing with strong quartering tailwinds, which aileron positions should be used?

A – Aileron down on the downwind side.

B – Ailerons neutral.

C – Aileron down on the side from which the wind is blowing.

4-43   PA.II.D.S3
Which aileron positions should a pilot generally use when taxiing in strong quartering headwinds?

A – Aileron up on the side from which the wind is blowing.

B – Aileron down on the side from which the wind is blowing.

C – Ailerons neutral.

4-44   PA.II.D.S3
Which wind condition would be most critical when taxiing a nosewheel equipped high-wing airplane?

A – Quartering tailwind.

B – Direct crosswind.

C – Quartering headwind.

4-40. Answer A. GFDPP 4A, FAR 91.119
Over a sparsely populated or open water area, you must remain at least 500 feet (horizontally or vertically) from any person, vessel, vehicle, or structure.

4-41. Answer B. GFDPP 4A, FAR 91.15
Objects can be dropped from an aircraft in flight, if reasonable precautions are taken to avoid injury or damage to persons or property.

4-42. Answer C. GFDPP 4A, AFH
With a quartering tailwind, the aileron should be down on the side from which the wind is blowing to prevent the wind from flowing under and lifting the wing.

4-43. Answer A. GFDPP 4A, AFH
To counteract the lifting tendency of a quartering headwind, the aileron should be up on the side from which the wind is blowing.

4-44. Answer A. GFDPP 4A, AFH
A tricycle-gear, high-wing airplane is most susceptible to a quartering tailwind because a strong airflow beneath the wing and horizontal stabilizer can lift the airplane and cause it to nose over and flip on its back.

SECTION A ■ Safety of Flight

4-45   PA.II.D.S3, PA.I.G.K1a

(Refer to Figure 9, area A.) How should the flight controls be held while taxiing a tricycle-gear equipped airplane into a left quartering headwind?

A – Left aileron up, elevator neutral.

B – Left aileron down, elevator neutral.

C – Left aileron up, elevator down.

4-45. Answer A. GFDPP 4A, AFH

While taxiing a tricycle-gear airplane in a quartering headwind, the aileron should be up on the side from which the wind is blowing, and the elevator neutral to prevent any lifting force on the tail. In this case, the wind is from the left, so the left aileron should be up.

4-46   PA.II.D.S3, PA.I.G.K1a

(Refer to Figure 9, area B.) How should the flight controls be held while taxiing a tailwheel airplane into a right quartering headwind?

A – Right aileron up, elevator up.

B – Right aileron down, elevator neutral.

C – Right aileron up, elevator down.

4-46. Answer A. GFDPP 4A, AFH

When taxiing a tailwheel airplane, position the ailerons the same as you do for a tricycle-gear airplane, However, to help keep the tailwheel on the ground, hold the elevator control aft (elevator up) in a headwind, and in a tailwind, hold the elevator control forward (elevator down). Because the tail of most tailwheel airplanes is lower than the nose while taxiing, a strong headwind with a neutral or down elevator could lift the tail.

4-47   PA.II.D.S3, PA.I.G.K1a

(Refer to Figure 9, area C.) How should the flight controls be held while taxiing a tailwheel airplane with a left quartering tailwind?

A – Left aileron up, elevator neutral.

B – Left aileron down, elevator neutral.

C – Left aileron down, elevator down.

4-47. Answer C. GFDPP 4A, AFH

For a quartering tailwind, the controls are held the same for both tailwheel and tricycle-gear airplanes. Ailerons are down on the side from which the wind is blowing. The elevator is down to prevent the wind from lifting the tail.

4-48   PA.II.B.K1, PA.II.F.K1

Pre-takeoff briefing of passengers about the use of seat belts for a flight is the responsibility of

A – all passengers.

B – the pilot in command.

C – the right seat pilot.

4-48. Answer B. GFDPP 4A, FAR 91.107, 91.519

As the pilot in command, you must ensure that each person on board is briefed on how to fasten and unfasten the safety belt and shoulder harness. You must also ensure that all persons on board are notified to fasten their safety belt (and shoulder harness, if installed) during taxi, takeoff, or landing.

4-49 PA.II.B.K1, PA.II.F.K1

Regarding passengers, what obligation, if any, does a pilot in command have concerning the use of safety belts?

A – The pilot in command must instruct the passengers to keep safety belts fastened for the entire flight.

B – The pilot in command must brief the passengers on the use of safety belts and notify them to fasten their safety belts during taxi, takeoff, and landing.

C – The pilot in command has no obligation concerning passengers' use of safety belts.

4-49. Answer B. GFDPP 4A, FAR 91.107, 91.519

As the pilot in command, you must ensure that each person on board is briefed on how to fasten and unfasten the safety belt and shoulder harness. You must also ensure that all persons on board are notified to fasten their safety belt (and shoulder harness, if installed) during taxi, takeoff, or landing.

4-50 PA.II.B.K1

Flight crewmembers are required to keep their safety belts and shoulder harnesses fastened during

A – takeoffs and landings.

B – all flight conditions.

C – flight in turbulent air.

4-50. Answer A. GFDPP 4A, FAR 91.105

According to the regulations, safety belts are required during takeoff and landing and while enroute. In addition, shoulder harnesses are required during takeoff and landing, unless the seat of the crewmembers' stations are not equipped with shoulder harnesses, or the crewmembers are not able to perform their duties with the shoulder harness fastened.

4-51 PA.II.B.K1

Safety belts are required to be properly secured about which persons in an aircraft and when?

A – Pilots only, during takeoffs and landings.

B – Passengers, during taxi, takeoffs, and landings only.

C – Each person on board the aircraft during the entire flight.

4-51. Answer B. GFDPP 4A, FAR 91.107

Passengers are required to have safety belts (and shoulder harnesses, if installed) fastened during taxi, takeoff, and landing.

4-52 PA.II.B.K1

Which best describes the flight conditions under which flight crewmembers are specifically required to keep their safety belts and shoulder harnesses fastened?

A – Safety belts during takeoff and landing; shoulder harnesses during takeoff and landing.

B – Safety belts during takeoff and landing; shoulder harnesses during takeoff and landing and while enroute.

C – Safety belts during takeoff and landing and while enroute; shoulder harnesses during takeoff and landing.

4-52. Answer C. GFDPP 4A, FAR 91.105

During takeoff and landing, and while enroute, each required flight crewmember shall keep the safety belt fastened while at the crewmember station. In addition, each required flight crewmember shall, during takeoff and landing, keep the shoulder harness fastened while at the crewmember station.

SECTION A ■ **Safety of Flight**

4-53   PA.II.B.K1

With certain exceptions, safety belts are required to be secured about passengers during

A – taxi, takeoffs, and landings.

B – all flight conditions.

C – flight in turbulent air.

4-53. Answer A. GFDPP 4A, FAR 91.107
Passengers are only required to have safety belts (and shoulder harnesses, if installed) fastened during taxi, takeoff, and landing.

4-54   PA.V.A.R4*

Your instructor has demonstrated a maneuver and wants you to try it. What steps should you complete to assume control of the airplane?

A – Tell the instructor that you are ready to try the maneuver and start performing the maneuver when the airplane is stabilized.

B – Start performing the maneuver after your instructor says, *"you have the flight controls,"* and let the instructor monitor your performance.

C – After your instructor says, *"you have the flight controls"*, say *"I have the flight controls"*, and watch for your instructor to confirm again, *"you have the flight controls."*

4-54. Answer C. GFDPP 4A, ACS
To ensure that it is clear who has control of the airplane, the FAA recommends the use of a three-step process when exchanging the flight controls.

- The pilot passing control says: *"You have the flight controls."*

- The pilot taking control says: *"I have the flight controls."*

- The pilot passing control says: *"You have the flight controls."*

4-55   PA.III.A.K8, PA.IX.C.K2e

Which incident requires an immediate notification to the nearest NTSB field office?

A – A forced landing due to engine failure.

B – Landing gear damage, due to a hard landing.

C – Flight control system malfunction or failure.

4-55. Answer C. GFDPP 4A, NTSB 830.5
The operator of an aircraft shall immediately, and by the most expeditious means available, notify the nearest NTSB field office when a flight control system malfunction or failure occurs.

4-56   PA.III.A.K8

Which incident would require an immediate notification to the nearest NTSB field office?

A – An in-flight generator or alternator failure.

B – An in-flight fire.

C – An in-flight loss of VOR receiver capability.

4-56. Answer B. GFDPP 4A, NTSB 830.5
The operator of an aircraft shall immediately, and by the most expeditious means available, notify the nearest NTSB field office when a fire in flight occurs.

4-57   PA.III.A.K8

Which incident requires an immediate notification be made to the nearest NTSB field office?

A – An overdue aircraft that is believed to be involved in an accident.

B – Inadvertent door or window opening in-flight.

C – An in-flight generator or alternator failure.

4-57. Answer A. GFDPP 4A, NTSB 830.5
The operator of an aircraft shall immediately, and by the most expeditious means available, notify the nearest NTSB field office when an overdue aircraft is believed to be involved in an accident.

4-58   PA.III.A.K8, PA.XIII.A.K1, K2

On a postflight inspection of your aircraft after an aborted takeoff due to an elevator malfunction, you find that the elevator control cable has broken. According to NTSB 830, you

A – must immediately notify the nearest NTSB office.

B – should notify the NTSB within 10 days.

C – must file a NASA report immediately

4-58. Answer A. GFDPP 4A, NTSB 830.5
NTSB 830.5 states that you must immediately notify the NTSB if an aircraft accident or serious incident occurs. Serious incidents include occurrences such as flight control system malfunction or failure; inability of any required flight crewmember to perform normal flight duties as a result of injury or illness; and in-flight fire.

4-59   PA.III.A.K8

The operator of an aircraft that has been involved in an incident is required to submit a report to the nearest field office of the NTSB

A – within 7 days.

B – within 10 days.

C – when requested.

4-59. Answer C. GFDPP 4A, NTSB 830.15
A report on an incident for which notification is required by 830.5(a) shall be filed only when requested by an authorized representative of the NTSB.

SECTION A ■ **Safety of Flight**

4-60   PA.III.A.K8

The operator of an aircraft that has been involved in an accident is required to file an accident report within how many days?

A – 5.

B – 7.

C – 10.

4-60. Answer C. GFDPP 4A, NTSB 830.15

The operator of an aircraft shall file a report within 10 days after an accident, or after 7 days if an overdue aircraft is still missing.

4-61   PA.III.A.K8

If an aircraft is involved in an accident that results in substantial damage to the aircraft, the nearest NTSB field office should be notified

A – immediately.

B – within 48 hours.

C – within 7 days.

4-61. Answer A. GFDPP 4A, NTSB 830.5

The operator of an aircraft shall immediately, and by the most expeditious means available, notify the nearest NTSB field office when an aircraft accident occurs.

4-62   PA.III.A.K8

May aircraft wreckage be moved prior to the time the NTSB takes custody?

A – Yes, but only if moved by a federal, state, or local law enforcement officer.

B – Yes, but only to protect the wreckage from further damage.

C – No, it may not be moved under any circumstances.

4-62. Answer B. GFDPP 4A, NTSB 830.10

Before the NTSB or its authorized representative takes custody of aircraft wreckage, mail, or cargo, such wreckage may not be disturbed or moved except to the extent necessary to protect the wreckage from further damage.

4-63   PA.IX.C.K5

Which is an appropriate action to take to manage a cabin door opening after departure?

A – Immediately attempt to close the door.

B – Perform a normal traffic pattern and landing and close the door on the ground.

C – Decrease the airspeed to just above a stall speed to make it easier to close the door.

4-63. Answer B. GFDPP 4A, AFH

Accidents have occurred on takeoff because pilots have stopped flying the airplane to concentrate on closing cabin or baggage doors. To manage an inadvertent door opening during flight:

- Stay calm and maintain control of the airplane.

- Do not release your shoulder harness in an attempt to reach the door.

- Close the door safely on the ground after landing as soon as practical by performing a normal traffic pattern and landing

4-64   PA.IX.D.K2, K3

Which is true regarding emergency equipment and survival gear?

A – FAR 91.205 requires that a fire extinguisher, ELT, and a survival kit be on board the airplane.

B – Because items might vary depending on the airplane, you must be familiar with using the specific emergency emergency equipment and survival equipment on board each airplane you fly.

C – You are required by FAR Part 91 to brief your passengers on the location and use of all emergency equipment and survival gear.

2-64. Answer B. GFDPP 4A, PHB

You should be familiar with the emergency equipment and survival gear in each airplane you fly. Depending on the airplane, emergency equipment can include items such as a fire extinguisher, egress hammer, or a parachute system.

Although not required by regulation, to enhance safety, you should:

- Pack your own gear or buy a commercial survival kit to keep in the airplane.

- Ensure the survival gear is appropriate for the terrain and weather along your route.

- Properly maintain and stow the equipment and/or gear.

- Brief your passengers on the location and use of all emergency equipment and survival gear.

4-65   PA.IX.D.K3a, K3b, K3c

When packing a survival kit to have on board the airplane, you should include

A – only the items required by regulation.

B – equipment/gear needed for the specific terrain and weather along your route.

C – no more than a few core items to limit the weight in the baggage area.

2-65. Answer B. GFDPP 4A, PHB

The only survival gear required by the FARs are flotation gear and a signaling device if the aircraft is operated for hire over water and beyond power-off gliding distance from shore. However, any complete survival kit should contain a basic core of survival supplies around which you can assemble the additional items that are appropriate to the terrain and weather you would encounter in an emergency, such as:

- Cold environments, including mountainous terrain—protective, warm clothing; a collapsible lightweight shelter.

- Overwater operations—a personal flotation device for you and each passenger.

- Hot desert conditions—sun block; clothing and shelter for protection from the sand and sun.

SECTION A ■ Safety of Flight

SECTION B
Airports

TRAFFIC PATTERN

- The correct traffic pattern procedure to use at a uncontrolled airport is to comply with any FAA traffic pattern established for the airport.

- Traffic pattern indicators on the segmented circle show the final and base legs to various runways on the airport.

- The wind cone or wind sock in the center of the segmented circle provides current wind direction.

- State your position on the airport when calling the tower for takeoff, particularly when at a runway intersection.

RUNWAY LAYOUT AND MARKINGS

- Runway numbers correspond to the magnetic direction of the runway and are rounded to the nearest 10 degrees, with the last zero dropped. For example, a runway that is oriented to 357 degrees is numbered 36.

- The area before a displaced threshold may be used for taxi and takeoff.

- You should land on the runway pavement after the displaced threshold.

- A closed runway is marked with Xs painted on its surface at each end.

AIRPORT SIGNS
There are six basic types of airport signs:

- **Direction Signs** indicate directions of taxiways leading out of an intersection. They have black inscriptions on a yellow background and always contain arrows that show the approximate direction of turn.

- **Mandatory Instruction Signs** denote an entrance to a runway, a critical area, or an area prohibited to aircraft. These signs are red with white letters or numbers. Examples of mandatory instruction signs include:

- **Runway Holding Position Signs** are located at the holding position on taxiways that intersect a runway or on runways that intersect other runways.

- **No Entry Signs** prohibit an aircraft from entering an area. Typically, this sign would be located on a taxiway intended to be used in only one direction or at the intersection of vehicle roadways with runways, taxiways, or aprons where the roadway may be mistaken as a taxiway or other aircraft movement surface.

- **Location Signs** identify either the taxiway or runway where your aircraft is located. These signs are black with yellow inscriptions and a yellow border. Location signs also identify the runway boundary or ILS critical area for aircraft exiting the runway. The runway boundary sign, which faces the runway and is visible to you when exiting the runway, is located adjacent to the holding position marking on the pavement, providing you with another visual cue to determine when you are clear of the runway.

- **Runway Distance Remaining Signs** provide distance remaining information to pilots during takeoff and landing operations. The signs are located along the sides of the runway, and the inscription consists of a white numeral on a black background. The signs indicate the distance remaining in thousands of feet. Runway distance remaining signs are recommended for runways used by turbojet aircraft.

- **Information Signs** advise you of such things as areas that cannot be seen from the control tower, applicable radio frequencies, and noise abatement procedures. These signs use yellow backgrounds with black inscriptions.

- **Destination Signs** indicate the general direction to a location on the airport, such as runways, aprons, terminals, military areas, civil aviation areas, cargo areas, international areas, and FBOs. They have black inscriptions on a yellow background and always have an arrow showing the direction of the taxiing route to that destination.

In addition to knowledge test questions about the purpose of runway and taxiway markings and airport signs, you might be required to identify specific signs and markings shown in illustrations. Refer to the FAA Airport Sign and Marking — Quick Reference Guide to review the various signs and markings.

AIRPORT SIGN AND MARKING – QUICK REFERENCE GUIDE

EXAMPLE	TYPE OF SIGN	PURPOSE	LOCATION/CONVENTION
4 - 22	Mandatory: Hold position for taxiway/runway intersection.	Denotes entrance to runway from a taxiway.	Located L side of taxiway within 10 feet of hold position markings.
22 - 4	Mandatory: Holding position for runway/runway intersection.	Denotes intersecting runway.	Located L side of rwy prior to intersection, & R side if rwy more than 150' wide, used as taxiway, or has "land & hold short" ops.
4 - APCH	Mandatory: Holding position for runway approach area.	Denotes area to be protected for aircraft approaching or departing a runway.	Located on taxiways crossing thru runway approach areas where an aircraft would enter an RSA or apch/departure airspace.
ILS	Mandatory: Holding position for ILS critical area/precision obstacle free zone.	Denotes entrance to area to be protected for an ILS signal or approach airspace.	Located on twys where the twys enter the NAVAID critical area or where aircraft on taxiway would violate ILS apch airspace (including POFZ).
⊖	Mandatory: No entry.	Denotes aircraft entry is prohibited.	Located on paved areas that aircraft should not enter.
B	Taxiway Location.	Identifies taxiway on which the aircraft is located.	Located along taxiway by itself, as part of an array of taxiway direction signs, or combined with a runway/taxiway hold sign.
22	Runway Location.	Identifies the runway on which the aircraft is located.	Normally located where the proximity of two rwys to one another could cause confusion.
≡≡≡	Runway Safety Area / OFZ and Runway Approach Area Boundary.	Identifies exit boundary for an RSA / OFZ or rwy approach.	Located on taxiways on back side of certain runway/taxiway holding position signs or runway approach area signs.
‖‖‖	ILS Critical Area/POFZ Boundary.	Identifies ILS critical area exit boundary.	Located on taxiways on back side of ILS critical area signs.
J →	Direction: Taxiway.	Defines designation/direction of intersecting taxiway(s).	Located on L side, prior to intersection, with an array L to R in clockwise manner.
↖ L	Runway Exit.	Defines designation/direction of exit taxiways from the rwy.	Located on same side of runway as exit, prior to exit.
22 ↑	Outbound Destination.	Defines directions to take-off runway(s).	Located on taxi routes to runway(s). Never collocated or combined with other signs.
FBO ↘	Inbound Destination.	Defines directions to airport destinations for arriving aircraft.	Located on taxi routes to airport destinations. Never collocated or combined with other types of signs.
NOISE ABATEMENT PROCEDURES IN EFFECT 2300 - 0500	Information.	Provides procedural or other specialized information.	Located along taxi routes or aircraft parking/staging areas. May not be lighted.
⬛	Taxiway Ending Marker.	Indicates taxiway does not continue beyond intersection.	Installed at taxiway end or far side of intersection, if visual cues are inadequate.
7	Distance Remaining.	Distance remaining info for take-off/landing.	Located along the sides of runways at 1000' increments.

EXAMPLE	TYPE OF MARKING	PURPOSE	LOCATION/CONVENTION
≡≡	Holding Position.	Denotes entrance to runway from a taxiway.	Located across centerline within 10 feet of hold sign on taxiways and on certain runways.
‖‖‖	ILS Critical Area/POFZ Boundary.	Denotes entrance to area to be protected for an ILS signal or approach airspace.	Located on twys where the twys enter the NAVAID critical area or where aircraft on taxiway would violate ILS apch airspace (including POFZ).
┄┄	Taxiway/Taxiway Holding Position.	Denotes location on taxiway or apron where aircraft hold short of another taxiway.	Used at ATCT airports where needed to hold traffic at a twy/twy intersection. Installed provides wing clearance.
═ ┄	Non-Movement Area Boundary.	Delineates movement area under control of ATCT, from non-movement area.	Located on boundary between movement and non-movement area. Located to ensure wing clearance for taxiing aircraft.
═══	Defines edge of usable, full strength taxiway.	Located along twy edge where contiguous shoulder or other paved surface NOT intended for use by aircraft.
═ ─ ═ ─ ═ Dashed Taxiway Edge.	Taxiway Edge.	Defines taxiway edge where adjoining pavement is usable.	Located along twy edge where contiguous paved surface or apron is intended for use by aircraft.
[4-22] [4-22] ↖T B	Surface Painted Holding Position.	Denotes entrance to runway from a taxiway.	Supplements elevated holding position signs. Required where hold line exceeds 200'. Also useful at complex intersections.
	Enhanced Taxiway Centerline.	Provides visual cue to help identify location of hold position.	Taxiway centerlines are enhanced 150' prior to a runway holding position marking.
	Surface Painted Taxiway Direction.	Defines designation/direction of intersecting taxiway(s).	Located L side for turns to left. R side for turns to right. Installed prior to intersection.
	Surface Painted Taxiway Location.	Identifies taxiway on which the aircraft is located.	Located R side. Can be installed on L side if combined with surface painted hold sign.

RUNWAY INCURSION AVOIDANCE

- A runway incursion is any occurrence at an airport involving the incorrect presence of an aircraft, vehicle or person on the protected area of a surface designated for the landing and takeoff of aircraft.

- Runway incursions are primarily caused by errors associated with clearances, communication, airport surface movement, and positional awareness.

- Actions to help prevent a runway incursion include but are not limited to:

    ○ Review the airport diagram and complete as many checklist items as possible before taxi.

    ○ Read back (in full) all active runway crossing, hold short, or line up and wait clearances.

    ○ Do not become absorbed in other tasks, or conversation, while your airplane is moving.If unsure of your position on the airport, stop and ask for assistance.

    ○ Monitor the radio to listen for other aircraft cleared onto your runway for takeoff or landing.

    ○ After landing, stay on the tower frequency until instructed to change frequencies.

    ○ To help others see your airplane during periods of reduced visibility or at night, use your exterior taxi/landing lights when practical.

LAHSO

- At controlled airports, air traffic control can clear an airplane to land and hold short. You may accept a land and hold short (LAHSO) clearance if you can determine that your airplane can safely land and stop within the available landing distance (ALD).

- Student pilots or pilots who are not familiar with LAHSO should not accept a LAHSO clearance.

- The pilot-in-command has the final authority to accept or decline any land and hold short clearance. Decline a LAHSO clearance if you do not believe the operation can be done safely.

AIRPORT LIGHTING

- A beacon at an airport that is on during the day usually means that the weather is below basic VFR minimums (ceiling less than 1,000 feet or visibility less than 3 miles).

- A military airport beacon alternates two quick flashes of white with one green flash.

- Taxiway edge lights are blue, and runway edge lights are white, except that on instrument runways, yellow replaces white on the last 2,000 feet or half the runway length, whichever is less.

- At airports with a three-step pilot-controlled lighting system, seven clicks of the microphone (or mic) sets the lights to high intensity, five clicks turns the lights to medium, and three clicks turns the lights to low. For each adjustment, you must key the mic the required number of times within a period of five seconds.

VISUAL GLIDEPATH INDICATORS

- If you are in Class D airspace approaching to land on a runway served by a VASI, FAR 91.129 requires you to maintain an altitude at or above the glide path until a lower altitude is necessary for a safe landing. The VASI glide path provides safe obstruction clearance to the runway.

- On a two-bar VASI, red over white indicates that you are on the glide path. White over white is above the glide path, and red over red is below the glide path.

- A precision approach path indicator (PAPI) has lights installed in a single row instead of far and near bars. On a four-light PAPI, an on-glide-path indication is two white lights and two red lights. Above the glide path, you see more white lights and below the glide path you see more red lights.

- A pulsating visual approach slope indicator (PVASI) provides a pulsating red light when you are below the glide path. Pulsating white indicates that you are above glide path, and steady white indicates that you are on the glide path.

SECTION B ■ Airports

AIRPORT SECURITY

- Report any suspicious activity at the airport to the Transportation Security Administration by calling 1-866-GA-SECURE (1-866-427-3287), and also notify airport management.

- If witnessing criminal activity at the airport, call 911 and talk to local law enforcement followed by calling 1-866-GA-SECURE.

NOTE: *An asterisk appearing after an ACS code (i.e. PA.VII.B.K1*) indicates that the question subject appears more than one time in the ACS. The code shown corresponds to the first instance of the subject in the ACS.*

4-66   PA.III.B.K1

Which is the correct traffic pattern departure procedure to use at a uncontrolled airport?

A – Depart in any direction consistent with safety, after crossing the airport boundary.

B – Make all turns to the left.

C – Comply with any FAA traffic pattern established for the airport.

4-66. Answer C. GFDPP 4B, FAR 91.126
Each person operating an aircraft to or from an airport without an operating control tower shall, in the case of an aircraft departing the airport, comply with any traffic patterns established for that airport in Part 93.

4-67   PA.II.D.K3, PA.III.B.S1

The numbers 9 and 27 on a runway indicate that the runway is oriented approximately

A – 009° and 027° true.

B – 090° and 270° true.

C – 090° and 270° magnetic.

4-67. Answer C. GFDPP 4B, AIM
Runway numbers correspond to the magnetic, not true, direction, and are rounded to the nearest 10°, with the last zero omitted.

4-68   PA.II.D.K3, PA.III.B.S1

The numbers 8 and 26 on the approach ends of a runway indicate that the runway is orientated approximately

A – 008° and 026° true.

B – 080° and 260° true.

C – 080° and 260° magnetic.

4-68. Answer C. GFDPP 4B, AIM
Runway numbers indicate the magnetic direction of a runway to the nearest 10 degrees. Runway 8 would have a magnetic direction of approximately 080°, and 26 would be approximately 260° magnetic.

SECTION B ■ **Airports**

4-69   PA.III.B.K1

The recommended entry position to an airport traffic pattern is

A – 45° to the base leg just below traffic pattern altitude.

B – 45° at the midpoint of the downwind leg at traffic pattern altitude.

C – to cross directly over the airport at traffic pattern altitude, and then join the downwind leg.

4-69. Answer B. GFDPP 4B, AIM

Always descend to traffic pattern altitude and level off before entering the traffic pattern. The recommended traffic pattern entry is 45° at the midpoint of the runway on the downwind leg. If it is necessary to cross over the airport to join a downwind leg on the opposite side, you should cross at 500 feet above the traffic pattern altitude.

4-70   PA.III.B.K2, S1

(Refer to Figure 49.) Select the proper traffic pattern and runway for landing.

A – Left-hand traffic and Runway 18.

B – Right-hand traffic and Runway 18.

C – Left-hand traffic and Runway 22.

4-70. Answer B. GFDPP 4B, AIM

The wind tetrahedron indicates that landing should be to the southwest, but Runway 22 is closed. Runway 18 is the next best choice, and the "L" mark at that end of the circle shows right-hand traffic for Runway 18.

4-71   PA.II.D.K4, PA.III.B.K2, S1

(Refer to Figure 49.) If the wind is as shown by the landing-direction indicator, the pilot should land on

A – Runway 18 and expect a crosswind from the right.

B – Runway 22 directly into the wind.

C – Runway 36 and expect a crosswind from the right.

4-71. Answer A. GFDPP 4B, AIM

The wind tetrahedron indicates that landing should be to the southwest, but Runway 22 is closed. Runway 18 is the next best choice, with a crosswind from the right.

4-72   PA.II.D.K4, PA.III.B.K2, S1

(Refer to Figure 50.) The segmented circle indicates that the airport traffic is

A – left hand for Runway 18 and right hand for Runway 36.

B – right hand for Runway 9 and left hand for Runway 27.

C – left hand for Runway 36 and right hand for Runway 18.

4-72. Answer C. GFDPP 4B, AIM

The segmented circle indicates left-hand traffic for Runways 9 and 36, and right-hand traffic for Runways 18 and 27.

SECTION B ■ Airports

4-73　PA.II.D.K4, PA.III.B.K2, S1

(Refer to Figure 50.) The traffic patterns indicated in the segmented circle have been arranged to avoid flights over an area to the

A – south of the airport.

B – north of the airport.

C – southeast of the airport.

4-73. Answer C. GFDPP 4B, AIM

Because the traffic pattern for the north-south runway is west of the field, and the pattern for the east-west runway is north of the field, no traffic patterns for landing should be flown southeast of the airport.

4-74　PA.II.D.K4, PA.III.B.K2, S1

(Refer to Figure 50.) The segmented circle indicates that a landing on Runway 26 is with a

A – right-quartering headwind.

B – left-quartering headwind.

C – right-quartering tailwind.

4-74. Answer A. GFDPP 4B, AIM

Because the wind cone shows wind from the northwest, a landing to the west experiences a right-quartering headwind.

4-75　PA.II.D.K4, PA.III.B.K2, S1

(Refer to Figure 50.) Which runway and traffic pattern should be used as indicated by the wind cone in the segmented circle?

A – Right-hand traffic on Runway 18.

B – Left-hand traffic on Runway 36.

C – Right-hand traffic on Runway 9.

4-75. Answer B. GFDPP 4B, AIM

With wind exactly from the northwest, you might choose either Runway 27 or 36, depending on runway length and other factors. Runway 27 is not an available answer choice, leaving Runway 36. A landing on Runway 36 would require a left-hand traffic pattern and would encounter a left-quartering headwind.

4-76　PA.II.D.K6c, PA.IV.A.S2

You should state your position on the airport when calling the tower for takeoff, especially when

A – visibility is less than 1 mile.

B – parallel runways are in use.

C – departing from a runway intersection.

4-76. Answer C. GFDPP 4B, AIM

It is always good practice to state your position, but the AIM specifically requests you to do so when calling from a runway intersection. For example: *"Centennial Tower, Cessna 5238-Kilo, at Alpha 3."*

4-77　PA.II.D.K3

The "yellow demarcation bar" marking indicates

A – runway with a displaced threshold from a blast pad, stopway, or taxiway that precedes the runway.

B – a hold line from a taxiway to a runway.

C – the beginning of available runway for landing on the approach side.

4-77. Answer A. GFDPP 4B, AIM

This double bar delineates a runway with a displaced threshold from a blast pad, stopway, or taxiway that precedes the runway.

SECTION B ■ **Airports**

4-78   PA.II.D.K3, PA.III.B.S1

(Refer to Figure 65, item E.) This sign is a visual clue that

A – confirms the aircraft's location to be on taxiway "B."

B – warns the pilot of approaching taxiway "B."

C – indicates "B" holding area is ahead.

4-78. Answer A. GFDPP 4B, AIM

A location sign has a black background with a yellow border and yellow numbers or letters inscribed in the center. A yellow letter designates a taxiway on which the aircraft is located.

4-79   PA.II.D.K3, PA.III.B.S1

(Refer to Figure 65, item F.) This sign confirms your position on

A – Runway 22.

B – routing to Runway 22.

C – Taxiway 22.

4-79. Answer A. GFDPP 4B, AIM

A location sign has a black background with a yellow border and yellow numbers or letters inscribed in the center. The yellow number designates a runway on which the aircraft is located.

4-80   PA.II.D.K3, PA.III.B.S1

(Refer to Figure 48.) The portion of the runway identified by the letter A may be used for

A – landing.

B – taxiing and takeoff.

C – taxiing and landing.

4-80. Answer B. GFDPP 4B, AIM

The area before a displaced threshold may be used for taxi and takeoff, and roll-out after landing in the opposite direction, but not for landing.

4-81   PA.II.D.K3, PA.III.B.S1

(Refer to Figure 65, item G.) From the cockpit, this marking confirms the aircraft to be

A – on a taxiway, about to enter runway zone.

B – on a runway, about to clear.

C – near an instrument approach clearance zone.

4-81. Answer B. GFDPP 4B, AIM

This sign is a runway boundary sign, which faces the runway and is visible to the pilot exiting the runway. It is located adjacent to the holding position marking on the pavement. The sign is intended to provide you with a visual cue you can use as a guide in deciding when you are "clear of the runway."

4-82   PA.II.D.K1, K3, PA.III.B.S1

(Refer to Figure 48.) According to the airport diagram, which statement is true?

A – Takeoffs may be started at position D on Runway 30, but the landing portion of this runway begins at position E.

B – Takeoffs may be started at position A on Runway 12, and the landing portion of this runway begins at position B.

C – The takeoff and landing portion of Runway 12 begins at position B.

4-82. Answer B. GFDPP 4B, AIM

The area before a displaced threshold may be used for taxi and takeoff, and roll-out after landing in the opposite direction. Landings may be made after the displaced threshold at position "B" on Runway 12.

4-83   PA.II.D.K1, K3, PA.III.B.S1

(Refer to Figure 48.) What is the difference between area A and area E on the airport depicted?

A – "A" may be used for taxi and takeoff; "E" may be used only as an overrun.

B – "A" may be used for all operations except heavy aircraft landings; "E" may be used only as an overrun.

C – "A" may be used only for taxiing; "E" may be used for all operations except landings.

4-83. Answer A. GFDPP 4B, AIM
The area before a displaced threshold may be used for taxi and takeoff, and roll-out after landing. Area "E" is a blastpad/stopway, and is not designed with the pavement strength to support continuous operations, but it may be used as an overrun.

4-84   PA.II.D.K1, K3, PA.III.B.S1

(Refer to Figure 48.) Area C on the airport depicted is a

A – stabilized area.

B – multiple heliport.

C – closed runway.

4-84. Answer C. GFDPP 4B, AIM
A closed runway is depicted by Xs.

4-85   PA.II.D.K3, PA.III.B.S1

(Refer to Figure 64.) Which marking indicates a vehicle lane?

A – A.

B – C.

C – E.

4-85. Answer B. GFDPP 4B, AIM
Lanes for ground vehicles look similar to a road painted on the airport surface.

4-86   PA.II.D.K3, PA.III.B.S1

(Refer to Figure 49.) The arrows that appear on the ends of the north/south runway indicate that these areas

A – may be used only for taxiing.

B – is usable for taxiing, takeoff, and landing.

C – cannot be used for landing, but may be used for taxiing and takeoff.

4-86. Answer C. GFDPP 4B, AIM
The area before a displaced threshold may be used for taxi and takeoff, and for roll-out after landing in the opposite direction, but not for landing.

SECTION B ■ **Airports**

4-87   PA.II.D.K3, K6b, PA.III.B.S1

At a towered airport, when approaching taxiway holding lines from the side with the continuous lines, you

A – may continue taxiing.

B – should not cross the lines without an ATC clearance.

C – should continue taxiing until all parts of the aircraft have crossed the lines.

4-87. Answer B. GFDPP 4B, AIM

When approaching a taxiway hold line from the side with the continuous (solid) line at a towered airport, you should not cross the hold line without ATC clearance. At a non-towered airport, stop and check for traffic before crossing any hold line.

4-88   PA.II.D.K6d, K6e

Which is true regarding runway incursions?

A – To help prevent a runway incursion during periods of reduced visibility or at night use your exterior taxi/landing lights when practical.

B – A runway incursion is defined as any occurrence at an airport involving an aircraft, vehicle, person, or object on the ground that results in a collision.

C – Runway incursions are primarily caused by controller errors.

4-88. Answer A. GFDPP 4B, AFH

The official definition of a runway incursion is "any occurrence at an airport involving an aircraft, vehicle, person, or object on the ground that creates a collision hazard or results in loss of separation with an aircraft taking off or intending to take off, landing, or intending to land."

Runway incursions are primarily caused by errors associated with clearances, communication, airport surface movement, and positional awareness.

In addition to other actions you can take to prevent runway incursion, to help others see your airplane during periods of reduced visibility or at night, use your exterior taxi/landing lights when practical.

4-89   PA.II.D.K6d, K6e

The runway status lights (RWSL) system is primarily used to

A – provide weather updates to pilots.

B – signal distress in case of an emergency.

C – indicate runway occupancy and help pilots avoid potential runway incursions.

4-89. Answer C. GFDPP 4B, AIM

Runway status lights (RWSL) are an automated system embedded in the pavement of runways and taxiways that are designed to automatically signal pilots when it is unsafe to enter, cross, or begin takeoff.

The primary purpose of the runway status lights (RWSL) system is to enhance runway safety by providing real-time information about runway occupancy and helping pilots avoid potential runway incursions. This system is particularly useful in poor visibility conditions or during night operations.

SECTION B ■ Airports

4-90   PA.II.D.K3, K6c, PA.III.B.S1

What is the purpose of the runway/runway hold position sign?

A – Denotes entrance to runway from a taxiway.

B – Denotes area protected for an aircraft approaching or departing a runway.

C – Denotes intersecting runways.

4-90. Answer C. GFDPP 4B, AIM
These signs are installed together with pavement markings only on runways that are used for land and hold short operations (LAHSO) or taxiing operations. Like other holding position signs, they consist of white runway numbers on a red background.

4-91   PA.II.D.K3, PA.III.B.S1

What does the outbound destination sign identify?

A – Identifies entrance to the runway from a taxiway.

B – Identifies direction to takeoff runways.

C – Identifies runway on which an aircraft is located.

4-91. Answer B. GFDPP 4B, AIM
An outbound destination sign displays black text on a yellow background, and a vertical black arrow. These signs always have an arrow showing the direction of the taxing route to that destination.

4-92   PA.IV.B.R3b*

How can you determine whether your destination airport uses land and hold short operations (LAHSO)?

A – The notation "L" preceding the runway length on a VFR chart.

B – LAHSO information published in the Airport/Facility Directory listing in the Chart Supplement.

C – NOTAMs from Flight Service during your preflight weather briefing.

4-92. Answer B. GFDPP 4B, AIM
LAHSO information is published in the Chart Supplement's Airport Facility Directory (A/FD) listing for that airport. The ATIS announces whether LAHSO is in use at the time you are preparing to land. An "L" preceding the runway length on a VFR chart indicates that runway lighting is available at that airport.

4-93   PA.IV.B.R3b*

Who has final authority to accept or decline any land and hold short (LAHSO) clearance?

A – Pilot in command.

B – Air traffic controller.

C – Second in command.

4-93. Answer A. GFDPP 4B, AIM
As the pilot in command (PIC), you have the final authority to accept or decline any LAHSO clearances. Decline a LAHSO clearance if you determine that it compromises safety.

SECTION B ■ Airports

4-94   PA.IV.B.R3b*

When should you decline a land and hold short (LAHSO) clearance?

A – When it compromises safety.

B – Only when the tower operator concurs.

C – Pilots may not decline a LAHSO clearance.

4-94. Answer A. GFDPP 4B, AIM
As the pilot in command (PIC), you have the final authority to accept or decline any LAHSO clearances. Decline a LAHSO clearance if you determine that it compromises safety.

4-95   PA.IV.B.R3b*, PA.II.D.K1

Where is the "available landing distance" (ALD) data published for an airport that utilizes land and hold short operations (LAHSO) published?

A – The special notices section of the Chart Supplement.

B – 14 CFR Part 91, General Operating and Flight Rules.

C – Aeronautical Information Manual (AIM).

4-95. Answer A. GFDPP 4B, AIM
ALD data is published in the listing for an airport that has LAHSO. Airport listings are in the A/FD section of the Chart Supplement.

4-96   PA.IV.B.R3b*

What is the minimum visibility for a pilot to receive a land and hold short (LAHSO) clearance?

A – 3 nautical miles.

B – 3 statute miles.

C – 1 statute mile.

4-96. Answer B. GFDPP 4B, AIM
You should only receive a LAHSO clearance when there is a minimum ceiling of 1,000 feet and 3 statute miles visibility.

4-97   PA.II.D.K3, PA.III.B.S1

An airport's rotating beacon operated during daylight hours indicates

A – there are obstructions on the airport.

B – that weather at the airport located in Class D airspace is below basic VFR weather minimums.

C – the air traffic control tower is not in operation.

4-97. Answer B. GFDPP 4B, AIM
When the airport beacon is on during the daytime, it means that the ceiling is less than 1,000 feet or the visibility is less than 3 miles, which is below basic VFR minimums.

4-98   PA.II.D.K3, PA.XI.A.K2

You can identify a military air station by a rotating beacon that emits

A – white and green alternating flashes.

B – two quick white flashes between each green flash.

C – green, yellow, and white flashes.

4-98. Answer B. GFDPP 4B, AIM

A military airport beacon has two quick flashes of white light between each green flash.

4-99   PA.II.D.K3, PA.XI.A.K2

How can a military airport be identified at night?

A – Alternate white and green light flashes.

B – Dual peaked (two quick) white flashes between each green flash.

C – White flashing lights with steady green at the same location.

4-99. Answer B. GFDPP 4B, AIM

A military airport beacon has two quick flashes of white light between each green flash.

4-100   PA.II.D.K3, PA.IV.B.K1*

While operating in Class D airspace, each pilot of an aircraft approaching to land on a runway served by a visual approach slope indicator (VASI) shall

A – maintain a 3° glide until approximately 1/2 mile to the runway before going below the VASI.

B – maintain an altitude at or above the glide slope until a lower altitude is necessary for a safe landing.

C – stay high until the runway can be reached in a power-off landing.

4-100. Answer B. GFDPP 4B, FAR 91.129, AIM

If you are in Class D airspace approaching to land on a runway served by a VASI, FAR 91.129 requires you to maintain an altitude at or above the glide path until a lower altitude is necessary for a safe landing. The VASI glide path provides safe obstruction clearance to the runway.

4-101   PA.II.D.K3, PA.IV.B.K1*

When approaching to land on a runway served by a visual approach slope indicator (VASI), the pilot shall

A – maintain an altitude that captures the glide slope at least 2 miles downwind from the runway threshold.

B – maintain an altitude at or above the glide slope.

C – remain on the glide slope and land between the two-light bar.

4-101. Answer B. GFDPP 4B, FAR 91.129, AIM

If you are in Class D airspace approaching to land on a runway served by a VASI, FAR 91.129 requires you to maintain an altitude at or above the glide path until a lower altitude is necessary for a safe landing. The VASI glide path provides safe obstruction clearance to the runway.

SECTION B ■ Airports

4-102   PA.II.D.K3, PA.IV.B.K1*

Which approach and landing objective is assured when the pilot remains on the proper glidepath of the VASI?

A – Runway identification and course guidance.

B – Safe obstruction clearance in the approach area.

C – Lateral course guidance to the runway.

4-102. Answer B. GFDPP 4B, AIM
The VASI glide path provides safe obstruction clearance to the runway.

4-103   PA.II.D.K3, PA.IV.B.K1*

A slightly high glide slope indication from a precision approach path indicator is

A – four white lights.

B – three white lights and one red light.

C – two white lights and two red lights.

4-103. Answer B. GFDPP 4B, AIM
A slightly high indication on a precision approach path indicator is three white lights and one red light.

4-104   PA.II.D.K3, PA.IV.B.K1*

A below glide slope indication from a pulsating visual approach slope indicator is a

A – pulsating white light.

B – steady white light.

C – pulsating red light.

4-104. Answer C. GFDPP 4B, AIM
A pulsating approach slope indicator provides a pulsating red light when below glide slope.

4-105   PA.II.D.K3, PA.IV.B.K1*

(Refer to Figure 47.) Illustration A indicates that the aircraft is

A – above the glide slope.

B – on the glide slope.

C – below the glide slope.

4-105. Answer B. GFDPP 4B, AIM
A red over white indication is on glide slope.

4-106   PA.II.D.K3, PA.IV.B.K1*

(Refer to Figure 47.) VASI lights as shown by illustration C indicate that the airplane is

A – above the glide slope.

B – on the glide slope.

C – below the glide slope.

4-106. Answer A. GFDPP 4B, AIM
A white over white indication is above glide slope.

4-107   PA.II.D.K3, PA.IV.B.K1*

(Refer to Figure 47.) While on final approach to a runway equipped with a standard two-bar VASI, the lights appear as shown by illustration D. This means that the aircraft is

A – above the glide path.

B – below the glide path.

C – on the glide path.

4-107. Answer B. GFDPP 4B, AIM

A red over red indication is below the glide slope.

4-108   PA.II.D.K3, PA.XI.A.K2

To set the high intensity runway lights on medium intensity, the pilot should click the microphone seven times, then click it

A – one time within four seconds.

B – three times within three seconds.

C – five times within five seconds.

4-108. Answer C. GFDPP 4B, AIM

At airports with a three-step pilot-controlled lighting system, five clicks turns them to medium intensity. For each adjustment, you must key the mic the required number of times within a period of five seconds.

4-109   PA.II.D.K6d, PA.XI.A.K2, K6

Airport taxiway edge lights are identified at night by

A – white directional lights.

B – blue omnidirectional lights.

C – alternate red and green lights.

4-109. Answer B. GFDPP 4B, AIM

Taxiway edge lights are blue.

4-110   PA.II.A.R5, PA.XII.A.R3

You observe someone breaking into an airplane. What should you do?

A – Call 1-866-GA SECURE and then call 911.

B – Call 911 and then call 1-866-GA SECURE.

C – Call 1-866-GA SECURE and then try to detain the suspected criminal.

4-110. Answer B. GFDPP 4B, TSA

As a pilot, you are responsible for following best practices to keep your airport and aircraft safe. Be able to recognize suspicious activity and know how to alert authorities. For criminal activity, such as someone breaking into an aircraft, do confront that person, but rather call 911 and then 1-866-GA-Secure (1-866-427-3287), and your airport, FBO, or flight school manager. Call 1-866-GA-Secure, but not 911, if you observe suspicious, but not criminal, activity.

SECTION C
Aeronautical Charts

You can answer many of the questions on the FAA knowledge test by referring to the figures and legends that are available during the test. Also refer to these resources during flight if you are unsure of the meaning of a symbol on the chart.

LATITUDE AND LONGITUDE

- In the northern hemisphere, latitude increases as you travel north.
- In the western hemisphere, longitude increases as you travel west.
- Each tick mark on the sectional chart represents one minute of latitude or longitude.

CHART SYMBOLOGY

- A blue segmented circle on a sectional chart depicts Class D airspace. A blue airport symbol depicts a tower-controlled airport.
- A common traffic advisory frequency (CTAF) is shown in the airport information followed by the circled letter "C".

NOTE: An asterisk appearing after an ACS code (i.e. PA.VII.B.K1*) indicates that the question subject appears more than one time in the ACS. The code shown corresponds to the first instance of the subject in the ACS.

4-111   PA.I.E.K2

(Refer to Figure 20, area 3.) Determine the approximate latitude and longitude of Currituck County Airport.

A – 36°24'N - 76°01'W.

B – 36°48'N - 76°01'W.

C – 47°24'N - 75°58'W.

4-111. Answer A. GFDPP 4C, PHB

This airport is located northeast of the number "3." Starting near the top of the chart excerpt near number "1," find the labels for the 37° latitude line and the 76° longitude line. That means the latitude line through the middle of the picture is 36°30'N. Count down the tick marks—one minute per tick mark—until abeam Currituck County Airport at 36°24'N. At one tick mark west of the 76° longitude line, the longitude of the airport is 76°01'W.

4-112   PA.I.E.K2

(Refer to Figure 21, area 3.) Which airport is located at approximately 47°21'30"N latitude and 101°01'30"W longitude?

A – Poleschook.

B – Washburn.

C – Johnson.

4-112. Answer B. GFDPP 4C, PHB

The 48° latitude line crosses the top third of the chart. The latitude line along the bottom third is 30' less, or 47°30'N. Count down 9-1/2 tick marks (minutes) for 47°21'30"N. Because the longitude of the airport is more than 101°W, move to the left of the 101° line 1-1/2 tick marks to arrive at 101°01'30"W. This intersection is at Washburn Airport (5C8).

4-113   PA.I.E.K2

(Refer to Figure 59, area 3.) Which airport is located at approximately 41°02'00"N latitude and 83°59'00"W longitude?

A – Ruhes.

B – Putnam.

C – Bluffton.

4-113. Answer B. GFDPP 4C, PHB

The 41° latitude line and the 84° longitude line are labeled near area 3. The airport is two tick marks above the 41° latitude line, which might be difficult to see, so count down from the 5-minute tick mark. At 83°59'00"W longitude, the airport is one tick mark east of the 84° longitude line. This intersection is at Putman Airport (OWX).

4-114   PA.I.E.K2

(Refer to Figure 21, area 2.) The CTAF/MULTICOM frequency for Garrison Airport is

A – 123.0 MHz.

B – 122.8 MHz.

C – 122.9 MHz.

4-114. Answer C. GFDPP 4C, AIM

The frequency next to the CTAF symbol (the letter "C" in a dark circle) is the multicom frequency of 122.9 MHz.

4-115   PA.I.E.K2

(Refer to Figure 59, area 3.) The CTAF/MULTICOM frequency for Wyandot County Airport (56D) is

A – 123.0 MHz.

B – 122.8 MHz.

C – 122.9 MHz.

4-115. Answer C. GFDPP 4C, AIM

The frequency next to the CTAF symbol (the letter "C" in a dark circle) is the multicom frequency of 122.9 MHz.

4-116   PA.I.E.K2, PA.II.D.K6b

(Refer to Figure 22, area 2 and Figure 31.) At Coeur D'Alene, which frequency should be used as a Common Traffic Advisory Frequency (CTAF) to self-announce position and intentions?

A – 122.05 MHz.

B – 108.8 MHz.

C – 122.8 MHz.

4-116. Answer C. GFDPP 4C, AIM

In this example, the airport data block located near Coeur D'Alene Airport lists 122.8 MHz as the CTAF frequency that should be used to self-announce your position and your intentions. The Chart Supplement excerpt also shows 122.8 MHz as the CTAF/UNICOM frequency. 122.05 MHz is the Flight Service frequency (Boise Radio) and 135.075 MHz is the AWOS frequency, also available at telephone number (208) 772-8215.

SECTION C ■ Aeronautical Charts

4-117   PA.I.E.K2, PA.II.D.K6b

(Refer to Figure 22, area 2 and Figure 31.) At Coeur D'Alene, which frequency should be used as a Common Traffic Advisory Frequency (CTAF) to monitor airport traffic?

A – 122.05 MHz.

B – 122.8 MHz.

C – 135.075 MHz.

4-117. Answer B. GFDPP 4C, AIM

At non-towered airports, the CTAF frequency is used to self-announce position or intentions. The Chart Supplement's A/FD section lists the CTAF/UNICOM frequency as 122.8 MHz. The CTAF symbol on the sectional chart is beside the frequency of 122.8 MHz.

4-118   PA.I.E.K2, PA.II.D.K6b

(Refer to Figure 22, area 2 and Figure 31.) What is the correct UNICOM frequency to be used at Coeur D'Alene to request fuel?

A – 108.8 MHz.

B – 122.8 MHz.

C – 135.075 MHz.

4-118. Answer B. GFDPP 4C, AIM

Use the UNICOM/CTAF frequency of 122.8 MHz to request fuel, transportation, or other airport information.

4-119   PA.I.E.K2, PA.VIII.F.K1

(Refer to Figure 59, area 1 and Figure 63.) What is the correct frequency to be used to contact Toledo Approach if approaching the Toledo Express Class C airspace from the east?

A – 123.975 MHz.

B – 126.1 MHz.

C – 134.35 MHz.

4-119. Answer B. GFDPP 4C, AIM

You can find the frequency in both the Chart Supplement's A/FD section and on the sectional chart. On the chart, a box east of the Class C airspace says to contact Toledo Approach within 20 NM on 126.1 MHz. In the A/FD, under Communications, The Approach/Departure control frequency is 126.1 MHz on radials 360°-179°, the east side of the airport. The frequency is 134.35 MHz on the west side (radials 180°-359°). A third frequency, 123.975 MHz, is also available, but not associated with any specific direction from the airport.

4-120   PA.I.E.K2, PA.VIII.F.K1

(Refer to Figure 25, area 3.) If Dallas Executive Tower is not in operation, which frequency should be used as a Common Traffic Advisory Frequency (CTAF) to monitor airport traffic?

A – 122.95 MHz.

B – 126.35 MHz.

C – 127.25 MHz.

4-120. Answer C. GFDPP 4C, AIM

The star next to the tower frequency "CT 127.25" MHz indicates that the tower operates part-time. You can find the hours of operation in the Chart Supplement's A/FD section. The circled "C" indicates that 127.25 MHz is the CTAF frequency when the tower is not operating.

4-121 PA.I.E.K2, PA.VIII.F.K1

(Refer to Figure 26, area 4.) The CTAF/UNICOM frequency at Jamestown Airport is

A – 118.425 MHz.

B – 122.2 MHz.

C – 123.0 MHz.

4-121. Answer C. GFDPP 4C, AIM
The CTAF symbol, the circled C, is next to the frequency 123.0 MHz. The frequency 118.425 MHz is for ASOS weather broadcasts and the frequency 122.2 MHz is for Flight Service.

4-122 PA.I.E.K2, PA.VIII.F.K1

(Refer to Figure 26, area 5.) What is the CTAF/UNICOM frequency at Barnes County Airport?

A – 118.725 MHz.

B – 122.8 MHz.

C – 1402 kHz.

4-122. Answer B. GFDPP 4C, AIM
The CTAF symbol, the circled C, is next to the frequency 122.8 MHz. The frequency 118.725 MHz is for ASOS weather broadcasts and 1402 is the field elevation.

4-123 PA.I.E.K2

(Refer to Figure 26, area 3.) When flying over Arrowwood National Wildlife Refuge, a pilot should fly no lower than

A – 2,000 feet AGL.

B – 2,500 feet AGL.

C – 3,000 feet AGL.

4-123. Answer A. GFDPP 4C, AIM
You are requested to maintain a minimum altitude of 2,000 feet above National Wildlife Refuges.

4-124 PA.I.E.K2

Pilots flying over a national wildlife refuge are requested to fly no lower than

A – 1,000 feet AGL.

B – 2,000 feet AGL.

C – 3,000 feet AGL.

4-124. Answer B. GFDPP 4C, AIM
You are requested to maintain a minimum altitude of 2,000 feet above National Wildlife Refuges.

4-125 PA.I.E.K2

(Refer to Figure 20, area 5.) The CAUTION box denotes what hazard to aircraft?

A – Unmarked balloon on cable to 3,008 feet MSL.

B – Unmarked balloon on cable to 3,008 feet AGL.

C – Unmarked blimp hangers at 308 feet MSL.

4-125. Answer A. GFDPP 4C, Appendix 1, Chart Legend
The CAUTION box indicates an unmarked balloon on a cable to 3,008 feet MSL.

SECTION C ■ **Aeronautical Charts**

4-126 PA.I.E.K2

(Refer to Figure 20, area 2.) The flag symbol at Lake Drummond represents a

A – compulsory reporting point for the Norfolk Class C Airspace.

B – compulsory reporting point for Hampton Roads Airport.

C – visual checkpoint used to identify position for initial callup to Norfolk Approach Control.

4-126. Answer C. GFDPP 4C, Appendix 1, Chart Legend

The flag represents a visual checkpoint used to identify your position for approach control. Because the flag is 22 nautical miles southwest of Norfolk International Airport, it can be assumed that the checkpoint is used when contacting Norfolk Approach.

4-127 PA.I.E.K2

(Refer to Figure 20, area 2.) The elevation of the Chesapeake Regional Airport is

A – 19 feet.

B – 23 feet.

C – 55 feet.

4-127. Answer A. GFDPP 4C, Appendix 1, Chart Legend

The elevation is the first number listed before the runway information. In this case, it is 19 feet.

4-128 PA.I.E.K2

(Refer to Figure 20, area 6.) The NALF Fentress (NFE) Airport is in what type of airspace?

A – Class C.

B – Class E.

C – Class G.

4-128. Answer B. GFDPP 4C, Appendix 1, Chart Legend

NFE is outside the solid magenta lines delineating the Norfolk Class C airspace, but inside a dashed magenta circle indicating Class E airspace at the surface.

4-129 PA.I.E.K2

(Refer to Figure 21.) The terrain elevation of the light tan area between Minot (area 1) and Audubon Lake (area 2) varies from

A – sea level to 2,000 feet MSL.

B – 2,000 feet to 2,500 feet MSL.

C – 2,000 feet to 2,700 feet MSL.

4-129. Answer B. GFDPP 4C, Appendix 1, Chart Legend

The colored scale shows that the tan area represents terrain above 2,000 feet MSL. In addition, the legend states that the contour interval is 500 feet. Between Minot and Audubon Lake, there are no contour lines in the tan area, which indicates that no terrain exists above 2,500 feet. Checking the airports in this area reveals that their elevations are all less than 2,500 feet. In addition, tower and windmill heights in MSL minus their AGL heights all yield base elevations less than 2,500 feet.

4-130 PA.I.E.K2

(Refer to Figure 21.) Which public use airports depicted are indicated as having fuel?

A – Minot Intl. (area 1).

B – Minot Intl. (area 1) and Mercer County Regional Airport (area 3).

C – Mercer County Regional Airport (area 3) and Garrison (area 2).

4-130. Answer A. GFDPP 4C, Appendix 1, Chart Legend

Tick marks around an airport symbol indicate that fuel is available.

4-131 PA.I.E.K2

(Refer to Figure 23, areas 2 and 3.) The flag symbols at Statesboro Bullock County Airport, Claxton-Evans County Airport, and Ridgeland Airport are

A – airports with special traffic patterns.

B – outer boundaries of Savannah Class C airspace.

C – visual checkpoints to identify position for initial callup before to entering Savannah Class C airspace.

4-131. Answer C. GFDPP 4C, Appendix 1, Chart Legend

The flag symbols represent checkpoints used to identify the aircraft position for approach control. In this case, they are visual checkpoints used when contacting Savannah Approach Control.

4-132 PA.I.E.K2

(Refer to Figure 23, area 3.) What altitude do you need to be flying at to clear the obstacle with high-intensity lights that is approximately 6 nautical miles southwest of Savannah International by 1,000 feet?

A – 1,531 feet MSL.

B – 1,534 feet AGL.

C – 2,548 feet MSL.

4-132. Answer C. GFDPP 4C, Appendix 1, Chart Legend, FAR 91.119

About 6 nautical miles southwest of the center of Savannah International Airport is an obstacle with a strobe light symbol. Its elevation is marked as 1,548 (1,534). The first number is the elevation in MSL, and the number in parentheses is height AGL. Other nearby numbers mark a group of obstacles about 10 miles away that do not have strobe lights.

4-133 PA.I.E.K2

(Refer to Figure 23, area 3.) The top of the group obstruction approximately 11 nautical miles from the Savannah VORTAC on the 010° radial is

A – 454 feet AGL.

B – 454 feet MSL.

C – 559 feet MSL.

4-133. Answer B. GFDPP 4C, Appendix 1, Chart Legend

At 11 NM, this obstruction is just outside the VOR compass rose. It is labeled "stacks" and has the number 454 next to it. When only one number is printed, not in parentheses, it is the MSL altitude. The nearby 559 (505) foot group is inside the edge of the Savannah Class C airspace, which is noticeably closer to the VOR than 11 NM. The Class C airspace does not line up with the Savannah VORTAC because the VORTAC is north of the center of Savannah/Hilton Head International Airport, as shown by the dot on the airport symbol.

SECTION C ■ Aeronautical Charts

4-134 PA.I.E.K2

(Refer to Figure 24, area 1.) What minimum altitude is required to vertically clear the obstacle in the uncongested area on the northeast side of Airpark East Airport?

A – 1,010 feet MSL.

B – 1,273 feet MSL.

C – 1,283 feet MSL.

4-134. Answer B. GFDPP 4C, FAR 91.119
Start by identifying Airpark East. Its symbol is below the label, and has an obstacle immediately northeast of it. In uncongested areas, the FARs require 500 feet of obstacle clearance. Add 500 feet to the obstacle elevation of 773 feet MSL.

4-135 PA.I.E.K2

(Refer to Figure 24, area 2.) What minimum altitude is necessary to vertically clear the obstacle in the uncongested area on the southeast side of Winnsboro Airport by 500 feet?

A – 823 feet MSL.

B – 1,013 feet MSL.

C – 1,403 feet MSL.

4-135. Answer C. GFDPP 4C, FAR 91.119
In uncongested areas, the FARs require 500 feet of obstacle clearance. Add 500 feet to the obstacle elevation of 903 feet MSL.

4-136 PA.I.E.K2

(Refer to Figure 25, area 2.) The control tower frequency for Addison Airport is

A – 122.95 MHz.

B – 126.0 MHz.

C – 133.4 MHz.

4-136. Answer B. GFDPP 4C, Appendix 1, Chart Legend
The tower frequency at Addison Airport is 126.0 MHz as indicated by the letters "CT."

4-137 PA.I.E.K2

(Refer to Figure 25, area 8.) What minimum altitude is required to fly over the Cedar Hill TV towers in the congested area east of Joe Pool Lake?

A – 2,731 feet MSL.

B – 3,049 feet MSL.

C – 3,549 feet MSL.

4-137. Answer C. GFDPP 4C, FAR 91.119
Because this area is a congested area, add 1,000 feet to the elevation of the highest obstacle. The highest elevation (height) of these towers is 2,549 feet MSL (1,731 feet AGL), so the minimum altitude would be 3,549 feet MSL. You would need an ATC clearance to fly at this altitude because the floor of Class B airspace at this location is 3,000 feet MSL.

4-138   PA.I.E.K2

(Refer to Figure 25, area 5.) The navigation facility at Dallas-Ft. Worth International (DFW) is a

A – VOR.

B – VORTAC.

C – VOR/DME.

4-138. Answer C. GFDPP 4C, Appendix 1, Chart Legend

This symbol appears immediately south of the runways at DFW. According to Legend 1, a hexagon surrounded by a square is a VOR/DME symbol.

4-139   PA.I.E.K2

Which is true concerning the blue and magenta colors used to depict airports on sectional aeronautical charts?

A – Airports with control towers underlying Class A, B, and C airspace are shown in blue, Class D and E airspace are magenta.

B – Airports with control towers underlying Class C, D, and E airspace are shown in magenta.

C – Airports with control towers underlying Class B, C, D, and E airspace are shown in blue.

4-139. Answer C. GFDPP 4C, Appendix 1, Chart Legend

Airports with control towers are depicted in blue on sectional charts.

4-140   PA.I.E.K2

Which statement about longitude and latitude is true?

A – Lines of longitude are parallel to the Equator.

B – Lines of longitude cross the Equator at right angles.

C – The 0° line of latitude passes through Greenwich, England.

4-140. Answer B. GFDPP 4C, PHB

Lines of longitude connect the poles and are perpendicular to the equator.

SECTION C ■ Aeronautical Charts

SECTION D

Airspace

CONTROLLED AND UNCONTROLLED AIRSPACE

- Airspace is divided into classes and categorized as either controlled or uncontrolled.

- Controlled airspace is a general term that describes five of the six airspace classes: Class A, Class B, Class C, Class D, and Class E. While operating in controlled airspace, you are subject to certain operating rules, as well as pilot qualification and aircraft equipment requirements.

- Class G airspace is referred to as uncontrolled airspace.

TRANSPONDERS AND ADS-B

- A transponder is an electronic device aboard the airplane that enhances your aircraft's identity on an ATC display. Transponders carry designations appropriate to their capabilities. A transponder with altitude encoding equipment is referred to as having Mode C capability. The Mode S transponder is required for use with ADS-B.

- The FARs require that you have an operating transponder with Mode C capability and ADS-B Out equipment:

    ◦ In Class A airspace.

    ◦ In Class B airspace and within 30 nautical miles of Class B primary airports.

    ◦ In and above Class C airspace.

    ◦ At and above 10,000 feet MSL (except at or below 2,500 feet AGL) over the 48 contiguous states.

- You also must have a Mode C transponder if you are in uncontrolled airspace above 10,000 feet MSL.

CLASS G AIRSPACE

- ATC does not exercise control of traffic and you are not required to communicate with controllers when operating in Class G airspace, unless a temporary control tower exists.

- Class G airspace typically starts at the surface—think G for Ground—and extends up to the base of the overlying controlled airspace (Class E), which is normally 700 or 1,200 feet AGL. In a few remote areas of the western U.S. and Alaska, Class G airspace can extend all the way up to 14,500 feet MSL, or to 1,500 feet AGL, whichever is higher.

CLASS E AIRSPACE

- Unless designated otherwise, Class E begins at 1,200 feet AGL and extends to the floor of Class A airspace at 18,000 feet MSL. Class E airspace also extends upward from FL600. Class E is the most common airspace in the United States—think E for Everywhere.

- Federal (or Victor) airways and T-routes (which are based on GPS) are Class E airspace. Airways and T-routes extend 4 nautical miles on each side of the airway centerline and are normally from 1,200 feet AGL up to 17,999 feet MSL.

- No communication requirements apply to operating within Class E airspace, but you can request traffic advisory services that ATC provides on a workload-permitting basis.

CLASS D AIRSPACE

- An airport that has an operating control tower, but does not provide Class B or C airspace ATC services, is surrounded by Class D airspace.

- The airspace at an airport with a part-time control tower is designated as Class D only when the tower is in operation. At airports where the control tower operates part-time, the airspace changes to Class E, or a combination of Class E and Class G, when the tower is closed.

- Two way radio communications are required for taking off and landing at an airport with an operating control tower.

- When approaching Class D airspace, you must contact the control tower of the primary airport.

- When departing a non-towered satellite airport, you must contact the control tower of the primary airport as soon as practicable after takeoff.

- Most Class D airspace is a circle with a radius of approximately 4 NM, however these areas can be various sizes and shapes, depending on the instrument procedures established for the airport.

CLASS C AIRSPACE

- To enter Class C airspace, you must have a Mode C transponder, ADS-B Out equipment, and establish two-way radio communications with approach control.

- Class C airspace usually consists of:

  ○ A 5 NM radius core surface area that extends from the surface up to 4,000 feet above the primary airport elevation.

  ○ A 10 NM radius shelf area that extends from 1,200 feet up to 4,000 feet above the airport elevation.

  ○ An "outer area" extends to 20 NM from the airport, where radar service is available, but contact with ATC is not required.

CLASS B AIRSPACE

- To enter Class B airspace, you must:

  ○ Hold either a private pilot certificate, or a student pilot certificate with a logbook endorsement from an appropriately rated instructor.

  ○ Have a Mode C transponder and ADS-B Out equipment.

  ○ Obtain an ATC clearance.

CLASS A AIRSPACE

- Class A airspace extends from 18,000 feet MSL up to and including FL600. Altitudes within Class A airspace are expressed to ATC by using the term flight level (FL). Class A covers the majority of the contiguous states and Alaska, as well as the area extending 12 nautical miles out from the U.S. coast.

- To operate within Class A airspace, you must:

  ○ Be instrument rated and current for instrument flight.

  ○ Operate under an IFR flight plan and in accordance with an ATC clearance at specified flight levels.

  ○ Set your altimeter to the standard setting of 29.92 inches Hg.

AIRSPACE CLASS REVIEW

You can use the following table as a quick reference for the VFR weather minimums and operating requirements of each airspace class. When trying to remember the weather minimums, it is easier to remember the minimums that are the same (highlighted) and know the few exceptions.

SECTION D ■ Airspace

SECTION D ■ Airspace

	VFR Minimum Visibility	VFR Minimum Distance from Clouds	Minimum Pilot Qualifications	VFR Entry and Equipment Requirements	ATC Services
CLASS A	N/A	N/A	Private Pilot Certificate Instrument Rating	IFR Flight Plan IFR Clearance	All Aircraft Separation
CLASS B	3 SM	Clear of Clouds	Private Pilot Certificate Student Pilot Logbook Endorsement	ATC Clearance Two-Way Radio Mode C Transponder/ADS-B Out	All Aircraft Separation
CLASS C	3 SM	500 ft Below 1,000 ft Above 2,000 ft Horizontal	Student Pilot Certificate	Establish Radio Communication Two-Way Radio Mode C Transponder/ADS-B Out	IFR/IFR Separation IFR/VFR Separation VFR Traffic Advisories (workload permitting)
CLASS D	3 SM	500 ft Below 1,000 ft Above 2,000 ft Horizontal	Student Pilot Certificate	Establish Radio Communication Two-Way Radio	IFR/IFR Separation VFR Traffic Advisories (workload permitting)
CLASS E	**Below 10,000 ft MSL:** 3 SM **At or Above 10,000 ft MSL:** 5 SM	**Below 10,000 ft MSL:** 500 ft Below 1,000 ft Above 2,000 ft Horizontal **At or Above 10,000 ft MSL:** 1,000 ft Below 1,000 ft Above 1 SM Horizontal	Student Pilot Certificate	**Below 10,000 ft MSL:** No Specific Equipment Required **At or Above 10,000 ft MSL and Above 2,500 ft AGL:** Mode C Transponder and ADB-B Out	IFR/IFR Separation VFR Traffic Advisories on Request (workload permitting)
CLASS G	**1,200 ft AGL and Below:** Day 1 SM; Night 3 SM **Below 10,000 ft MSL:** Day 1 SM; Night 3 SM **At or Above 10,000 MSL :** 5 SM (above 1,200 ft AGL)	**1,200 ft AGL and Below:** Day; Clear of Clouds Night; 500 ft Below, 1,000 ft Above, 2,000 ft Horizontal **Below 10,000 ft MSL:** 500 ft Below 1,000 ft Above 2,000 ft Horizontal (above 1,200 ft AGL) **At or Above 10,000 ft MSL:** 1,000 ft Below 1,000 ft Above 1 SM Horizontal (above 1,200 ft AGL)	Student Pilot Certificate	**Below 10,000 ft MSL:** No Specific Equipment Required **At or Above 10,000 ft MSL and Above 2,500 ft AGL:** Mode C Transponder	VFR Traffic Advisories on Request (workload permitting)

SPECIAL VFR

- You may only operate within the areas of Class B, C, D, or E airspace that extend to the surface around an airport, when the ground visibility is at least 3 statute miles and the cloud ceiling is at least 1,000 feet AGL. If ground visibility is not reported, you can use flight visibility.

- When the visibility is below 3 statute miles and/or the ceiling is below 1,000 feet AGL, you may obtain a special VFR clearance from the ATC facility having jurisdiction over the affected airspace.

- A special VFR clearance allows you to enter, leave, or operate within most Class D and Class E surface areas and in some Class B and Class C surface areas if the flight visibility is at least 1 statute mile and you can remain clear of clouds. At least 1 statute mile of ground visibility is required for takeoff and landing. However, if ground visibility is not reported, you must have at least 1 statute mile flight visibility.

- Special VFR is not permitted between sunset and sunrise unless you have a current instrument rating and the aircraft is equipped for instrument flight.

- The phrase NO SVFR included with the airport information on a VFR chart indicates that you cannot obtain a special VFR clearance to operate at the airport.

AIRSPEED LIMITATIONS

The following indicated airspeed limitations apply.

- 250 knots:

  - Below 10,000 feet MSL.

  - Within Class B airspace.

- 200 knots:

  - Under Class B airspace, or in a VFR corridor through a Class B area.

  - In Class C or D airspace, at or below 2,500 feet above the surface, and within 4 NM of the primary airport.

SPECIAL USE AIRSPACE

- Alert area—designated for unusual types of aerial activities, such as parachute jumping, glider towing, or high concentrations of student pilot training. All pilots flying within an alert area are equally responsible for collision avoidance.

- Military operations area (MOA)—a block of airspace in which military training and other military maneuvers are conducted. Avoid these areas or exercise extreme caution when military activity is being conducted. When operating VFR in an MOA, you should exercise extreme caution when military training is being conducted.

- Warning area—extends from three nautical miles outward from the coast of the United States; contains activity that can be hazardous to nonparticipating aircraft.

- Restricted area—often has invisible hazards to aircraft, such as artillery firing, aerial gunnery, or guided missiles. Permission to fly through a restricted area must be granted by the controlling agency.

- Prohibited area—established for security or other reasons associated with national welfare and contains airspace within which the flight of aircraft is prohibited.

- National security area (NSA)—established at locations where there is a requirement for increased security and safety of ground facilities. You are requested to voluntarily avoid flying through an NSA. At times, flight through an NSA may be prohibited by a NOTAM to provide a greater level of security and safety.

- Controlled firing area—not depicted on aeronautical charts; activities are discontinued immediately when a spotter aircraft, radar, or ground lookout personnel determines an aircraft might be approaching the area.

OTHER AIRSPACE AREAS

- Local airport advisory (LAA)—a service only available in Alaska and extends 10 statute miles from airports where a flight service station is located on the field, but there is no operating control tower.

- Military training routes (MTR)—established below 10,000 feet MSL for operations at speeds in excess of 250 knots.

  - Routes are classified as VR for VFR operations and IR for IFR operations.

  - Flights on routes marked IR are under ATC control regardless of the weather.

  - Routes at or below 1,500 feet AGL are identified by 4-digit numbers; routes with segments above 1,500 feet AGL are identified by 3-digit numbers.

SECTION D ■ Airspace

- Parachute jump aircraft operations—parachute jumping sites that have been used on a frequent basis and have been in use for at least one year. These areas are depicted on aeronautical charts by a magenta parachute symbol.

- Terminal radar service area (TRSA)—do not fit any of the U.S. airspace classes. TRSAs have never been established as controlled airspace and, therefore, FAR Part 91 does not contain any rules for TRSA operations. By contacting approach control, you can receive radar services within a TRSA, but participation is not mandatory.

- The primary airport within the TRSA is surrounded by Class D airspace and the outer portion of a TRSA normally overlies Class E airspace beginning at 700 or 1,200 feet AGL.

TEMPORARY FLIGHT RESTRICTIONS

- Temporary flight restrictions (TFRs) are regulatory actions that temporarily restrict certain aircraft from operating within a defined area to protect persons or property in the air or on the ground. You are responsible for knowing about and avoiding TFRs. Several types of TFRs are defined by the FARs:

  - Disaster/hazard TFRs.

  - Space flight operations.

  - High barometric pressure.

  - VIP TFRs—flight restrictions in the proximity of the President, Vice President, or other government officials. TFRs are the most restrictive in the vicinity of the President and typically prohibit all flight training activity within 30 miles of the president and create 10-mile radius no-fly zones that ban almost all general aviation activity..

  - Emergency air traffic rules—established by the FAA immediately after determining that, without such action, the air traffic control system could not operate at the required level of safety and efficiency.

  - Air shows and sporting events.

  - Special Security Instructions—an example of this type of TFR is described by a standing NOTAM that applies to major sporting events.

- TFRs are issued in NOTAMs that specify the dimensions, restrictions, and effective times. To determine if a TFR affects your flight, obtain NOTAMs from Flight Service during your online or phone briefing. You can also obtain a list of TFR NOTAMs with graphic depictions at tfr.faa.gov. The FAA cautions that the depicted TFR data might not be a complete listing, so always follow up with Flight Service during flight planning.

ADIZ

- An Air Defense Identification Zone (ADIZ) facilitates early identification of all aircraft in the vicinity of a nation's airspace boundaries. The Alaskan ADIZ, which lies along the coastal waters of Alaska has different operating rules than the contiguous U.S. ADIZ. The AIM and FAR Part 99 specify requirements to enter a United States ADIZ.

- To operate within the Contiguous U.S. ADIZ, you must file an IFR, or defense VFR (DVFR) flight plan containing the time and point at which you plan to enter the ADIZ. Set your Mode C transponder to the assigned code prior to entering the ADIZ and maintain two-way communication with the appropriate ATC facility.

WASHINGTON DC SPECIAL FLIGHT RULES AREA

- The Washington DC Special Flight Rules Area (SFRA) is airspace where the ready identification, location, and control of aircraft is required in the interests of national security. Depicted on charts, the SFRA includes all airspace within a 30 nautical mile radius of the Washington DC VOR (DCA) from the surface up to but not including flight level 180 (FL180).

- Areas with additional requirements include: the Leesburg Maneuvering Area with its own special procedures, and the flight restricted zone (FRZ)—a highly-restricted ring of airspace within 13 to 15 nautical miles of the Washington DC VOR, which is directly over the nation's capital. Only specially authorized aircraft may fly in the FRZ under IFR flight plans. Flight under VFR and general aviation aircraft operations are prohibited.

- If you are planning to fly under VFR within 60 nautical miles of the Washington DC VOR you must complete the FAA Special Awareness Training course at faasafety.gov. After you finish the training, print the completion certificate and carry it with you. You must present it at the request of an FAA, NTSB, law enforcement, or TSA authority.

INTERCEPT PROCEDURES

- If you penetrate an area with security-related flight restrictions, you risk being intercepted by U.S. military or law enforcement aircraft.

- Aircraft that are within security-related restricted airspace without authorization may be intercepted by law enforcement or military aircraft. If you are intercepted:

    ◦ Do not adjust your altitude, heading, or airspeed until directed by the intercepting aircraft.

    ◦ Follow instructions from the intercepting aircraft by given visual signals or radio communications until positively released.

    ◦ Attempt to contact the intercepting aircraft or ATC on 121.5 and provide your aircraft identity, position, and nature of the flight.

    ◦ Squawk 7700 on your transponder unless otherwise instructed by ATC

NOTE: An asterisk appearing after an ACS code (i.e. PA.VII.B.K1*) indicates that the question subject appears more than one time in the ACS. The code shown corresponds to the first instance of the subject in the ACS.

SECTION D ■ Airspace

4-141   PA.I.E.K1, K2

(Refer to Figure 26, area 2.) The day VFR visibility and cloud clearance requirements to operate over the town of Cooperstown, after departing and climbing out of the Cooperstown Airport at or below 700 feet AGL are

A – 1 mile and clear of clouds.

B – 1 mile 1,000 feet above, 500 feet below, and 2,000 feet horizontally from clouds.

C – 3 miles and clear of clouds.

4-141. Answer A. GFDPP 4D, FAR 91.155
This area is under the magenta shaded ring of Class E airspace, which begins at 700 feet AGL. Class G is below that, beginning at the surface and up to 700 feet AGL. Therefore, the visibility and cloud clearance at or below 700 feet AGL are 1 statue mile and clear of clouds in the day.

4-142   PA.I.E.K1

What minimum visibility and clearance from clouds are required for VFR operations in Class G airspace at 700 feet AGL or below during daylight hours?

A – 1-mile visibility and clear of clouds.

B – 1-mile visibility, 500 feet below, 1,000 feet above, and 2,000 feet horizontal clearance from clouds.

C – 3-miles visibility and clear of clouds.

4-142. Answer A. GFDPP 4D, FAR 91.155
For VFR flight in uncontrolled or class G airspace below 1,200 feet during daytime, you are only required to have 1 mile visibility and remain clear of clouds.

4-143   PA.I.E.K1, K2

(Refer to Figure 78.) What are the basic VFR weather minima required to takeoff from the Onawa, IA (K36) airport during the day?

A – 3 statute miles visibility, 500 feet below the clouds, 1,000 feet above the clouds, and 2,000 feet horizontally from the clouds.

B – 0 statute miles, clear of clouds.

C – 1 statute mile, clear of clouds.

4-143. Answer C. GFDPP 4D, FAR 91.155

This airport is in Class G airspace (uncontrolled) from the surface up to 1,200 feet AGL. For VFR flight in uncontrolled airspace below 1,200 feet AGL during daytime, you are required to have 1 mile visibility and remain clear of clouds.

4-144   PA.I.E.K1, K2

(Refer to Figure 26, area 1.) Identify the airspace over the town of McHenry.

A – Class G airspace — surface up to but not including 700 feet MSL, Class E airspace — 700 feet to 14,500 feet MSL.

B – Class G airspace — surface up to but not including 1,200 feet AGL, Class E airspace — 1,200 feet AGL up to but not including 18,000 feet MSL.

C – Class G airspace — surface up to but not including 18,000 feet MSL.

4-144. Answer B. GFDPP 4D, Appendix 1, Chart Legend

The town of McHenry is outside of the hard-edged side of the magenta bands, thus Class G airspace—from the surface, up to but not including 1,200 feet AGL. Class E airspace extends from 1,200 feet AGL to 18,000 MSL, unless otherwise marked.

4-145   PA.I.E.K1

During operations outside controlled airspace at altitudes of more than 1,200 feet AGL, but less than 10,000 feet MSL, the minimum flight visibility for VFR flight at night is

A – 1 mile.

B – 3 miles.

C – 5 miles.

4-145. Answer B. GFDPP 4D, FAR 91.155

In Class G airspace at these altitudes, night VFR operations require 3 statute miles visibility.

4-146   PA.I.E.K1

During operations outside controlled airspace at altitudes of more than 1,200 feet AGL, but less than 10,000 feet MSL, the minimum distance below clouds requirement for VFR flight at night is

A – 500 feet.

B – 1,000 feet.

C – 1,500 feet.

4-146. Answer A. GFDPP 4D, FAR 91.155

At night, in uncontrolled airspace below 10,000 feet MSL (both above and below 1,200 feet AGL), the VFR cloud clearance is 500 feet below.

4-147   PA.I.E.K1

During operations outside controlled airspace at altitudes of more than 1,200 feet AGL, but less than 10,000 feet MSL, the minimum flight visibility for day VFR is

A – 1 mile.

B – 3 miles.

C – 5 miles.

4-147. Answer A. GFDPP 4D, FAR 91.155

In uncontrolled airspace below 10,000 feet MSL and above 1,200 feet AGL, required daytime visibility is 1 statute mile.

4-148   PA.I.E.K1

During operations at altitudes of more than 1,200 feet AGL and at or above 10,000 feet MSL, the minimum distance above clouds requirement for VFR flight is

A – 500 feet.

B – 1,000 feet.

C – 1,500 feet.

4-148. Answer B. GFDPP 4D, FAR 91.155

For VFR flights at these altitudes, whether in controlled airspace or not, you are required to remain 1,000 feet above clouds. The only exception is for daytime operations below 1,200 feet AGL in uncontrolled airspace. In this case, it is clear of clouds.

4-149   PA.I.E.K1

VFR flight in controlled airspace above 1,200 feet AGL and below 10,000 feet MSL requires a minimum visibility and vertical cloud clearance of

A – 3 miles, and 500 feet below or 1,000 feet above the clouds in controlled airspace.

B – 5 miles, and 1,000 feet below or 1,000 feet above the clouds at all altitudes.

C – 5 miles, and 1,000 feet below or 1,000 feet above the clouds only in Class A airspace.

4-149. Answer A. GFDPP 4D, FAR 91.155

Below 10,000 feet MSL in Class C and D airspace, the cloud clearances are 500 feet below, 1,000 feet above, and 2,000 feet horizontal. In Class B airspace, however, the cloud clearance is just "clear of clouds."

4-150   PA.I.E.K1

During operations within controlled airspace at altitudes of less than 1,200 feet AGL, the minimum horizontal distance from clouds requirement for VFR flight is

A – 1,000 feet.

B – 1,500 feet.

C – 2,000 feet.

4-150. Answer C. GFDPP 4D, FAR 91.155

In controlled airspace, other than Class B, below 10,000 feet, it does not matter whether you are above or below 1,200 feet AGL. The VFR cloud clearance is 2,000 feet horizontal.

SECTION D ■ **Airspace**

4-151   PA.I.E.K1

The minimum flight visibility required for VFR flights above 10,000 feet MSL and more than 1,200 feet AGL in controlled airspace is

A – 1 mile.

B – 3 miles.

C – 5 miles.

4-151. Answer C. GFDPP 4D, FAR 91.155
At or above 10,000 feet MSL and above 1,200 feet AGL, the required visibility is 5 statute miles, regardless of whether you're in controlled or uncontrolled airspace.

4-152   PA.I.E.K1

During operations within controlled airspace at altitudes of more than 1,200 feet AGL, but less than 10,000 feet MSL, the minimum distance above clouds requirement for VFR flight is

A – 500 feet.

B – 1,000 feet.

C – 1,500 feet.

4-152. Answer B. GFDPP 4D, FAR 91.155
Below 10,000 feet MSL in Class C and D airspace, the cloud clearances are 500 feet below, 1,000 feet above, and 2,000 feet horizontal. In Class B airspace, however, the cloud clearance is just "clear of clouds."

4-153   PA.I.E.K1

A Mode C transponder and ADS-B Out equipment are required in which airspace?

A – Class A, Class B (and within 30 miles of the Class B primary airport), and Class C.

B – Class D and Class E (below 10,000 feet MSL).

C – Class D and Class G (below 10,000 feet MSL).

4-153. Answer A. GFDPP 4D, AIM, FAR 91.215
A Mode C transponder and ADS-B Out equipment are required in Class A, Class B, and Class C airspace.

4-154   PA.I.D.K1, PA.I.E.K1

A Mode C transponder and ADS-B Out equipment are required in

A – Class B airspace and within 30 miles of the Class B primary airport.

B – Class D airspace.

C – Class E airspace below 10,000 feet MSL.

4-154. Answer A. GFDPP 4D, AIM, FAR 91.131
For airspace within 30 nautical miles of a primary airport within a Class B airspace area, from the surface upward to 10,000 feet MSL, unless otherwise authorized by ATC, aircraft operating within this airspace must be equipped with a Mode C transponder and ADS-B Out equipment.

This equipment is not required in Class D airspace, or in Class E airspace below 10,000 feet MSL.

4-155 PA.I.E.K1

For VFR flight operations above 10,000 feet MSL and more than 1,200 feet AGL, the minimum horizontal distance from clouds required is

A – 1,000 feet.

B – 2,000 feet.

C – 1 mile.

4-155. Answer C. GFDPP 4D, FAR 91.155
Whether in controlled or uncontrolled airspace at these altitudes, the minimum VFR horizontal distance from clouds is 1 statute mile.

4-156 PA.I.E.K1, K2

(Refer to Figure 26, area 2.) The visibility and cloud clearance requirements to operate VFR during daylight hours over the town of Cooperstown between 1,200 feet AGL and 10,000 feet MSL are

A – 3 miles and 1,000 feet above, 500 feet below, and 2,000 feet horizontally from clouds.

B – 1 mile and clear of clouds.

C – 1 mile and 1,000 feet above, 500 feet below, and 2,000 feet horizontally from clouds.

4-156. Answer A. GFDPP 4D, FAR 91.155
Cooperstown is within the Class E airspace of the Cooperstown Airport. The floor of this airspace is 700 feet AGL. Visibility of 3 miles and 1,000 feet above, 500 feet below, and 2,000 feet horizontally from clouds is required to operate VFR in controlled airspace below 10,000 feet MSL.

4-157 PA.I.E.K1

Unless otherwise specified, Federal airways include that Class E airspace extending upward from

A – 700 feet above the surface up to and including 17,999 feet MSL.

B – 1,200 feet above the surface up to and including 17,999 feet MSL.

C – the surface up to and including 18,000 feet MSL.

4-157. Answer B. GFDPP 4D, AIM
Federal (or Victor) airways normally begin at 1,200 feet AGL and extend up to 17,999 feet MSL.

4-158 PA.I.E.K1, K2

(Refer to Figure 22, area 1.) The visibility and cloud clearance requirements to operate VFR during daylight hours over Sandpoint Airport at 1,200 feet AGL are

A – 1 mile and 1,000 feet above, 500 feet below, and 2,000 feet horizontally from each cloud.

B – 1 mile and clear of clouds.

C – 3 miles and 1,000 feet above, 500 feet below, and 2,000 feet horizontally from each cloud.

4-158. Answer C. GFDPP 4D, FAR 91.155
The airspace is Class E above 700 feet AGL. The day VFR minimums are 3 miles and 1,000 feet above, 500 feet below, and 2,000 feet horizontally from all clouds.

SECTION D ■ Airspace

4-159 PA.I.E.K1, K2

(Refer to Figure 26, area 5.) The airspace overlying and within 5 miles of Barnes County Airport is

A – Class D airspace from the surface to the floor of the overlying Class E airspace.

B – Class E airspace from the surface to 1,200 feet MSL.

C – Class G airspace from the surface to 700 feet AGL.

4-159. Answer C. GFDPP 4D, Appendix 1, Chart Legend
The magenta shading around the airport indicates that Class E airspace begins at 700 feet AGL. Below 700 feet the airspace is uncontrolled (Class G) airspace. Class D airspace, marked with a blue dashed line around an airport, is not shown anywhere on this chart.

4-160 PA.I.E.K1

What minimum visibility and clearance from clouds are required for VFR operations on an airway below 10,000 feet MSL is

A – 1-mile visibility and clear of clouds.

B – 3-miles visibility, 500 feet below, 1,000 feet above, and 2,000 feet horizontally from clouds.

C – 3-miles visibility, 500 feet above, 1,000 feet below, and 2,000 feet horizontally from clouds.

4-160. Answer B. GFDPP 4D, FAR 91.155
The VFR cloud clearances for Class E airspace apply. Below 10,000 feet MSL, you must maintain 3 miles visibility and 500 feet below, 1,000 feet above, and 2,000 feet horizontally from clouds.

4-161 PA.I.E.K1, K2

(Refer to Figure 22, area 3.) The vertical limits of that portion of Class E airspace designated as a Federal airway over Magee Airport are

A – 1,200 feet AGL to 17,999 feet MSL.

B – 7,500 feet MSL to 17,999 feet MSL.

C – 700 feet MSL to 12,500 feet MSL.

4-161. Answer A. GFDPP 4D, Appendix 1, Chart Legend
Class E airspace includes Federal (or Victor) airways that usually extend to 4 nautical miles on each side of the airway centerline and, unless otherwise indicated, extends from 1,200 feet AGL up to, but not including, 18,000 feet MSL.

4-162 PA.I.E.K1

The width of a Federal airway from either side of the centerline is

A – 4 nautical miles.

B – 6 nautical miles.

C – 8 nautical miles.

4-162. Answer A. GFDPP 4D, AIM
A Federal (or Victor) airway is normally 8 NM wide (4 NM each side of the centerline).

4-163   PA.I.E.K1

With certain exceptions, Class E airspace extends upward from either 700 feet or 1,200 feet AGL to, but does not include,

A – 10,000 feet MSL.

B – 14,500 feet MSL.

C – 18,000 feet MSL.

4-163. Answer C. GFDPP 4D, AIM

Unless otherwise indicated, E airspace begins at 700 feet or 1,200 feet AGL and continues up to, but not including 18,000 MSL. Notice that the upper limit is defined by MSL, while the lower limit is typically defined by AGL.

4-164   PA.I.E.K1

T-routes are GPS based and are Class E airspace extending upward from

A – 700 feet above the surface up to and including 17,999 feet MSL.

B – 1,200 feet above the surface up to and including 17,999 feet MSL.

C – the surface up to and including 18,000 feet MSL.

4-164. Answer B. GFDPP 4D, AIM

T-routes normally begin at 1,200 feet AGL and extend up to 17,999 feet MSL. They are considered Class E airspace.

4-165   PA.I.E.K1

Normal VFR operations in Class D airspace with an operating control tower require the ceiling and visibility to be at least

A – 1,000 feet and 1 mile.

B – 1,000 feet and 3 miles.

C – 2,500 feet and 3 miles.

4-165. Answer B. GFDPP 4D, FAR 91.155

To operate in Class D airspace, the VFR visibility minimum is three statute miles. In addition, the ceiling must be at least 1,000 feet.

4-166   PA.I.E.K1, K2

(Refer to Figure 25, area 7.) The airspace overlying Collin County Mc Kinney Airport (TKI) is controlled from the surface to

A – 2,900 feet MSL.

B – 2,500 feet MSL.

C – 700 feet AGL.

4-166. Answer A. GFDPP 4D, Appendix 1, Chart Legend

The controlled airspace overlying TKI airport is Class D airspace. The top of Class D airspace (MSL) for this airport is shown as [29] within the square dashed box, indicating 2,900 feet MSL. Class B airspace begins at 4,000 feet MSL above TKI airport.

4-167   PA.I.E.K1, K2

(Refer to Figure 25, area 4.) The airspace directly overlying Fort Worth Meacham is

A – Class B airspace to 10,000 feet MSL.

B – Class C airspace to 5,000 feet MSL.

C – Class D airspace to 3,200 feet MSL.

4-167. Answer C. GFDPP 4D, Appendix 1, Chart Legend

The blue segmented circle indicates that Fort Worth Meacham is located in Class D airspace. The [32] indicates that the ceiling of the Class D airspace is 3,200 feet MSL. Class B airspace begins at 4,000 feet MSL above Fort Worth Meacham airport.

SECTION D ■ Airspace

4-168 PA.I.E.K1

At what altitude shall the altimeter be set to 29.92, when climbing to cruising flight level?

A – 14,500 feet MSL.

B – 18,000 feet MSL.

C – 24,000 feet MSL.

4-168. Answer B. GFDPP 4D, FAR 91.121

To standardize altimeter settings in Class A airspace, you are required to set your altimeters to 29.92 at and above 18,000 feet MSL.

4-169 PA.I.E.K1, K2

A blue segmented circle on a sectional chart depicts which class airspace?

A – Class B.

B – Class C.

C – Class D.

4-169. Answer C. GFDPP 4D, Appendix 1, Chart Legend

Class D airspace is designated on sectional charts by a blue segmented circle. Class B airspace is indicated by a solid blue line. Class C airspace is designated by a solid magenta line.

4-170 PA.I.E.K1

Airspace at an airport with a part-time control tower is classified as Class D airspace only

A – when the weather minimums are below basic VFR.

B – when the associated control tower is in operation.

C – when the associated Flight Service Station is in operation.

4-170. Answer B. GFDPP 4D, AIM

In order for airspace to be classified as Class D, a control tower must be operating.

4-171 PA.I.E.K1

The basic VFR weather minimums for operating an aircraft within Class D airspace are

A – 500-foot ceiling and 1-mile visibility.

B – 1,000-foot ceiling and 3-miles visibility.

C – clear of clouds and 2-miles visibility.

4-171. Answer B. GFDPP 4D, FAR 91.155

To take off or land under VFR in Class D airspace, the ceiling must be at least 1,000 feet and the ground visibility must be at least 3 statute miles. Flight visibility may be used if ground visibility is not available.

4-172 PA.I.E.K1

When a control tower on an airport within Class D airspace ceases operation for the day, what happens to the airspace designation?

A – The airspace designation normally does not change.

B – The airspace remains Class D airspace as long as a weather observer or automated weather system is available.

C – The airspace reverts to Class E or a combination of Class E and G airspace during the hours the tower is not in operation.

4-172. Answer C. GFDPP 4D, AIM
Class D Airspace exists only when the control tower is operating. When the tower closes, the airspace becomes Class E (or occasional Class G) and you normally continue to use the tower frequency as a common traffic advisory frequency (CTAF).

4-173 PA.I.E.K1

No person may take off or land an aircraft under basic VFR at an airport that lies within Class D airspace unless the

A – flight visibility at that airport is at least 1 mile.

B – ground visibility at that airport is at least 1 mile.

C – ground visibility at that airport is at least 3 miles.

4-173. Answer C. GFDPP 4D, FAR 91.155
To take off or land under VFR in a Class D airspace area, the ceiling must be at least 1,000 feet and the ground visibility must be at least 3 statute miles. Flight visibility may be used if ground visibility is not available.

4-174 PA.I.E.K1, PA.II.D.K6b

A non-towered satellite airport, within the same Class D airspace as that designated for the primary airport, requires two-way radio communications be established and maintained with the

A – satellite airport's UNICOM.

B – associated Flight Service facility.

C – primary airport's control tower.

4-174. Answer C. GFDPP 4D, FAR 91.129
When approaching Class D airspace, you must contact the control tower of the primary airport before entering the airspace. When departing from a non-towered satellite airport, you must establish and maintain two-way radio communications with the ATC facility having jurisdiction over the Class D airspace area as soon as practical after departing.

4-175 PA.I.E.K1

The lateral dimensions of Class D airspace are based on

A – the number of airports that lie within the Class D airspace.

B – 5 statute miles from the geographical center of the primary airport.

C – the instrument procedures for which the controlled airspace is established.

4-175. Answer C. GFDPP 4D, AIM
Generally, Class D airspace extends upward from the surface to 2,500 feet above the airport elevation (charted in MSL) surrounding those airports that have an operational control tower. The configuration of each Class D airspace area is individually tailored, and when instrument procedures are published, the airspace will normally be designed to contain the procedures.

SECTION D ■ Airspace

4-176   PA.III.A.K1

Unless otherwise authorized, two-way radio communications with air traffic control are required for landings or takeoffs at all towered airports

A – regardless of weather conditions.

B – only when weather conditions are less than VFR.

C – within Class D airspace only when weather conditions are less than VFR.

4-176. Answer A. GFDPP 4D, AIM

When operating at an airport where a control tower is in operation, you must be in radio contact with ATC whether or not VFR conditions exist.

4-177   PA.III.A.K2, K4

Two-way radio communication must be established with the air traffic control facility having jurisdiction over the area before entering which class airspace?

A – Class C.

B – Class E.

C – Class G.

4-177. Answer A. GFDPP 4D, AIM, FAR 91.130

You must establish two-way radio communications before entering a Class C airspace area, and maintain it while operating within the Class C airspace.

4-178   PA.I.E.K1

Under what condition may an aircraft operate from a satellite airport within Class C airspace?

A – The pilot must file a flight plan before departure.

B – The pilot must monitor ATC until clear of the Class C airspace.

C – The pilot must contact ATC as soon as practicable after takeoff.

4-178. Answer C. GFDPP 4D, FAR 91.130

You must establish two-way communications with ATC as soon as practical after takeoff.

4-179   PA.I.E.K1

What minimum radio equipment is required for operation within Class C airspace?

A – Two-way radio communications equipment and a Mode C transponder.

B – Two-way radio communications equipment, a Mode C transponder, and DME.

C – Two-way radio communications equipment, a Mode C transponder, and ADS-B Out equipment.

4-179. Answer C. GFDPP 4D, AIM, FAR 91.130, 91.215

To operate in a Class C airspace area, you are required to have a two-way radio, a transponder with Mode C capability and ADS-B Out equipment.

4-180   PA.I.E.K1

Which initial action should a pilot take before entering Class C airspace?

A – Contact approach control on the appropriate frequency.

B – Contact the tower and request permission to enter.

C – Contact Flight Service for traffic advisories.

4-180. Answer A. GFDPP 4D, AIM

Before entering Class C airspace, you must establish contact with approach control.

4-181   PA.I.E.K1, K2

(Refer to Figure 20, area 1.) What minimum radio equipment is required to land and take off at Norfolk International?

A – Mode C transponder, ADS-B Out equipment, and VOR.

B – A two-way radio, a Mode C transponder, and ADS-B Out equipment.

C – Mode C transponder, VOR, and DME.

4-181. Answer B. GFDPP 4D, FAR 91.130

The area depicted is Class C airspace. Aircraft operating in Class C airspace must be equipped with a two-way radio, and unless otherwise authorized by ATC, a Mode C transponder, and ADS-B Out equipment. A VOR or DME is not required to operate in Class C airspace.

4-182   PA.I.E.K1

The vertical limit of Class C airspace above the primary airport is normally

A – 1,200 feet AGL.

B – 3,000 feet AGL.

C – 4,000 feet AGL.

4-182. Answer C. GFDPP 4D, AIM

Although the configuration of each Class C airspace area is individually tailored, the airspace usually consists of a 5 NM radius core surface area that extends from the surface up to 4,000 feet above the airport elevation, and a 10 NM radius shelf area that extends no lower than 1,200 feet up to 4,000 feet above the airport elevation. This altitude is rounded to an even 100-foot MSL altitude and shown as an MSL altitude on charts.

4-183   PA.I.E.K1

The radius of the procedural outer area of Class C airspace is normally

A – 10 NM.

B – 20 NM.

C – 30 NM.

4-183. Answer B. GFDPP 4D, AIM

Class C airspace areas have a procedural outer area not shown on charts. Normally this area extends 20 NM from the primary Class C airspace airport. You may obtain radar services in this area, however contact with ATC is not required.

SECTION D ■ Airspace

4-184 PA.I.E.K1, K2

(Refer to Figure 23, area 3.) What is the floor of the Savannah Class C airspace at the shelf area (outer circle)?

A – 1,300 feet AGL.

B – 1,300 feet MSL.

C – 1,700 feet MSL.

4-184. Answer B. GFDPP 4D, AIM

The floor of the shelf area of most Class C airspace is around 1,200 feet AGL, rounded to an exact MSL altitude. At Savannah, it is 1,300 feet MSL, about 1,250 feet above the airport elevation. The exact upper and lower limits of the shelf area of the Class C airspace depicted on the chart are 4,100 and 1,300 feet (41/13) MSL.

4-185 PA.I.D.K1, PA.I.E.K1

All operations within Class C airspace must be in

A – accordance with instrument flight rules.

B – compliance with ATC clearances and instructions.

C – an aircraft equipped a Mode C transponder and ADS-B Out equipment.

4-185. Answer C. GFDPP 4D, AIM, FAR 91.130, 91.215

All aircraft operating within Class C airspace must be equipped with a two-way radio, and unless authorized by ATC, a Mode C transponder, and ADS-B Out equipment.

4-186 PA.I.E.K1, K2

(Refer to Figure 69, area 4.) You are in the uncongested area about 7 NM SW of Corpus Christi International Airport, and unable to reach Corpus Christi Approach because of radio congestion. What altitude or course should you use to avoid the tall TV towers in this area?

A – Maintain at least 1,604 feet MSL; it is not necessary to contact ATC for obstacle clearance.

B – Fly over the TV towers at 1,499 feet MSL to remain under the floor of the Class C airspace.

C – Remain under the floor of Class C airspace and maintain a safe lateral distance from all the TV towers.

4-186. Answer C. GFDPP 4D, FAR 91.119

Because this is an uncongested area, you should be 500 feet above the elevation of the highest obstacle, which is 1,104 feet MSL. However, the minimum safe altitude of 1,604 feet MSL is above the floor of the Class C airspace, and two-way radio communications with ATC is required. If you cannot climb to the minimum safe altitude, you must maintain a safe horizontal clearance from these towers. If you are unable to reach ATC and climb into Class C airspace, deviating to the south and remaining below 1,200 feet MSL could be the best course of action, particularly in lower visibility, or if you are unfamiliar with the TV towers in the area.

4-187 PA.I.E.K2

(Refer to Figure 25, area 2.) The floor of Class B airspace at Addison Airport is

A – at the surface.

B – 3,000 feet MSL.

C – 3,100 feet MSL.

4-187. Answer B. GFDPP 4D, Appendix 1, Chart Legend

The altitudes of this portion of the Class B airspace are indicated by "110" over "30." This means the Class B airspace extends from a floor of 3,000 feet MSL up to 11,000 feet MSL. In addition, the [–30] indicates that the Class D airspace at Addison extends up to, but does not include 3,000 feet, which is the floor of the overlying Class B airspace.

4-188   PA.I.D.K1, PA.I.E.K1

What minimum radio equipment is required for VFR operation within Class B airspace?

A – Two-way radio communications equipment and a Mode C or Mode S transponder.

B – Two-way radio communications equipment, a Mode C transponder, and ADS-B Out equipment.

C – Two-way radio communications equipment, a Mode C transponder, and a VOR or TACAN receiver.

4-188. Answer B. GFDPP 4D, AIM, FAR 91.131

For all operations, a two-way radio capable of communications with ATC on appropriate frequencies for that area; and unless otherwise authorized by ATC, a Mode C transponder, and ADS-B Out equipment.

4-189   PA.I.E.K1

What minimum pilot certification is required for operation within Class B airspace?

A – Sport pilot certificate.

B – Private pilot certificate or student pilot certificate with appropriate logbook endorsements.

C – Private Pilot Certificate with an instrument rating.

4-189. Answer B. GFDPP 4D, FAR 91.131

To operate in a Class B airspace area, you must hold a private pilot certificate. However, within certain Class B airspace areas, student pilot operations may be conducted after receiving specific training and a logbook endorsement from an authorized flight instructor.

4-190   PA.I.E.K2

(Refer to Figure 25, area 4.) The floor of Class B airspace overlying Hicks Airport (T67) north-northwest of Fort Worth Meacham Field is

A – at the surface.

B – 3,200 feet MSL.

C – 4,000 feet MSL.

4-190. Answer C. GFDPP 4D, Appendix 1, Chart Legend

The altitudes of this sector of the Class B airspace are indicated by "110" over "40." This means the Class B airspace extends from a floor of 4,000 feet MSL up to 11,000 feet MSL.

4-191   PA.I.E.K1

With certain exceptions, all aircraft within 30 miles of a Class B primary airport from the surface upward to 10,000 feet MSL must be equipped with

A – an operable VOR or TACAN receiver and an ADF receiver.

B – instruments and equipment required for IFR operations.

C – a Mode C transponder and ADS-B Out equipment.

4-191. Answer C. GFDPP 4D, AIM, FAR 91.215

A Mode C transponder and ADS-B Out equipment are required to be in use when within 30 miles of a Class B primary airport.

SECTION D ■ Airspace

SECTION D ■ Airspace

4-192   PA.I.E.K1

In which type of airspace are VFR flights prohibited?

A – Class A.

B – Class B.

C – Class C.

4-192. Answer A. GFDPP 4D, FAR 91.135
Only IFR operations are allowed in Class A airspace. VFR flights are allowed in Class B and C airspace if authorized by ATC.

4-193   PA.I.E.K1, K4

What is the minimum weather condition required for airplanes operating under special VFR in Class D airspace?

A – 1-mile flight visibility.

B – 1-mile flight visibility and 1,000-foot ceiling.

C – 3-miles flight visibility and 1,000-foot ceiling.

4-193. Answer A. GFDPP 4D, FAR 91.157
When authorized by ATC, special VFR allows you to operate with one statute mile visibility as long as you can remain clear of clouds.

4-194   PA.I.E.K1, K4

What are the minimum requirements for airplane operations under special VFR in Class D airspace at night?

A – The airplane must be under radar surveillance at all times while in Class D airspace.

B – The airplane must be equipped for IFR and with an altitude reporting transponder.

C – The pilot must be instrument rated, and the airplane must be IFR equipped.

4-194. Answer C. GFDPP 4D, FAR 91.157
For special VFR at night, you must have a current instrument rating, and the airplane must be equipped for IFR operations.

4-195   PA.I.E.K1, K4

No person may operate an airplane within Class D airspace at night under special VFR unless the

A – flight can be conducted 500 feet below the clouds.

B – airplane is equipped for instrument flight.

C – flight visibility is at least 3 miles.

4-195. Answer B. GFDPP 4D, FAR 91.157
For special VFR at night, you must have a current instrument rating, and the airplane must be equipped for IFR operations.

4-196   PA.I.E.K1, K4

A special VFR clearance authorizes the pilot of an aircraft to operate VFR while within Class D airspace when the visibility is

A – less than 1 mile and the ceiling is less than 1,000 feet.

B – at least 1 mile and the aircraft can remain clear of clouds.

C – at least 3 miles and the aircraft can remain clear of clouds.

4-196. Answer B. GFDPP 4D, FAR 91.157
When authorized by ATC, special VFR allows you to operate with one statute mile visibility as long as you can remain clear of clouds.

4-197   PA.I.E.K1, K2, K4

(Refer to Figure 25.) At which airports is fixed-wing Special VFR not authorized?

A – Fort Worth Meacham and Fort Worth Spinks.

B – Dallas-Fort Worth International and Dallas Love Field.

C – Addison and Dallas Executive.

4-197. Answer B. GFDPP 4D, Appendix 1, Chart Legend
The "NO SVFR" above the airport name indicates that fixed-wing special VFR is not authorized.

4-198   PA.I.E.K1, K4

What ATC facility should the pilot contact to receive a special VFR departure clearance in Class D airspace?

A – Flight Service.

B – Air traffic control tower.

C – Air route traffic control center.

4-198. Answer B. GFDPP 4D, AIM, FAR 91.157
When a control tower is located within the Class B, Class C, or Class D surface area, requests for clearances should be to the tower. In a Class E surface area, a clearance may be obtained from the nearest tower, Flight Service, or center.

4-199   PA.I.E.K1

Unless otherwise authorized, what is the maximum indicated airspeed at which a person may operate an aircraft below 10,000 feet MSL?

A – 200 knots.

B – 250 knots.

C – 288 knots.

4-199. Answer B. GFDPP 4D, FAR 91.117
Unless otherwise authorized by the Administrator, you not may operate an aircraft below 10,000 feet MSL at an indicated airspeed of more than 250 knots.

SECTION D ■ **Airspace**

4-200   PA.I.E.K1

When flying in the airspace underlying Class B airspace, the maximum speed authorized is

A – 200 knots.

B – 230 knots.

C – 250 knots.

4-200. Answer A. GFDPP 4D, FAR 91.117
You may not operate an aircraft in the airspace underlying a Class B airspace area, or in a VFR corridor designated through a Class B airspace area, at an indicated airspeed of more than 200 knots.

4-201   PA.I.E.K1

When flying in a VFR corridor designated through Class B airspace, the maximum speed authorized is

A – 180 knots.

B – 200 knots.

C – 250 knots.

4-201. Answer B. GFDPP 4D, FAR 91.117
You may not operate an aircraft in a VFR corridor designated through a Class B airspace area, or in the airspace underlying a Class B airspace area, at an indicated airspeed of more than 200 knots.

4-202   PA.I.E.K1

Unless otherwise authorized, the maximum indicated airspeed at which aircraft may be flown when at or below 2,500 feet AGL and within 4 nautical miles of the primary airport of Class C airspace is

A – 200 knots.

B – 230 knots.

C – 250 knots.

4-202. Answer A. GFDPP 4D, FAR 91.117
The maximum indicated airspeed inside or within 4 NM of the primary airport in Class C airspace is 200 knots.

4-203   PA.I.E.K3

Under what condition, if any, may pilots fly through a restricted area?

A – When flying on airways with an ATC clearance.

B – With authorization from the controlling agency.

C – Regulations do not allow this.

4-203. Answer B. GFDPP 4D, FAR 91.133
The controlling agency may grant permission to fly through a restricted area.

4-204   PA.I.E.K3

Responsibility for collision avoidance in an alert area rests with

A – the controlling agency.

B – all pilots.

C – Air traffic control.

4-204. Answer B. GFDPP 4D, AIM
All pilots flying in an alert area, whether participating in activities or transitioning the area, are equally responsible for collision avoidance.

4-205 PA.I.E.K3

What action should a pilot take when operating under VFR in a military operations area (MOA)?

A – Obtain a clearance from the controlling agency before entering the MOA.

B – Operate only on the airways that transverse the MOA.

C – Exercise extreme caution when military activity is being conducted.

4-205. Answer C. GFDPP 4D, AIM
Due to the possibility of military training activities, you should use extra caution and be vigilant for military traffic when operating in an MOA.

4-206 PA.I.E.K3

(Refer to Figure 21, area 2.) What hazards to aircraft may exist in areas such as Devils Lake West MOA?

A – Military training activities that require acrobatic or abrupt flight maneuvers.

B – High volume of pilot training or an unusual type of aerial activity.

C – Unusual, often invisible, hazards to aircraft such as artillery firing, aerial gunnery, or guided missiles.

4-206. Answer A. GFDPP 4D, AIM
Most training activities in a military operations area (MOA) involve acrobatic or abrupt flight maneuvers.

4-207 PA.I.E.K3

You should not fly through a restricted area unless you have

A – filed an IFR flight plan.

B – received prior authorization from the controlling agency.

C – received prior permission from the commanding officer of the nearest military base.

4-207. Answer B. GFDPP 4D, AIM, FAR 91.133
You must receive prior authorization from the controlling agency before operating in a restricted area. If you are operating on an IFR flight plan and ATC clears you through that airspace, then they are acting as the controlling agency and providing you the needed authorization. The FAA is not the controlling agency when the restricted area is "hot" and may not clear you through the airspace at those times.

4-208 PA.I.D.K1

(Refer to Figure 21, area 3.) What type of military flight operations should you expect along IR644?

A – VFR training flights above 1,500 feet AGL at speeds less than 250 knots.

B – IFR training flights above 1,500 feet AGL at speeds more than 250 knots.

C – Instrument training flights below 1,500 feet AGL at speeds more than 150 knots.

4-208. Answer B. GFDPP 4D, AIM
Generally, military training routes (MTR) are established below 10,000 feet MSL for operations at speeds in excess of 250 knots. MTRs are classified as VR for VFR operations and IR for IFR operations. IR routes are designed to be flown by military aircraft at speeds often more than 250 knots. An IR route with three letters in the designator (IR644) indicates one or more segments are above 1,500 feet AGL. MTRs that are entirely at or below 1,500 feet AGL are identified by a four-digit designator (VR1175).

SECTION D ■ Airspace

4-209   PA.I.E.K3

(Refer to Figure 20, area 4.) What hazards to aircraft may exist in restricted areas such as R-5302B?

A – Military training activities that require acrobatic or abrupt flight maneuvers.

B – Unusual, often invisible, hazards such as aerial gunnery or guided missiles.

C – High volume of pilot training or an unusual type of aerial activity.

4-209. Answer B. GFDPP 4D, AIM
Restricted areas can have invisible hazards to aircraft, such as artillery firing, aerial gunnery, or guided missiles.

4-210   PA.I.E.K3

Information concerning regularly used parachute jump aircraft operations areas may be found in the

A – NOTAMs publications.

B – Aeronautical Information Manual.

C – General Information section of the Chart Supplement.

4-210. Answer C. GFDPP 4D, Chart Supplement, PHB
Established parachute jump aircraft operations areas are listed in the General Information section of the Chart Supplement. Frequently used sites are also depicted on aeronautical charts, and single events or infrequently used sites are listed in NOTAMs.

4-211   PA.I.E.K3

How can you learn the time and location of TFRs?

A – Obtain a weather briefing from Flight Service and ask for TFRs.

B – Obtain a list of TFR NOTAMs with graphical depictions at tfr.tsa.com.

C – Check TFRs and their effective times at tfr.gov or in your flight planning application.

4-211. Answer A. GFDPP 4D, AIM
Although the graphical TFR listings at faa.gov and in flight planning programs and applications provide detailed information regarding TFRs, official dissemination is through Flight Service by telephone at 1-800-wx-brief or at website 1800wxbrief.com.

4-212   PA.I.E.K3

If you are planning to fly under VFR within 60 nautical miles of the Washington DC SFRA

A – squawk 7700 and announce your intentions on 121.5.

B – remain clear of the Class B airspace until receiving clearance to enter the airspace.

C – you must complete the FAA Special Awareness Training course.

4-212. Answer C. GFDPP 4D, AIM
If you are planning to fly under VFR within 60 nautical miles of the Washington DC VOR you must complete the FAA Special Awareness Training course at faasafety.gov.

4-213 PA.I.E.K3

What must you do before flying anywhere within 60 miles of the Washington (DCA) VOR?

A – Take special awareness training from the FAA and have a certificate of completion in your possession.

B – Take an orientation flight with an authorized flight instructor, obtain a logbook endorsement, and carry your logbook.

C – Obtain ground instruction from an authorized ground or flight instructor, obtain a logbook endorsement, and carry your logbook.

4-213. Answer A. GFDPP 4D, FAR 91.161

You may not act as a pilot of an aircraft while flying within a 60-nautical mile radius of the DCA VOR under VFR unless you have completed Special Awareness Training and hold a certificate of training completion.

This training is available at the FAAsafety.gov website. Upon completion of the training, print a copy of the completion certificate and be able to present it to an authorized representative of the FAA, NTSB, TSA, or to a law enforcement officer.   .

4-214 PA.I.D.K6

If you penetrate an area with security related flight restrictions, you risk being intercepted by U.S military or law enforcement aircraft. If you are intercepted, you should

A – turn 180 degrees from your present heading and return to your departure airport.

B – follow instructions from the intercepting aircraft, squawk 7700 on your transponder, and attempt to contact the intercepting aircraft on 121.5.

C – squawk 7500 on your transponder and attempt to contact the intercepting aircraft on 123.0.

4-214. Answer B. GFDPP 4D, AIM

Do not adjust your altitude, heading, or airspeed until directed by the intercepting aircraft. Follow instructions from the intercepting aircraft by given visual signals or radio communications until positively released. Attempt to contact the intercepting aircraft or ATC on 121.5 and provide your aircraft identity, position, and nature of the flight. Squawk 7700 on your transponder unless otherwise instructed by ATC

4-215 PA.I.E.K3

You are flying VFR and not talking to ATC and observe a military jet fighter off your left wingtip rocking its wings and then making a slow left turn away from you. What action should you take?

A – Land at the nearest airport and wait for authorities to meet you.

B – Follow the aircraft, squawk 7700, and contact the VHF guard frequency on 121.5 MHz.

C – Proceed on course, you are free to go because the intercepting aircraft has turned away from you.

4-215. Answer B. GFDPP 4D, AIM

When the intercepting aircraft rocks its wings and makes a *slow* turn away from you, it means that you are intercepted and must follow it. An *abrupt* breakaway turn that does not cross your flight path means that you are free to proceed on course.

SECTION D ■ **Airspace**

CHAPTER 5

Communication and Flight Information

SECTION A
ATC Services

ADS-B SYSTEM

- The automatic dependent surveillance-broadcast (ADS-B) system incorporates GPS, aircraft transmitters and receivers, and ground stations to provide pilots and ATC with specific data about the position and speed of aircraft. Two forms of ADS-B equipment apply to aircraft—ADS-B Out and ADS-B In.

- The ADS-B system relies on the GPS in each aircraft for position information.

- ADS-B Out transmits line-of-sight signals from the aircraft to ATC ground receivers and to receivers in other aircraft.

- ADS-B In receives the lateral position, altitude, and velocity of transmitting aircraft and presents this data on a cockpit display of traffic information (CDTI), which can be a dedicated display or integrated into an existing display, such as a GPS moving map or multi-function display (MFD).

- ADS-B provides precise real-time data to controllers that immediately indicates when an aircraft deviates from its assigned flight path.

- ADS-B offers an effective range of 100 nautical miles, providing ATC a large area in which to implement traffic conflict detection and resolution.

- In addition to traffic information, ADS-B can provide the flight information service broadcast (FIS-B) to suitably equipped aircraft. FIS-B delivers textual and graphical weather products, as well as TFR locations and special use airspace status.

TRANSPONDERS

- A Mode S two-way data link enables the transponder to exchange information with ATC and with other Mode S-equipped aircraft. Mode S is required for ADS-B.

- Your ATC transponder must have been tested and inspected within the preceding 24 calendar months, or its use is not permitted.

- Unless otherwise required, set the transponder to squawk 1200 when operating under VFR.

- When leaving Class B, C or D airspace and being advised that radar service is terminated, squawk 1200.

- Avoid inadvertently selecting the transponder codes 7500—hijacking; 7600—communication failure; or 7700—emergency.

FLIGHT SERVICE

- Flight Service, sometimes referred to as a flight service station (FSS), provides weather briefings, NOTAMs, and flight plan filing at 1-800-WX-BRIEF (1-800-992-7433) or 1800wxbrief.com.

- Flight Service can provide updated weather information and receive your pilot reports (PIREPs) during flight. The call sign for a flight service station is its name, followed by the word "radio." On initial call-up, provide the full call sign of your aircraft, using the phonetic alphabet.

CONTROL TOWER SERVICES—ATIS

- Automatic terminal information service (ATIS) is the continuous broadcast of recorded information concerning non-control data in selected high-activity terminal areas.

- Absence of the sky condition and visibility on an ATIS broadcast indicates that the ceiling is at least 5,000 feet and the visibility is 5 miles or more.

TRACON AND ARTCC SERVICES

- A terminal radar approach control (TRACON) uses the ADS-B system and radar equipment to provide ATC services in the terminal area of airports typically in Class B, Class C, or TRSA airspace.

- To receive TRACON services at busy airports and airports in Class B and Class C airspace, obtain a transponder code and a departure frequency from clearance delivery prior to taxi. You must utilize Class B and C services.

- At airports that are not in Class B or C airspace, you can request optional flight following—a service to alert you to relevant traffic—by obtaining the appropriate frequency from the control tower, Flight Service, or airport information sources.

- While enroute, air route traffic control centers (ARTCCs) are the facilities that provide flight following.

- Basic radar service for VFR aircraft provides traffic advisories and limited vectoring on a workload permitting basis.

- Terminal radar service areas (TRSAs) provide sequencing and separation for participating VFR aircraft.

INTERPRETING TRAFFIC ADVISORIES

- When giving traffic advisories and safety alerts, ATC references traffic from your airplane as if it were a clock with 12 o'clock at the nose. *"Cessna 78 Juliet Romeo, traffic at 11 o'clock, 2 miles, southbound, Archer at 6,500."* This means that traffic appears to the controller to be 30 degrees left of the airplane nose. You should look for the traffic anywhere between the nose and the left wing of your airplane or between 12 and 9 o'clock.

- Wind correction angles do not show up on the ATC display so the position of the traffic is based on your ground track, not your heading.

NOTE: An asterisk appearing after an ACS code (i.e. PA.VII.B.K1*) indicates that the question subject appears more than one time in the ACS. The code shown corresponds to the first instance of the subject in the ACS.

5-1   PA.VI.B.K3

The automatic dependent surveillance-broadcast (ADS-B) system includes what primary components?

A – ADS-B radar transmitter; aircraft radar receiver, ADS-B Out transmitter, and ADS-B In receiver.

B – ADS-B ground station; aircraft GPS receiver, ADS-B Out transmitter, and ADS-B In receiver.

C – ADS-B ground station; aircraft VOR/DME or GPS receiver, and ADS-B In receiver.

5-1. Answer B. GFDPP 5A, AIM
The ADS-B system depends on accurate GPS positions transmitted from each aircraft—GPS is a core component. The ADS-B Out transmitter in your aircraft sends your GPS position to ATC through the ADS-B ground station, which transmits information back to aircraft through the ADS-B In receiver.

SECTION A ■ **ATC Services**

5-2   PA.VI.B.K3

Automatic dependent surveillance-broadcast (ADS-B) provides what services?

A – Weather products on FIS-B through your ADS-B Out transceiver.

B – Traffic alert and collision avoidance system (TCAS) with conflict resolution through your ADS-B In receiver.

C – Traffic information service-broadcast (TIS-B) through your ADS-B Out transmitter and ADS-B In receiver.

5-2. Answer C. GFDPP 5A, AIM

The ADS-B ground station receives the position that each aircraft transmits through ADS-B out, integrates it with non-ADS-B radar targets, and transmits the complete traffic picture back to aircraft. Properly equipped aircraft receive the traffic information broadcast through their ADS-B In receivers. Aircraft equipped with ADS-B In and Out can also receive traffic information directly from other aircraft, but this capability is not the same as the TCAS system that is required on airliners.

5-3   PA.III.A.K4, PA.VI.B.K4

When making routine transponder code changes, pilots should avoid inadvertent selection of which code?

A – 7200.

B – 7000.

C – 7500.

5-3. Answer C. GFDPP 5A AIM

Avoid inadvertent selection of transponder codes that set off alarms at ATC facilities. These codes are: 7500 for hijacking, 7600 for two-way radio communications failure, and 7700 for other emergencies.

5-4   PA.III.A.K4, PA.VI.B.K4

Unless otherwise authorized, if flying a transponder equipped aircraft, you should squawk which VFR code?

A – 1200.

B – 7600.

C – 7700.

5-4. Answer A. GFDPP 5A, AIM

The transponder code for standard VFR is 1200. Code 7700 is to communicate an emergency and 7600 is for two-way radio communications failure.

5-5   PA.III.A.K4, PA.VI.B.K4

What happens when you activate the IDENT function on your transponder?

A – A data block appears on the ATC display that contains your aircraft call sign and type.

B – The transponder return from your aircraft momentarily blossoms on the ATC display to enable easy identification.

C – The ATC display momentarily zooms in on the transponder return from your aircraft to enable easy identification.

5-5. Answer B. GFDPP 5A, AIM

Pressing the IDENT button or softkey on your transponder system causes the transponder return to blossom on the ATC display and enables easy identification of your aircraft. It does not cause any change to the data block or zoom level on the ATC display.

5-6   PA.III.A.K4, PA.VI.B.K4

When should you activate the IDENT function on your transponder?

A – On initial callup.

B – On ATC request.

C – When setting or changing a transponder code.

5-6. Answer B. GFDPP 5A, AIM
Press the IDENT button or softkey on your transponder system only when ATC asks you to "Ident."

5-7   PA.I.B.K1b

An ATC transponder was tested and inspected on March 15, 2024. The next test is due on

A – September 30, 2024.

B – March 31, 2025.

C – March 31, 2026.

5-7. Answer C. GFDPP 5A, FAR 91.413
An ATC transponder must have been tested and inspected within the preceding 24 calendar months, or its use is not permitted.

5-8   PA.I.C.K2

You can obtain pilot reports and updated weather information along your route of flight by contacting

A – CTAF.

B – UNICOM.

C – Flight Service.

5-8. Answer C. GFDPP 5A, AIM
Flight Service provides all non-ATC services to pilots before and during flight. UNICOM is a privately owned air/ground communication station. CTAF stands for common traffic advisory frequency, designated for the purpose of carrying out airport advisory practices while operating to or from an airport without an operating control tower. UNICOM advisories usually include wind direction and speed, favored runway, and known traffic.

5-9   PA.I.C.K2, R2c

Absence of the sky condition and visibility on an ATIS broadcast indicates that

A – weather conditions are at or above VFR minimums.

B – the sky condition is clear and visibility is unrestricted.

C – the ceiling is at least 5,000 feet and visibility is 5 miles or more.

5-9. Answer C. GFDPP 5A, AIM
ATIS is broadcast at certain busy airports, and provides noncontrol weather and runway information. If the ceiling is at least 5,000 feet and visibility is 5 miles or more, reporting of the ceiling/sky condition, visibility, and obstructions to vision is optional.

SECTION A ■ **ATC Services**

5-10   PA.VIII.F.K3

Automatic terminal information service (ATIS) is the continuous broadcast of recorded information concerning

A – pilots of radar-identified aircraft whose aircraft is in dangerous proximity to terrain or to an obstruction.

B – nonessential information to reduce frequency congestion.

C – noncontrol information in selected high-activity terminal areas.

5-10. Answer C. GFDPP 5A, AIM
ATIS is broadcast at certain busy airports, and provides noncontrol weather and runway information.

5-11   PA.VI.B.K3

During ground operations, from whom should a departing VFR aircraft request radar traffic information?

A – Flight Service.

B – Ground control, on initial contact.

C – Tower, just before takeoff.

5-11. Answer B. GFDPP 5A, AIM
Request radar traffic information by notifying ground control on initial contact with your request and proposed direction of flight. At airports in Class B or C airspace, you might make this request from clearance delivery. Flight Service does not provide radar services and requesting radar service from the tower just before takeoff could delay your departure.

5-12   PA.III.A.K7, PA.VI.B.K3, PA.VIII.F.K3

What basic terminal radar services does ATC provide to VFR aircraft?

A – Traffic advisories, safety alerts, limited vectoring, and, in Class B airspace, separation between all aircraft.

B – Sequencing and separation between all aircraft operating in Class B and C airspace.

C – Wind shear warning at participating airports.

5-12. Answer A. GFDPP 5A, AIM
Basic radar service for VFR aircraft provides safety alerts, traffic advisories, and limited vectoring on a workload-permitting basis. In Class B airspace, ATC provides sequencing and separation for all aircraft. In Class C airspace, ATC provides sequencing for all aircraft and separation of VFR aircraft only from IFR aircraft; not other VFR aircraft.

5-13   PA.VI.B.K3

TRSA service in the terminal radar program provides

A – sequencing and separation for all VFR aircraft.

B – IFR separation (1,000 feet vertical and 3 miles lateral) between all aircraft.

C – warning to pilots when their aircraft are in unsafe proximity to terrain, obstructions, or other aircraft.

5-13. Answer C. GFDPP 4C, 5A, AIM
Terminal radar service areas (TRSAs) are airspace areas surrounding about 30 U.S. airports that offer radar services, but where contact with ATC is not mandatory outside of the Class D airspace. TRSA provides basic radar service, including safety alerts—warnings to pilots when their aircraft are in unsafe proximity to terrain, obstructions, or other aircraft. In addition, TRSA service provides sequencing for *participating* VFR aircraft to the primary airport. Separation is provided between *IFR* aircraft, not all aircraft.

5-14   PA.VI.B.K3, PA.III.A.K4

If air traffic control advises that radar service is terminated when you are departing Class C airspace, you should set the transponder code to

A – 0000.

B – 1200.

C – 4096.

5-14. Answer B. GFDPP 5A, AIM

When radar service is terminated, you no longer squawk a discrete code and should switch to the standard VFR code of 1200.

5-15   PA.VI.B.K3

ATC issues a safety alert to aircraft under their control

A – that are at an altitude believed to place the aircraft in unsafe proximity to terrain or obstructions.

B – when another aircraft under their control is at an altitude that places the aircraft in unsafe proximity to each other.

C – if radar shows that the aircraft is headed into level 5 or 6 convective activity.

5-15. Answer A. GFDPP 5A, AIM

The types of safety alerts are:

- Terrain or obstruction alert—immediately issued to an aircraft under ATC control if ATC is aware that the aircraft is at an altitude believed to place the aircraft in unsafe proximity to terrain or obstructions.

- Aircraft conflict alert—immediately issued to an aircraft under ATC control if ATC is aware of an aircraft *not under their control* at an altitude believed to place the aircraft in unsafe proximity to each other.

ATC *does not* provide convective thunderstorm avoidance assistance unless a pilot requests it and ATC agrees to provide the service based on workload.

5-16   PA.VI.B.K3

(Refer to Figure 52.) Which type radar service is provided to VFR aircraft at Lincoln Municipal?

A – Sequencing to the primary Class C airport and standard separation.

B – Sequencing to the primary Class C airport and conflict resolution so that radar targets do not touch, or 1,000 feet vertical separation.

C – Sequencing to the primary Class C airport, traffic advisories, conflict resolution, and safety alerts.

5-16. Answer C. GFDPP 5A, AIM

The VFR services provided within a Class C airspace area include:

- Sequencing for all arriving aircraft to the primary Class C airport.

- Traffic advisories and safety alerts between VFR aircraft.

- Traffic advisories and conflict resolution (radar targets do not touch or 500 feet vertical separation) between IFR and VFR aircraft.

5-17   PA.VI.B.K3

An ATC radar facility issues the following advisory to a pilot flying on a heading of 090°: "TRAFFIC 3 O'CLOCK, 2 MILES, WESTBOUND..." Where should the pilot look for this traffic?

A – East.

B – South.

C – West.

5-17. Answer B. GFDPP 5A, AIM

Because the pilot is heading east, the 3 o'clock position is to the right, which is south.

SECTION A ■ **ATC Services**

5-18   PA.VI.B.K3

An ATC radar facility issues the following advisory to a pilot flying on a heading of 360°: "TRAFFIC 10 O'CLOCK, 2 MILES, SOUTHBOUND..." Where should the pilot look for this traffic?

A – Northwest.

B – Northeast.

C – Southwest.

5-18. Answer A. GFDPP 5A, AIM
Because the pilot's 12 o'clock position is north, the 10 o'clock position is northwest.

5-19   PA.VI.B.K3

An ATC radar facility issues the following advisory to a pilot during a local flight: "TRAFFIC 2 O'CLOCK, 5 MILES, NORTHBOUND..." Where should the pilot look for this traffic?

A – Between directly ahead and 90° to the left.

B – Between directly behind and 90° to the right.

C – Between directly ahead and 90° to the right.

5-19. Answer C. GFDPP 5A, AIM
The pilot's 12 o'clock position is directly ahead, and the 3 o'clock position is 90° to the right. The 2 o'clock position is approximately 60° right.

5-20   PA.VI.B.K3

An ATC radar facility issues the following advisory to a pilot flying north in a calm wind: "TRAFFIC 9 O'CLOCK, 2 MILES, SOUTHBOUND..." Where should the pilot look for this traffic?

A – South.

B – North.

C – West.

5-20. Answer C. GFDPP 5A, AIM
The pilot's 12 o'clock position is north, so the 9 o'clock position is west.

5-21   PA.VI.B.K3

ATC advises, *"traffic at your 12 o'clock."* This advisory is relative to your

A – true course.

B – ground track.

C – magnetic heading.

5-21. Answer B. GFDPP 5A, AIM
Controllers can see the ground track of an aircraft on radar, but cannot factor in any crab angle applied for wind correction. You must consider this when interpreting ATC traffic advisories.

SECTION B
Radio Procedures

USING NUMBERS ON THE RADIO

- State altitudes as individual numbers, with the word "thousand" included as appropriate.

- At altitudes of 10,000 feet and above, each digit of the thousands is pronounced, for example "one zero thousand."

COORDINATED UNIVERSAL TIME

- Aviation uses the 24-hour clock system, along with an international standard called coordinated universal time (UTC), or Zulu time (Z).

- When a time is in UTC, it is the time at the 0° line of longitude, which passes through Greenwich, England. In the United States, you add hours to convert local time to Zulu time and, to convert Zulu time to local time, you subtract hours.

- To convert the local departure or arrival time to UTC, add the hours of difference from the number on a time conversion table.

COMMON TRAFFIC ADVISORY FREQUENCY

- Broadcast your position and intentions to other aircraft on the common traffic advisory frequency (CTAF) at airports without operating control towers or when the tower is closed.

- A UNICOM is a privately owned air/ground communication station that can provide airport advisory information.

- At an airport that does not have a tower, an FSS, or a UNICOM, the CTAF is the MULTICOM frequency, 122.9 MHz.

- The self-announce procedure is broadcasting your position or intended flight activity or ground operation on the designated CTAF.

- You should monitor the CTAF and communicate your intentions within 10 miles of the airport.

ATC FACILITIES

- If instructed by ground control to taxi to a runway, you should read back the assigned route and proceed to the next intersecting runway where further clearance is required to continue.

- After landing at a tower-controlled airport, contact ground control when advised by the tower to do so.

- When transmitting to Flight Service in flight, use the name of the Flight Service facility in the area followed by the word "radio." Give your full call sign of your aircraft, using the phonetic alphabet.

SECTION B ■ **Radio Procedures**

LOST COMMUNICATION PROCEDURES

- If you believe that your radio has failed, set your transponder to code 7600 to alert an ATC facility of your radio facility.

- If you experience two-way communications failure, and need to land at a tower-controlled airport, determine the direction and flow of the traffic, join the pattern, and look for the following light signals from the tower.

COLOR AND TYPE OF SIGNAL	MEANING	
	On the Ground	In Flight
Steady Green	Cleared for takeoff	Cleared to land
Flashing Green	Cleared to taxi	Return for landing (to be followed by steady green at proper time)
Steady Red	Stop	Give way to other aircraft and continue circling
Flashing Red	Taxi clear of landing area (runway) in use	Airport unsafe—do not land
Flashing White	Return to starting point on airport	(No assigned meaning)
Alternating Red and Green	Exercise extreme caution	Exercise extreme caution

EMERGENCY PROCEDURES

- If you are not already in contact with ATC, use the frequency 121.5 MHz.

- An emergency locator transmitter (ELT) may be tested during the first 5 minutes after the hour.

NOTE: An asterisk appearing after an ACS code (i.e. PA.VII.B.K1*) indicates that the question subject appears more than one time in the ACS. The code shown corresponds to the first instance of the subject in the ACS.

5-22   PA.III.A.K2

When flying HAWK N666CB, the proper phraseology for initial contact with McAlester FSS is

A – "MCALESTER RADIO, HAWK SIX SIX SIX CHARLIE BRAVO, RECEIVING ARDMORE VORTAC, OVER."

B – "MCALESTER STATION, HAWK SIX SIX SIX CEE BEE, RECEIVING ARDMORE VORTAC, OVER."

C – "MCALESTER FLIGHT SERVICE STATION, HAWK NOVEMBER SIX CHARLIE BRAVO, RECEIVING ARDMORE VORTAC, OVER."

5-22. Answer A. GFDPP 5B, AIM

When transmitting to Flight Service in flight, use the name of the Flight Service facility in the area followed by the word "radio." Give your full call sign of your aircraft, using the phonetic alphabet.

5-23   PA.III.A.K2

The correct method of stating 5,500 feet MSL to ATC is

A – "FIVE POINT FIVE."

B – "FIFTY-FIVE HUNDRED FEET MSL."

C – "FIVE THOUSAND FIVE HUNDRED."

5-23. Answer C. GFDPP 5B, AIM

Altitudes should be stated as individual numbers with the word hundreds or thousands added as appropriate. In this case, 5,500 feet should be read as "FIVE THOUSAND FIVE HUNDRED."

5-24   PA.III.A.K2

The correct method of stating 10,500 feet MSL to ATC is

A – "TEN THOUSAND, FIVE HUNDRED FEET."

B – "TEN POINT FIVE."

C – "ONE ZERO THOUSAND, FIVE HUNDRED."

5-24. Answer C. GFDPP 5B, AIM

For altitudes at 10,000 feet MSL and above, pronounce each digit of the thousands, so that 10,500 becomes "ONE ZERO THOUSAND FIVE HUNDRED."

5-25   PA.I.D.K3b

(Refer to Figure 27.) An aircraft departs an airport in the eastern daylight time zone at 0945 EDT for a 2-hour flight to an airport located in the central daylight time zone. The landing should be at what coordinated universal time?

A – 1345Z.

B – 1445Z.

C – 1545Z.

5-25. Answer C. GFDPP 5B, AIM

To convert the local departure time to UTC, add four hours (0945 + 4 = 1345). Two hours later is 1545Z.

5-26   PA.I.D.K3b

(Refer to Figure 27.) An aircraft departs an airport in the central standard time zone at 0930 CST for a two-hour flight to an airport located in the mountain standard time zone. The landing should be at what time?

A – 0930 MST.

B – 1030 MST.

C – 1130 MST.

5-26. Answer B. GFDPP 5B, AIM

To find the arrival time, add two hours to the 0930 departure time to get 1130 CST. Because mountain time is one hour earlier than central time, subtract one hour, for a landing time of 1030 MST.

SECTION B ■ **Radio Procedures**

5-27   PA.I.D.K3b

(Refer to Figure 27.) An aircraft departs an airport in the central standard time zone at 0845 CST for a 2-hour flight to an airport located in the mountain standard time zone. The landing should be at what coordinated universal time?

Λ – 1345Z.

B – 1445Z.

C – 1645Z.

5-27. Answer C. GFDPP 5B, AIM
Departure time (0845) plus two hours is 1045 CST. Convert CST to UTC by adding six hours, for a landing time of 1645Z.

5-28   PA.I.D.K3b

(Refer to Figure 27.) An aircraft departs an airport in the mountain standard time zone at 1615 MST for a 2-hour 15-minute flight to an airport located in the Pacific standard time zone. The estimated time of arrival at the destination airport should be

A – 1630 PST.

B – 1730 PST.

C – 1830 PST.

5-28. Answer B. GFDPP 5B, AIM
Add 2:15 to 1615 MST to find the arrival time of 1830 MST. Because PST is one hour earlier than MST, the arrival time is 1730 PST.

5-29   PA.I.D.K3b

(Refer to Figure 27.) An aircraft departs an airport in the Pacific standard time zone at 1030 PST for a 4-hour flight to an airport located in the central standard time zone. The landing should be at what coordinated universal time?

A – 2030Z.

B – 2130Z.

C – 2230Z.

5-29. Answer C. GFDPP 5B, AIM
Add four hours to 1030 PST to find the arrival time of 1430 PST. To convert PST to UTC, add eight hours. The landing time is 2230Z.

5-30   PA.I.D.K3b

(Refer to Figure 27.) An aircraft departs an airport in the mountain standard time zone at 1515 MST for a 2-hour 30-minute flight to an airport located in the Pacific standard time zone. What is the estimated time of arrival at the destination airport?

A – 1645 PST.

B – 1745 PST.

C – 1845 PST.

5-30. Answer A. GFDPP 5B, AIM
Add 2:30 to 1515 MST to find the arrival time of 1745 MST. Convert MST to PST by subtracting one hour. The answer is 1645 PST.

5-31   PA.III.A.K2

(Refer to Figure 52.) What is the recommended communication procedure for landing at Lincoln Municipal during the hours when the tower is not in operation?

A – Monitor airport traffic and announce your position and intentions on 118.5 MHz.

B – Contact UNICOM on 122.95 MHz for traffic advisories.

C – Monitor ATIS for airport conditions, then announce your position on 122.95 MHz.

5-31. Answer A. GFDPP 5B, Chart Supplement
The CTAF frequency is listed as 118.5, and is used when the tower is not in operation. Standard procedures are to monitor airport traffic and announce your position on CTAF.

5-32   PA.III.A.K2

As standard operating practice, all inbound traffic to an airport without a control tower should continuously monitor the appropriate facility from a distance of

A – 25 miles.

B – 20 miles.

C – 10 miles.

5-32. Answer C. GFDPP 5B, AIM
In addition to monitoring a CTAF within 10 miles when inbound or outbound from a non-towered airport, you should also make an initial call to announce your intention if you plan to land at the airport. You should then report on each leg of the pattern.

5-33   PA.III.A.K1, K2

(Refer to Figure 20, area 3.) What is the recommended communication procedure for a landing at Currituck County Airport?

A – Transmit intentions on 122.9 MHz when 10 miles out and give position reports in the traffic pattern.

B – Contact Elizabeth City Tower on 120.5 MHz for airport and traffic advisories.

C – Contact FSS for area traffic information.

5-33. Answer A. GFDPP 5B, AIM
The CTAF symbol is next to the frequency of the MULTICOM frequency of 122.9. The normal procedure is to transmit intentions when 10 miles out and give position reports in the pattern.

5-34   PA.III.A.K1, K2

(Refer to Figure 26, area 2.) What is the recommended communication procedure when inbound to land at Cooperstown Airport?

A – Broadcast intentions when 10 miles out on the CTAF/MULTICOM frequency, 122.9 MHz.

B – Contact UNICOM when 10 miles out on 122.8 MHz.

C – Circle the airport in a left turn before entering traffic.

5-34. Answer A. GFDPP 5B, AIM
The CTAF symbol is next to the frequency of the MULTICOM frequency of 122.9. The normal procedure is to transmit intentions when 10 miles out and give position reports in the pattern.

SECTION B ■ **Radio Procedures**

5-35   PA.III.A.K2

After landing at a tower-controlled airport, when should the pilot contact ground control?

A – When advised by the tower to do so.

B – Before turning off the runway.

C – After reaching a taxiway that leads directly to the parking area.

5-35. Answer A. GFDPP 5B, AIM
The tower normally instructs you to exit the runway and contact ground control.

5-36   PA.III.A.K2

If instructed by ground control to taxi to Runway 9, the pilot may proceed

A – via taxiways and across runways to, but not onto, Runway 9.

B – to the next intersecting runway where further clearance is required.

C – via taxiways and across runways to Runway 9, where an immediate takeoff may be made.

5-36. Answer B. GFDPP 5B, AIM
You must have an explicit clearance to cross any runway to taxi to your departure runway, and you must have a clearance to taxi onto the departure runway.

5-37   PA.II.D.S7

An ATC clearance provides

A – priority over all other traffic.

B – adequate separation from all traffic.

C – authorization to proceed under specified traffic conditions in controlled airspace.

5-37. Answer C. GFDPP 5B, AIM
A clearance is authorization from ATC to operate under specific conditions in controlled airspace.

5-38   PA.III.A.K6

Which is an action you should take if you are unable to contact ATC as you are nearing an airport with a control tower?

A – Immediately squawk 7600 on your transponder.

B – Troubleshoot the problem by taking actions such as trying a different frequency; checking the radio volume and squelch; or switching to an alternate transceiver.

C – Enter the traffic pattern and land as soon as possible.

5-38. Answer B. GFDPP 5B, PHB

If you are unable to contact ATC:

- Ensure that you are using the correct frequency. Try a different frequency for the ATC facility, if available.

- Check the volume and squelch on your transceiver.

- Check the switch position on your audio control panel.

- Ensure that the headset mic and speaker plugs (or the handheld mic plug, if applicable) are properly inserted into the jacks.

- Try the handheld mic if you are using headsets.

- If your aircraft is equipped with more than one radio, try the alternate transceiver.

- If it is within range, try requesting assistance from the last ATC facility with which you had contact.

- If after taking these steps, you still are unable to contact ATC, follow the lost communication procedures.

5-39   PA.III.A.K3, K5

If the radio fails in an aircraft, what is the recommended procedure when landing at a controlled airport?

A – Observe the traffic flow, enter the pattern, and look for a light signal from the tower.

B – Enter a crosswind leg and rock the wings.

C – Flash the landing lights and cycle the landing gear while circling the airport.

5-39. Answer A. GFDPP 5B, AIM, FAR 91.125

To avoid conflicts and cause the least disruption in the traffic flow, determine the landing direction, and enter the pattern. Watch the tower for a light signal and acknowledge by rocking the wings. At night, acknowledge by flashing the landing or navigation lights.

5-40   PA.III.A.K3, K5

A steady green light signal directed from the control tower to an aircraft in flight is a signal that the pilot

A – is cleared to land.

B – should give way to other aircraft and continue circling.

C – should return for landing.

5-40. Answer A. GFDPP 5B, FAR 91.125

A steady green light while in flight means that you are cleared to land.

SECTION B ■ **Radio Procedures**

5-41   PA.III.A.K3, K5

Which light signal from the control tower clears a pilot to taxi?

A – Flashing green.

B – Steady green.

C – Flashing white.

5-41. Answer A. GFDPP 5B, FAR 91.125

While on the ground, a flashing green light means cleared to taxi.

5-42   PA.III.A.K3, K5

If the control tower uses a light signal to direct a pilot to give way to other aircraft and continue circling, the light is

A – flashing red.

B – steady red.

C – alternating red and green.

5-42. Answer B. GFDPP 5B, FAR 91.125

While in flight, a steady red light means give way and continue circling.

5-43   PA.II.D.K2, PA.III.A.K3, K5

A flashing white light signal from the control tower to a taxiing aircraft is an indication to

A – taxi at a faster speed.

B – taxi only on taxiways and not cross runways.

C – return to the starting point on the airport.

5-43. Answer C. GFDPP 5B, FAR 91.125

A flashing white light while operating on the ground means return to your starting point on the airport.

5-44   PA.III.A.K3, K5

An alternating red and green light signal directed from the control tower to an aircraft in flight is a signal to

A – hold position.

B – exercise extreme caution.

C – not land; the airport is unsafe.

5-44. Answer B. GFDPP 5B, FAR 91.125

An alternating red and green signal means the same whether you are in flight or on the ground — exercise extreme caution.

5-45   PA.III.A.K3, K5

While on final approach for landing, an alternating green and red light followed by a flashing red light is received from the control tower. Under these circumstances, the pilot should

A – discontinue the approach, fly the same traffic pattern and approach again, and land.

B – exercise extreme caution and abandon the approach, realizing the airport is unsafe for landing.

C – abandon the approach, circle the airport to the right, and expect a flashing white light when the airport is safe for landing.

5-45. Answer B. GFDPP 5B, FAR 91.125
An alternating red and green signal means exercise extreme caution. The flashing red signal that follows means that the airport is unsafe.

5-46   PA.III.A.K1, PA.IX.A.K2, PA.IX.B.K6

While on a VFR cross-country flight and not in contact with ATC, what frequency would you use in the event of an emergency?

A – 121.5 MHz.

B – 122.5 MHz.

C – 128.725 MHz.

5-46. Answer A. GFDPP 5B, AIM
If you are not already talking to ATC on another frequency, squawk 7700 and use the emergency frequency of 121.5 to obtain assistance.

5-47   PA.IX.B.K5, PA.IX.D.K1

When activated, an emergency locator transmitter (ELT) transmits on

A – 118.0 MHz, 118.8 MHz., and 119.1 MHz

B – 121.5 MHz, 243.0 MHz, and 406 MHz.

C – 119.0 MHz, 123.0 MHz, and 124.0 MHz.

5-47. Answer B. GFDPP 5B, AIM
The operating frequencies are 121.5 MHz, 243.0 MHz, and the newer 406 MHz. ELTs operating on 121.5 MHz and 243.0 MHz are analog devices. The newer 406 MHz ELT is a digital transmitter that can be encoded with the owner's contact information or aircraft data. The latest 406 MHz ELT models can also be encoded with the aircraft's position data which can help search and rescue (SAR) forces locate the aircraft much more quickly after a crash. The 406 MHz ELTs also transmits a stronger signal when activated than the older 121.5 MHz ELTs.

SECTION B ■ **Radio Procedures**

5-48   PA.IX.B.K5, PA.IX.D.K1

When must the battery in an emergency locator transmitter (ELT) be replaced (or recharged if the battery is rechargeable)?

A – After one-half of the battery's useful life.

B – During each annual and 100-hour inspection.

C – Every 24 calendar months.

5-48. Answer A. GFDPP 5B, FAR 91.207
The ELT battery must be replaced or recharged after one-half of the useful life of the battery has expired.

5-49   PA.IX.B.K5

When may an emergency locator transmitter (ELT) be tested?

A – Any time.

B – At 15 and 45 minutes past the hour.

C – During the first 5 minutes after the hour.

5-49. Answer C. GFDPP 5B, AIM
ELTs should be tested in accordance with the manufacturer's instructions, preferably in a shielded or screened room. When this cannot be done, aircraft operational testing is authorized as follows:

- Analog 121.5/243 MHz ELTs should only be tested during the first 5 minutes after any hour. If operational tests must be made outside of this period, they should be coordinated with the nearest FAA control tower. Tests should be no longer than three audible sweeps. If the antenna is removable, a dummy load should be substituted during test procedures.

- Digital 406 MHz ELTs should only be tested in accordance with the unit's manufacturer's instructions.

- Airborne tests are not authorized.

5-50   PA.IX.B.K5

Which procedure is recommended to ensure that the emergency locator transmitter (ELT) has not been activated?

A – Turn off the aircraft ELT after landing.

B – Ask the airport tower if they are receiving an ELT signal.

C – Monitor 121.5 before engine shutdown.

5-50. Answer C. GFDPP 5B, AIM
By monitoring 121.5, you can hear the ELT signal if it is activated.

SECTION C
Sources of Flight Information

CHART SUPPLEMENT

- Chart Supplements include data that cannot be readily depicted in graphic form on charts. This data applies to public and joint-use airports, seaplane bases, and heliports, as well as navaids and airspace.

- Each of the seven volumes of Chart Supplements covers a specific region of the contiguous United States, Puerto Rico, and the U.S. Virgin Islands. Two additional volumes cover Alaska and the Pacific region, including Hawaii.

- The Chart Supplements are updated every 56 days.

- Each Chart Supplement contains five primary sections:

  - The Airport/Facility Directory Legend
  - The Airport/Facility Directory
  - Notices
  - Associated Data
  - Airport Diagrams

ADVISORY CIRCULARS

- You can obtain FAA advisory circulars from the FAA website (**FAA.gov**).

- FAA advisory circulars (ACs) that relate to certain subject matters are identified by numeric codes. ACs pertaining to airmen are issued under subject number 60 (example AC 60-22). Airspace ACs are issued under subject number 70. ATC and General Operations ACs are issued under subject number 90.

NOTICES TO AIR MISSIONS

- Notices to Air Missions (NOTAMs) provide time-critical flight planning information regarding a facility, service, procedure, or hazard that is temporary or not known far enough in advance to be included in the most recent aeronautical charts or Chart Supplements.

- Two primary NOTAM types are:

  - NOTAM(D) (distant NOTAM)—disseminated for all navigational facilities that are part of the U.S. airspace system, and all public use airports, seaplane bases, and heliports listed in the Chart Supplements.

  - FDC NOTAMs—issued by the National Flight Data Center and contain regulatory information. For example, FDC NOTAMs may advise of changes in flight data which affect instrument approach procedures (IAP), aeronautical charts, and flight restrictions prior to normal publication.

NOTE: An asterisk appearing after an ACS code (i.e. PA.VII.B.K1*) indicates that the question subject appears more than one time in the ACS. The code shown corresponds to the first instance of the subject in the ACS.

SECTION C ■ Sources of Flight Information

SECTION C ■ Sources of Flight Information

5-51   PA.I.D.K1a

When planning for a cross-country, you can find airport information for your destination, such as communication frequencies and runway lengths in flight information sources, such as:

A – Advisory circulars and NOTAMs.

B – Electronic flight bags and the Notices section of the Chart Supplement.

C – Electronic flight bags and the Airport/Facility Directory in the Chart Supplement.

5-51. Answer C. GFDPP 5C, PHB

The Airport/Facility Directory (A/FD) in the Chart Supplement contains a descriptive listing of all airports that are open to the public. Electronic flight bags also contain airport information and might also provide access to A/FD pages. The Notices of the Chart Supplement describes unique restrictions or operating procedures and summarizes special air traffic rules established in FAR Part 93.

Notices to Air Missions (NOTAMs) provide time-critical flight planning information regarding a facility, service, procedure, or hazard. The FAA issues advisory circulars (ACs) to provide nonregulatory guidance and information in a variety of subject areas.

5-52   PA.I.D.S2, PA.III.A.K2, PA.VIII.F.K1

(Refer to Figure 52.) When approaching Lincoln Municipal from the west at noon for the purpose of landing, initial communication should be with

A – Lincoln Approach Control on 124.0 MHz.

B – Minneapolis Center on 128.75 MHz.

C – Lincoln Tower on 118.5 MHz.

5-52. Answer A. GFDPP 5C, A/FD Legend

Lincoln Municipal Airport is in Class C airspace. The approach frequency is under Communications in the Airport/Facility Directory (A/FD) excerpt from the Chart Supplement. When west of the airport (180°–359°), use the approach frequency of 124.0. When east of the airport (360°–179°), use the approach frequency of 124.8. Lincoln Class C airspace is in effect from 1130Z to 0600Z (5:30 a.m. to 12:00 a.m. local time).

5-53   PA.I.D.K, S2

(Refer to Figure 52.) Where is Loup City Municipal located in relation to the city?

A – Northeast approximately 3 miles.

B – Northwest approximately 1 mile.

C – East approximately miles.

5-53. Answer B. GFDPP 5C, A/FD Legend

Each Airport/Facility Directory (A/FD) listing in the Chart Supplement begins with airport identification and location information, including the distance and direction from the associated city. The entry 1 NW indicates that the airport is 1 mile northwest of the city.

5-54   PA.I.D.K1, S2

(Refer to Figure 52.) The landing distance available on Rwy 17 at Lincoln Airport is

A – 5,400 feet.

B – 5,800 feet.

C – 8,286 feet.

5-54. Answer A. GFDPP 5C, A/FD Legend

The landing distance available (LDA) appears under Runway Declared Distance Information section. For Rwy 17, the LDA is 5,400 feet.

5-55　PA.I.D.K1, PA.I.D.S2

(Refer to Figure 52.) The landing distance available on Rwy 32 at Lincoln Airport is

A – 7,816 feet.

B – 8,286 feet.

C – 8,649 feet.

5-55. Answer A. GFDPP 5C, A/FD Legend

The landing distance available (LDA) appears under Runway Declared Distance Information section. For Rwy 32, the LDA Is 7,816 feet.

5-56　PA.I.D.S2

You can obtain FAA advisory circulars by

A – distribution from the nearest FAA district office.

B – searching for them at FAA.gov.

C – subscribing to the Federal Register.

5-56. Answer B. GFDPP 5C, PHB

Search for the current advisory circulars (ACs) by name, number, or subject at faa.gov. The Federal Register includes changes to regulations, but not the content of ACs.

5-57　PA.I.D.S2

FAA advisory circulars containing subject matter related to Airmen are issued under which subject number?

A – 60.

B – 70.

C – 90.

5-57. Answer A. GFDPP 5C, AC 60

Advisory circulars relating to Airmen are issued under subject number 60.

5-58　PA.I.D.S2

FAA advisory circulars containing subject matter related to Air Traffic Control and General Operations are issued under which subject number?

A – 60.

B – 70.

C – 90.

5-58. Answer C. GFDPP 5C, AC 90

Advisory circulars relating to Air Traffic Control and General Operations are issued under subject number 90.

SECTION C ■ **Sources of Flight Information**

5-59   PA.I.D.S2

Which is true about NOTAMs? An example of a NOTAM(D) is an permanent change in flight data on an aeronautical chart.

A – An example of a NOTAM(D) is a temporary runway closure.

B – An example of an FDC NOTAM is an alert about snow and ice on runways and taxiways.

5-59. Answer B. GFDPP 5C, AIM

Notices to Air Missions (NOTAMs) provide time-critical flight planning information regarding a facility, service, procedure, or hazard that is temporary or not known far enough in advance to be included in the most recent aeronautical charts or Chart Supplements. NOTAM(D)s are disseminated for all navigational facilities that are part of the U.S. airspace system, and all public use airports, seaplane bases, and heliports listed in the Chart Supplements. FDC NOTAMs are issued by the National Flight Data Center and contain regulatory information. For example, FDC NOTAMs may advise of changes in flight data which affect instrument approach procedures (IAP), aeronautical charts, and flight restrictions prior to normal publication.

CHAPTER 6

Meteorology For Pilots

SECTION A
Basic Weather Theory

ATMOSPHERIC CIRCULATION

- Atmospheric circulation refers to the movement of air relative to the surface of the earth.

- Every physical process of weather is accompanied by, or is the result of, a heat exchange.

- Unequal heating of the surface of the earth leads to variations in pressure, which is why altimeter settings differ between weather reporting points.

- Coriolis force deflects winds to the right in the northern hemisphere as they flow from high-pressure to low-pressure areas. However, below 2,000 feet AGL, friction with the surface of the earth reduces the effect of Coriolis force, so winds at the surface generally are different in direction from the winds aloft.

- Convective circulation patterns associated with sea breezes are caused by cool, dense air moving inland from over the water.

NOTE: An asterisk appearing after an ACS code (i.e. PA.VII.B.K1*) indicates that the question subject appears more than one time in the ACS. The code shown corresponds to the first instance of the subject in the ACS.

6-1   PA.I.C.K3c

What causes variations in altimeter settings between weather reporting points?

A – Unequal heating of the earth's surface.

B – Variation of terrain elevation.

C – Coriolis force.

6-1. Answer A. GFDPP 6A, AW

The unequal heating of the surface not only modifies air density and creates circulation patterns, it also causes changes in pressure. This is one of the main reasons for differences in altimeter settings between weather reporting stations.

6-2   PA.I.C.K3b

The wind at 5,000 feet AGL is southwesterly while the surface wind is southerly. This difference in direction is primarily due to

A – stronger pressure gradient at higher altitudes.

B – friction between the wind and the surface.

C – stronger Coriolis force at the surface.

6-2. Answer B. GFDPP 6A, AW

Above 2,000 feet AGL, wind flows along isobars because of Coriolis force. Below that altitude, friction with the surface of the earth weakens Coriolis force and allows the wind to flow more directly toward the low pressure area. In the northern hemisphere, this action results in a counterclockwise shift in wind direction closer to the surface.

6-3   PA.I.C.K3b

Convective circulation patterns associated with sea breezes are caused by

A – warm, dense air moving inland from over the water.

B – water absorbing and radiating heat faster than the land.

C – cool, dense air moving inland from over the water.

6-3. Answer C. GFDPP 6A, AW

During the day, land surfaces become warmer than the adjacent water surfaces—warming the air above the land and causing it to rise. The rising air is replaced by the inland flow of cooler, denser air located over the water. The warm air then flows out over the water where it cools and descends. This action starts the cycle all over again. During the night, the process is reversed as the land cools off faster than the water.

6-4   PA.I.C.K3c

Every physical process of weather is accompanied by, or is the result of, a

A – movement of air.

B – pressure differential.

C – heat exchange.

6-4. Answer C. GFDPP 6A, AW

Every physical process of weather such as heating, cooling, evaporation, and condensation, is caused by, or is the result of, a heat exchange.

SECTION A ■ **Basic Weather Theory**

SECTION B
Weather Patterns

ATMOSPHERIC STABILITY

- The rate at which temperature decreases with an increase in altitude is referred to as the lapse rate.

- The stability of air can be determined by its actual lapse rate. A high lapse rate tends to indicate unstable air, and a low lapse rate is an indicator of stability in the atmosphere.

TEMPERATURE INVERSIONS

- A temperature inversion means that temperature increases as altitude increases. An inversion is associated with a stable layer of air.

- The most frequent type of ground or surface-based temperature inversion is produced by terrestrial radiation on clear and relatively still nights.

- The weather conditions that can be expected beneath a low-level temperature inversion layer when the relative humidity is high are smooth air, poor visibility, fog, haze, or low clouds.

MOISTURE

- The processes by which moisture is added to unsaturated air are evaporation and sublimation.

- The dewpoint is the temperature to which the air must be cooled to become saturated.

- The amount of water vapor that air can hold depends on the air temperature.

- If the temperature of the collecting surface is at or below the dewpoint of the adjacent air, and the dewpoint is below freezing, frost forms.

- Frost on the wings affects takeoff performance by disrupting the smooth flow of air over the airfoil, adversely affecting the airfoil's lifting capacity. Frost is a hazard that can prevent the airplane from becoming airborne at normal takeoff speed.

CLOUDS AND FOG

- Clouds are divided into four families according to their height range—low clouds, middle clouds, high clouds, and clouds with vertical development.

- Clouds, fog, or dew will always form when water vapor condenses. Air cools when lifted and moisture condenses out of the air as the relative humidity reaches 100 percent.

- You can estimate the bases of cumulus clouds using the convergence rate of temperature and the dewpoint of lifted air—4.4°F per 1,000 feet. Divide the temperature/dewpoint spread at the surface by 4.4 to find the height of the cloud base above the surface, in thousands of feet.

- The suffix "nimbus," used in naming clouds, means a rain cloud.

- Stratus clouds form when moist, stable air is lifted.

- A characteristic of stable air is the presence of stratiform clouds.

- When moist, stable air flows upslope, you can expect the formation of stratus type clouds.

- If the temperature/dewpoint spread is small and decreasing, and the temperature is above freezing, fog or low clouds are likely to develop.

- Radiation fog forms as warm, moist air lies over flatland areas on clear, calm nights.

- Advection fog is caused when a low layer of warm, moist air moves over a cooler surface. It is most common under cloudy skies along coastlines in the winter—wind transports air from the warm water to the cooler land.

- Upslope fog can form in moderate to strong winds when moist, stable air is forced up a sloping land mass.

- Advection fog and upslope fog depend upon wind to exist.

- In steam fog, low-level turbulence can occur and icing can become hazardous.

- You can expect clouds with extensive vertical development and associated turbulence when an unstable air mass is forced upward.

PRECIPITATION

- Precipitation can be defined as any form of particles, whether liquid or solid, that fall from the atmosphere.

- Precipitation can reduce visibility, affect engine performance, increase braking distance, and cause dramatic shifts in wind direction and velocity.

- Freezing drizzle and freezing rain freeze upon contact with the ground or other objects such as power lines, trees, or aircraft. Freezing rain poses a serious hazard to all aircraft, even aircraft with deicing and anti-icing equipment.

- The presence of ice pellets at the surface is evidence that a temperature inversion exists with freezing rain at a higher altitude.

AIR MASSES

- Air masses are large-scale parcels of air that have a set of characteristics (i.e. moist, unstable) that distinguishes them from one another.

- Characteristics of a moist, unstable air mass are cumuliform clouds and showery precipitation.

- Characteristics of unstable air include turbulence and good surface visibility.

- A stable air mass generally contains smooth air.

FRONTS

- The boundary between two different air masses is referred to as a front.

- One of the most easily recognizable discontinuities across a front is a change in temperature.

- One weather phenomenon that always occurs when flying across a front is a change in the wind direction.

- Steady precipitation preceding a front is an indication of stratiform clouds with little or no turbulence.

SECTION B ■ Weather Patterns

NOTE: *An asterisk appearing after an ACS code (i.e. PA.VII.B.K1*) indicates that the question subject appears more than one time in the ACS. The code shown corresponds to the first instance of the subject in the ACS.*

6-5   PA.I.C.K3a

What measurement can be used to determine the stability of the atmosphere?

A – Atmospheric pressure.

B – Actual lapse rate.

C – Surface temperature.

6-5. Answer B. GFDPP 6B, AW

Stability is the atmosphere's resistance to vertical motion. Lapse rate generally refers to the decrease in temperature with an increase in altitude. A high lapse rate tends to indicate unstable air, and a low lapse rate is an indicator of stability in the atmosphere.

6-6 PA.I.C.K3a

What would decrease the stability of an air mass?

A – Warming from below.

B – Cooling from below.

C – Decrease in water vapor.

6-6. Answer A. GFDPP 6B, AW
Stability is affected by a change in the lapse rate of an air mass. Warming from below or cooling from above increases the lapse rate and makes the air less stable.

6-7 PA.I.C.K3c

What feature is associated with a temperature inversion?

A – A stable layer of air.

B – An unstable layer of air.

C – Chinook winds on mountain slopes.

6-7. Answer A. GFDPP 6B, AW
Temperature inversions occur in stable air. Inversions cannot form in unstable air. As Chinook winds descend, the temperature rises. This phenomenon is the opposite of an inversion (cooler air under a warmer layer). In the U.S., the typical example of Chinook winds is the downslope, easterly flow from the Rocky Mountains.

6-8 PA.I.C.K3c

When there is a temperature inversion, you would expect to experience

A – clouds with extensive vertical development above an inversion aloft.

B – good visibility in the lower levels of the atmosphere and poor visibility above an inversion aloft.

C – an increase in temperature as altitude increases.

6-8. Answer C. GFDPP 6B, AW
Normally, temperature decreases with altitude. During an inversion, cooler air is trapped beneath a warmer layer of air. Therefore, temperature increases with altitude.

6-9 PA.I.C.K3c

The most frequent type of ground or surface-based temperature inversion is that which is produced by

A – terrestrial radiation on a clear, relatively still night.

B – warm air being lifted rapidly aloft in the vicinity of mountainous terrain.

C – the movement of colder air under warm air, or the movement of warm air over cold air.

6-9. Answer A. GFDPP 6B, AW
An inversion commonly forms on clear, cool nights when the ground radiates heat and cools faster than the overlying air.

6-10   PA.I.C.K3c, K3f

Which weather conditions should be expected beneath a low-level temperature inversion layer when the relative humidity is high?

A – Smooth air, poor visibility, fog, haze, or low clouds.

B – Light wind shear, poor visibility, haze, and light rain.

C – Turbulent air, poor visibility, fog, low stratus type clouds, and showery precipitation.

6-10. Answer A. GFDPP 6B, AW
Low-level temperature inversions normally occur in stable, smooth air, with poor visibility due to trapped pollutants, which are commonly referred to as condensation nuclei. In addition, high humidity tends to cause formation of fog and low clouds.

6-11   PA.I.C.K3c

What is meant by the term "dewpoint"?

A – The temperature at which condensation and evaporation are equal.

B – The temperature at which dew always forms.

C – The temperature to which air must be cooled to become saturated.

6-11. Answer C. GFDPP 6B, AW
When air is cooled to its dewpoint, it can hold no more moisture, and is said to be saturated.

6-12   PA.I.C.K3c

The amount of water vapor which air can hold depends on the

A – dewpoint.

B – air temperature.

C – stability of the air.

6-12. Answer B. GFDPP 6B, AW
The amount of moisture in the air primarily depends on the temperature. For example, warm air can hold more moisture than cool air.

6-13   PA.I.C.K3d, K3f, K3j

Clouds, fog, or dew will always form when

A – water vapor condenses.

B – water vapor is present.

C – relative humidity reaches 100 percent.

6-13. Answer A. GFDPP 6B, AW
Condensation occurs when water vapor changes to liquid form. Examples are when water vapor changes to clouds, fog, or dew.

6-14   PA.I.C.K3d

What are the processes by which moisture is added to unsaturated air?

A – Evaporation and sublimation.

B – Heating and condensation.

C – Supersaturation and evaporation.

6-14. Answer A. GFDPP 6B, AW
Evaporation occurs when liquid water changes to water vapor. Sublimation is the changing of ice directly to water vapor. Both processes add moisture to the air.

SECTION B ■ **Weather Patterns**

6-15   PA.I.C.K3k

Which conditions result in the formation of frost?

A – The temperature of the collecting surface is at or below freezing when small droplets of moisture fall on the surface.

B – The temperature of the collecting surface is at or below the dewpoint of the adjacent air and the dewpoint is below freezing.

C – The temperature of the surrounding air is at or below freezing when small drops of moisture fall on the collecting surface.

6-15. Answer B. GFDPP 6B, AW
When the dewpoint of the surrounding air is below freezing, and the collecting surface is at or below the dewpoint, water vapor sublimates directly into ice crystals or frost instead of condensing into dew.

6-16   PA.I.C.K3k

Why is frost considered hazardous to flight?

A – Frost changes the basic aerodynamic shape of the airfoils, thereby increasing lift.

B – Frost slows the airflow over the airfoils, thereby increasing control effectiveness.

C – Frost spoils the smooth flow of air over the wings, thereby decreasing lifting capability.

6-16. Answer C. GFDPP 6B, AW
Frost disrupts the smooth airflow over the wing and can cause early separation of the airflow, resulting in a loss of lift.

6-17   PA.I.C.K3k, PA.I.G.K1i

How does frost affect the lifting surfaces of an airplane on takeoff?

A – Frost may prevent the airplane from becoming airborne at normal takeoff speed.

B – Frost changes the camber of the wing, increasing lift during takeoff.

C – Frost may cause the airplane to become airborne with a lower angle of attack at a lower indicated airspeed.

6-17. Answer A. GFDPP 6B, PHB
Frost disrupts the smooth airflow over the wing and can cause early separation of the airflow, resulting in a loss of lift. By disrupting the airflow over the wings, frost can prevent an airplane from becoming airborne at the normal takeoff speed.

6-18   PA.I.C.K3k

How does frost on the wings of an airplane affect takeoff performance?

A – Frost disrupts the smooth flow of air over the wing, adversely affecting its lifting capability.

B – Frost changes the camber of the wing, increasing its lifting capability.

C – Frost causes the airplane to become airborne with a higher angle of attack, decreasing the stall speed.

6-18. Answer A. GFDPP 6B, PHB
Frost disrupts the smooth airflow over the wing and can cause early separation of the airflow, resulting in a loss of lift.

6-19　PA.I.C.K3f

What is the approximate base of the cumulus clouds if the surface air temperature at 1,000 feet MSL is 70°F and the dewpoint is 48°F?

A – 4,000 feet MSL.

B – 5,000 feet MSL.

C – 6,000 feet MSL.

6-19. Answer C. GFDPP 6B, AW

When warm, moist air rises in a convective current, the temperature and dewpoint converge at 4.4°F per 1,000 feet. You can estimate the cloud base by dividing the temperature/dewpoint spread at the surface by 4.4°F per 1,000 feet.

In this case: (70°F − 48°F) ÷ 4.4°F × 1,000 feet = 5,000 feet AGL. Because the surface is 1,000 feet MSL, the cloud base should be approximately 6,000 feet MSL.

6-20　PA.I.C.K3f

At approximately what altitude above the surface would the pilot expect the base of cumuliform clouds if the surface air temperature is 82°F and the dewpoint is 38°F?

A – 9,000 feet AGL.

B – 10,000 feet AGL.

C – 11,000 feet AGL.

6-20. Answer B. GFDPP 6B, AW

When warm, moist air rises in a convective current, the temperature and dewpoint converge at 4.4°F per 1,000 feet. You can estimate the cloud base by dividing the temperature/dewpoint spread at the surface by 4.4°F per 1,000 feet.

In this case: (82°F − 38°F) ÷ 4.4°F × 1,000 feet = 10,000 feet AGL.

6-21　PA.I.C.K3a, K3f

What is a characteristic of stable air?

A – Stratiform clouds.

B – Unlimited visibility.

C – Cumulus clouds.

6-21. Answer A. GFDPP 6B, AW

Very little vertical development of clouds occurs in stable air, and stratiform clouds and poor visibility are typical. Cumulus clouds and good visibility are indicators of unstable air.

6-22　PA.I.C.K3f

The suffix "nimbus," used in naming clouds, means

A – a cloud with extensive vertical development.

B – a rain cloud.

C – a middle cloud containing ice pellets.

6-22. Answer B. GFDPP 6B, AW

The word "nimbus" is the Latin word for rainstorm or cloud, and is used today to designate rain clouds, such as cumulonimbus or nimbostratus.

6-23　PA.I.C.K3f

Clouds are divided into four families according to their

A – outward shape.

B – height range.

C – composition.

6-23. Answer B. GFDPP 6B, AW

Clouds are also grouped by families according to their altitudes (height range). The four families are low, middle, high, and clouds with extensive vertical development.

SECTION B ■ **Weather Patterns**

6-24   PA.I.C.K3f

When warm, moist, stable air flows upslope, it

A – produces stratus type clouds.

B – causes showers and thunderstorms.

C – develops convective turbulence.

6-24. Answer A. GFDPP 6B, AW
Stratus clouds are produced in stable air. When moist air flows upslope it cools to its saturation point, and clouds form.

6-25   PA.I.C.K3f

If an unstable air mass is forced upward, what type clouds can be expected?

A – Stratus clouds with little vertical development.

B – Stratus clouds with considerable associated turbulence.

C – Clouds with considerable vertical development and associated turbulence.

6-25. Answer C. GFDPP 6B, AW
Clouds with extensive vertical development are formed when unstable air is lifted. These cumulus type clouds are associated with moderate to severe turbulence.

6-26   PA.I.C.K3j

What situation is most conducive to the formation of radiation fog?

A – Warm, moist air over low, flatland areas on clear, calm nights.

B – Moist, tropical air moving over cold, offshore water.

C – The movement of cold air over much warmer water.

6-26. Answer A. GFDPP 6B, AW
On clear, calm nights in flat areas, radiation fog forms when moist air cools to its dewpoint. Ground fog is a form of radiation fog.

6-27   PA.I.C.K3j

If the temperature/dewpoint spread is small and decreasing, and the temperature is 62°F, what type weather is most likely to develop?

A – Freezing precipitation.

B – Thunderstorms.

C – Fog or low clouds.

6-27. Answer C. GFDPP 6B, AW
When the temperature/dewpoint spread decreases to zero, the likely result is the condensation of water vapor into visible moisture, such as fog or low clouds.

6-28   PA.I.C.K3j

In which situation is advection fog most likely to form?

A – A warm, moist air mass on the windward side of mountains.

B – An air mass moving inland from the coast in winter.

C – A light breeze blowing colder air out to sea.

6-28. Answer B. GFDPP 6B, AW
Advection fog is caused when a low layer of warm, moist air moves over a cooler surface. It is most common under cloudy skies along coastlines where wind transports air from the warm water to the cooler land.

6-29   PA.I.C.K3j

What types of fog depend upon wind to exist?

A – Radiation fog and ice fog.

B – Steam fog and ground fog.

C – Advection fog and upslope fog.

6-29. Answer C. GFDPP 6B, AW
Advection fog is caused when a low layer of warm, moist air moves over a cooler surface. It is most common under cloudy skies along coastlines where wind transports air from the warm water to the cooler land. Upslope fog can form in moderate to strong winds when moist, stable air is forced up a sloping land mass.

Radiation fog forms when moist air cools to its dewpoint on *calm* nights. Ground fog is a form of radiation fog.

6-30   PA.I.C.K3j

Low-level turbulence can occur and icing can become hazardous in which type of fog?

A – Rain-induced fog.

B – Upslope fog.

C – Steam fog.

6-30. Answer C. GFDPP 6B, AW
Steam fog is formed by cold, dry air moving over warmer water. As the water particles evaporate and rise, they often freeze and fall back into the water. Icing and low-level turbulence can result.

6-31   PA.I.C.K3d

The presence of ice pellets at the surface is evidence that there

A – are thunderstorms in the area.

B – has been cold frontal passage.

C – is a temperature inversion with freezing rain at a higher altitude.

6-31. Answer C. GFDPP 6B, AW
Due to a temperature inversion, a warm layer of air is aloft and keeps the rain in liquid form. As the rain falls through colder air, it begins to freeze, finally turning into ice pellets. Ice pellets always indicate freezing rain at a higher altitude. Ice pellets can form under various conditions, and do not necessarily indicate thunderstorms.

6-32   PA.I.C.K3e

What are characteristics of a moist, unstable air mass?

A – Cumuliform clouds and showery precipitation.

B – Poor visibility and smooth air.

C – Stratiform clouds and showery precipitation.

6-32. Answer A. GFDPP 6B, AW
Cumuliform clouds are indicative of unstable air. These clouds normally produce showery, not continuous, precipitation. Poor visibility, smooth air, and stratiform clouds are characteristic of stable air.

6-33   PA.I.C.K3e

What are characteristics of unstable air?

A – Turbulence and good surface visibility.

B – Turbulence and poor surface visibility.

C – Nimbostratus clouds and good surface visibility.

6-33. Answer A. GFDPP 6B, AW
The lifting motion of unstable air produces turbulence. Clouds and pollutants are not trapped as they are in stable layers of air, and good visibility is typical with unstable air. Poor surface visibility and nimbostratus clouds are typical of stable air masses.

SECTION B ■ **Weather Patterns**

6-34   PA.I.C.K3e

A stable air mass is most likely to have which characteristic?

A – Showery precipitation.

B – Turbulent air.

C – Smooth air.

6-34. Answer C. GFDPP 6B, AW

Stable air resists the lifting motion that is associated with turbulence, and is typically smooth. Showery precipitation and turbulent air are characteristics of unstable air.

6-35   PA.I.C.K3e

One of the most easily recognized discontinuities across a front is

A – a change in temperature.

B – an increase in cloud coverage.

C – an increase in relative humidity.

6-35. Answer A. GFDPP 6B, AW

Because a front is the boundary between air masses of differing temperatures, one of the easiest ways to recognize frontal passage is the change in temperature.

6-36   PA.I.C.K3e

One weather phenomenon that always occurs when flying across a front is a change in the

A – wind direction.

B – type of precipitation.

C – stability of the air mass.

6-36. Answer A. GFDPP 6B, AW

A shift in wind direction always occurs across a front.

6-37   PA.I.C.K3e

Steady precipitation preceding a front is an indication of

A – stratiform clouds with moderate turbulence.

B – cumuliform clouds with little or no turbulence.

C – stratiform clouds with little or no turbulence.

6-37. Answer C. GFDPP 6B, AW

Steady precipitation, stratiform clouds, and little or no turbulence are all typical of stable air.

SECTION C
Weather Hazards

THUNDERSTORMS

- Cumulonimbus clouds have the greatest turbulence.
- The conditions necessary for the formation of cumulonimbus clouds are a lifting action and unstable, moist air.
- Thunderstorms are formed when high humidity, a lifting force, and unstable conditions combine.
- A non-frontal, narrow band of active thunderstorms that often develops ahead of a cold front is known as a squall line. Squall line thunderstorms generally produce the most intense hazards to aircraft.

LIFE CYCLE

- Certain characteristics, such as cloud shape, air current direction, and precipitation intensity, are associated with each stage of a thunderstorm.
- The three stages of a thunderstorm are:
    - Cumulus stage—a lifting action initiates the vertical movement of air causing a continuous updraft.
    - Mature stage— water drops in the cloud grow too large to be supported by the updrafts and rain falls from the cloud. Thunderstorms reach their greatest intensity during this stage, producing strong updrafts and downdrafts, severe turbulence, lightning, heavy rain, hail, strong surface winds, and gust front.
    - Dissipating stage—characterized predominantly by downdrafts that spread out within the cell, taking the place of the weakening updrafts. Because upward movement is necessary for condensation and the release of the latent energy, the entire thunderstorm begins to weaken.

HAZARDS

- The weather hazards associated with thunderstorms are not confined to the cloud itself. You can encounter turbulence in clear conditions as far as 20 miles from the storm.
- If thunderstorm activity exists in the vicinity of an airport at which you plan to land, you can expect to encounter wind-shear turbulence during the landing approach.
- Lightning is always associated with thunderstorms.

TURBULENCE

- Upon encountering severe turbulence, the pilot should attempt to maintain a level flight attitude.
- Towering cumulus clouds indicate convective turbulence.

WAKE TURBULENCE

- Whenever an airplane generates lift, air spills over the wingtips from the high pressure areas below the wings to the low pressure areas above them. This flow causes rapidly rotating whirlpools of air called wingtip vortices, or wake turbulence.
- The intensity of wake turbulence depends on aircraft weight, speed, and configuration.
- The greatest vortex strength occurs when the generating aircraft is heavy, clean, and slow.
- Wingtip vortices created by a large aircraft tend to sink below the aircraft that is generating the turbulence.

SECTION C ■ Weather Hazards

- When taking off or landing at an airport where heavy aircraft are operating, be particularly alert to the hazards of wingtip vortices because this turbulence tends to sink into the flight path of the aircraft operating below the aircraft generating the turbulence.

- The wind condition that requires maximum caution when avoiding wake turbulence on landing is a light, quartering tailwind.

MOUNTAIN WAVE TURBULENCE

- An almond or lens-shaped cloud that appears stationary, but that can contain winds of up to 50 knots or more, is referred to as a lenticular cloud.

- Crests of standing mountain waves are often marked by stationary, lens-shaped clouds known as standing lenticular clouds.

- A pilot can expect possible mountain wave turbulence when winds of 40 knots or greater blow across a mountain ridge, when the air is stable.

WIND SHEAR

- Wind shear can occur at any altitude, in any direction.

- Hazardous wind shear can be expected in areas of low-level temperature inversion, frontal zones, and clear air turbulence.

- A pilot can expect a wind shear zone in a temperature inversion whenever the wind speed at 2,000–4,000 feet above the surface is at least 25 knots.

ICING

- Visible moisture is necessary for the formation of in-flight structural icing.

- Areas of freezing rain create the environment in which structural icing is most likely to have the highest accumulation rate.

VOLCANIC ASH

- Ash from a volcanic eruption can affect a widespread area and cause severe damage, especially if the volcanic cloud is only a few hours old.

- If you inadvertently enter a volcanic ash cloud you should not attempt to fly straight through or climb out of the cloud because ash cloud can be hundreds of miles wide and extend to great heights.

SECTION C ■ Weather Hazards

NOTE: *An asterisk appearing after an ACS code (i.e. PA.VII.B.K1*) indicates that the question subject appears more than one time in the ACS. The code shown corresponds to the first instance of the subject in the ACS.*

6-38   PA.I.C.K3f, K3h

The conditions necessary for the formation of cumulonimbus clouds are a lifting action and

A – unstable air containing an excess of condensation nuclei.

B – unstable, moist air.

C – either stable or unstable air.

6-38. Answer B. GFDPP 6C, AW
Three conditions are normally required for the formation of cumulonimbus clouds—lifting action, instability, and moisture.

6-39   PA.I.C.K3h

What feature is normally associated with the cumulus stage of a thunderstorm?

A – Roll cloud.

B – Continuous updraft.

C – Frequent lightning.

6-39. Answer B. GFDPP 6C, AW

In the early, or cumulus, stage of a thunderstorm, continuous updrafts cause the cloud to build upwards.

6-40   PA.I.C.K3h

The mature stage of a thunderstorm begins with

A – formation of the anvil top.

B – the start of precipitation.

C – continuous downdrafts.

6-40. Answer B. GFDPP 6C, AW

The mature stage of a thunderstorm begins when the rain drops grow too large to be supported by the updrafts, and precipitation begins to fall.

6-41   PA.I.C.K3h

What conditions are necessary for the formation of thunderstorms?

A – High humidity, lifting force, and unstable conditions.

B – High humidity, high temperature, and cumulus clouds.

C – Lifting force, moist air, and extensive cloud cover.

6-41. Answer A. GFDPP 6C, AW

Three conditions are normally required for the formation of cumulonimbus clouds—lifting action, instability, and moisture. As moist, unstable air is lifted, it builds cumulonimbus clouds, which form thunderstorms.

6-42   PA.I.C.K3h

During the life cycle of a thunderstorm, which stage is characterized predominately by downdrafts?

A – Cumulus.

B – Dissipating.

C – Mature.

6-42. Answer B. GFDPP 6C, AW

The mature stage of a thunderstorm begins when the rain drops grow too large to be supported by the updrafts, and precipitation begins to fall. As a thunderstorm dissipates, updrafts weaken and downdrafts become predominate.

6-43   PA.I.C.K3h

Thunderstorms reach their greatest intensity during the

A – mature stage.

B – downdraft stage.

C – cumulus stage.

6-43. Answer A. GFDPP 6C, AW

Thunderstorms reach their greatest intensity during the mature stage, producing strong updrafts and downdrafts, severe turbulence, lightning, heavy rain, hail, strong surface winds, and gust fronts.

SECTION C ■ Weather Hazards

6-44   **PA.I.C.K3h**

Thunderstorms that generally produce the most intense hazard to aircraft are

A – squall line thunderstorms.

B – steady-state thunderstorms.

C – warm front thunderstorms.

6-44. Answer A. GFDPP 6C, AW

Squall lines are a narrow band of thunderstorms that often develop ahead of a cold front. Squall lines often contain severe steady-state thunderstorms and present the most hazardous conditions to aircraft.

6-45   **PA.I.C.K3h**

A nonfrontal, narrow band of active thunderstorms that often develop ahead of a cold front is known as a

A – prefrontal system.

B – squall line.

C – dry line.

6-45. Answer B. GFDPP 6C, AW

Squall lines are a narrow band of thunderstorms that often develop ahead of a cold front. Squall lines often contain severe steady-state thunderstorms and present the most hazardous conditions to aircraft.

6-46   **PA.I.C.K3h**

If thunderstorm activity is in the vicinity of an airport at which you plan to land, which hazardous atmospheric phenomenon might be expected on the landing approach?

A – Precipitation static.

B – Wind-shear turbulence.

C – Steady rain.

6-46. Answer B. GFDPP 6C, AW

In the vicinity of thunderstorms, expect strong surface winds with wind-shear turbulence. Showery precipitation, not steady rain, is associated with thunderstorms. Static wicks typically prevent excessive precipitation static, which is not considered an atmospheric phenomenon.

6-47   **PA.I.C.K3h**

Which weather phenomenon is always associated with a thunderstorm?

A – Lightning.

B – Heavy rain.

C – Hail.

6-47. Answer A. GFDPP 6C, AW

Because lightning causes thunder, lightning is always associated with a thunderstorm.

6-48   **PA.I.C.K3f, K3g**

What cloud types would indicate convective turbulence?

A – Cirrus clouds.

B – Nimbostratus clouds.

C – Towering cumulus clouds.

6-48. Answer C. GFDPP 6C, AW

Towering cumulus clouds are formed by convective currents, caused by rising heated air. These rising air currents cause convective turbulence.

6-49   PA.I.C.K3f, K3g

What clouds have the greatest turbulence?

A – Towering cumulus.

B – Cumulonimbus.

C – Nimbostratus.

6-49. Answer B. GFDPP 6C, AW

Cumulonimbus clouds, which form thunderstorms and tornadoes, produce the most severe turbulence.

6-50   PA.I.C.K3g

Upon encountering severe turbulence, which flight condition should the pilot attempt to maintain?

A – Constant altitude and airspeed.

B – Constant angle of attack.

C – Level flight attitude.

6-50. Answer C. GFDPP 6C, AW

If entering severe turbulence, the best procedure is to slow to at or below maneuvering airspeed and maintain a constant level flight attitude. Variations in airspeed and altitude should be expected and tolerated.

6-51   PA.II.F.R3*

Wingtip vortices are created only when an aircraft is

A – operating at high airspeeds.

B – heavily loaded.

C – developing lift.

6-51. Answer C. GFDPP 6C, AFH

Whenever an airplane generates lift, air spills over the wingtips from the high pressure areas below the wings to the low pressure areas above them. This flow causes rapidly rotating whirlpools of air called wingtip vortices, or wake turbulence.

6-52   PA.II.F.R3*

The greatest vortex strength occurs when the generating aircraft is

A – light, dirty, and fast.

B – heavy, dirty, and fast.

C – heavy, clean, and slow.

6-52. Answer C. GFDPP 6C, AFH

Heavy aircraft, in a clean configuration, flying at low airspeeds with high angles of attack, generate the strongest wingtip vortices.

6-53   PA.II.F.R3*

Wingtip vortices created by large aircraft tend to

A – sink below the aircraft generating turbulence.

B – rise into the traffic pattern.

C – rise into the takeoff or landing path of a crossing runway.

6-53. Answer A. GFDPP 6C AFH

Wingtip vortices tend to sink below the flight path of the aircraft that generated them.

SECTION C ■ Weather Hazards

6-54   PA.II.F.R3*

When taking off or landing at an airport where heavy aircraft are operating, pilots should be alert to the hazards of wingtip vortices because this turbulence tends to

A – rise from a crossing runway into the takeoff or landing path.

B – rise into the traffic pattern area surrounding the airport.

C – sink into the flight path of aircraft operating below the aircraft generating the turbulence.

6-54. Answer C. GFDPP 6C, AFH
Wingtip vortices tend to sink below the flight path of the aircraft that generated them.

6-55   PA.IV.B.R2d*

The wind condition that requires maximum caution when avoiding wake turbulence on landing is a

A – light, quartering headwind.

B – light, quartering tailwind.

C – strong headwind.

6-55. Answer B. GFDPP 6C, AFH
A light, quartering tailwind is the most hazardous because it can move the upwind vortex over the runway and forward into the landing zone.

6-56   PA.IV.A.R2d*

When departing behind a heavy aircraft, the pilot should avoid wake turbulence by maneuvering the aircraft

A – below and downwind from the heavy aircraft.

B – above and upwind from the heavy aircraft.

C – below and upwind from the heavy aircraft.

6-56. Answer B. GFDPP 6C, AFH
Because wake turbulence tends to sink and drift downwind, when you are departing behind a large aircraft, stay above and upwind of the preceding aircraft.

6-57   PA.IV.B.R2d*

When landing behind a large aircraft, which procedure should be followed for vortex avoidance?

A – Stay above its final approach flight path all the way to touchdown.

B – Stay below and to one side of its final approach flight path.

C – Stay well below its final approach flight path and land at least 2,000 feet behind.

6-57. Answer A. GFDPP 6C AFH
Because wake turbulence tends to sink, when you are landing behind a large aircraft stay above the large aircraft's flight path and land beyond its touchdown point.

6-58 PA.III.B.R3

How does the wake turbulence vortex circulate around each wingtip?

A – Inward, upward, and around each tip.

B – Inward, upward, and counterclockwise.

C – Outward, upward, and around each tip.

6-58. Answer C. GFDPP 6C, AFH
Whenever an airplane generates lift, air spills over the wingtips from the high pressure areas below the wings to the low pressure areas above them. The vortices move outward, upward and around each wingtip.

6-59 PA.IV.B.R2d*

When landing behind a large aircraft, the pilot should avoid wake turbulence by staying

A – above the large aircraft's final approach path and landing beyond the large aircraft's touchdown point.

B – below the large aircraft's final approach path and landing before the large aircraft's touchdown point.

C – above the large aircraft's final approach path and landing before the large aircraft's touchdown point.

6-59. Answer A. GFDPP 6C, AFH
Because wake turbulence tends to sink, when you are landing behind a large aircraft stay above the large aircraft's flight path and land beyond its touchdown point.

6-60 PA.I.C.K3f

An almond or lens-shaped cloud that appears stationary, but that can contain winds of 50 knots or more, is referred to as

A – an inactive frontal cloud.

B – a funnel cloud.

C – a lenticular cloud.

6-60. Answer C. GFDPP 6C, AW
Lenticular clouds are the lens-shaped clouds that form at the crests of mountain waves.

6-61 PA.I.C.K3f

Crests of standing mountain waves may be marked by stationary, lens-shaped clouds known as

A – mammatocumulus clouds.

B – standing lenticular clouds.

C – roll clouds.

6-61. Answer B. GFDPP 6C, AW
Lenticular clouds are the lens-shaped clouds that form at the crests of mountain waves.

SECTION C ■ **Weather Hazards**

6-62   PA.I.C.K3b, K3g

Possible mountain wave turbulence could be anticipated when winds of 40 knots or greater blow

A – across a mountain ridge, and the air is stable.

B – down a mountain valley, and the air is unstable.

C – parallel to a mountain peak, and the air is stable.

6-62. Answer A. GFDPP 6C, AW

Mountain waves are formed when strong winds (40 knots or greater) flow across a barrier, such as a mountain ridge. When the air is stable, the flow is laminar, or layered, and creates a series of waves. Unstable air that is forced upward tends to continue rising, often creating thunderstorms.

6-63   PA.I.C.K3b, K3g

Where does wind shear occur?

A – Only at higher altitudes.

B – Only at lower altitudes.

C – At all altitudes, in all directions.

6-63. Answer C. GFDPP 6C, AW

Wind shear can occur at middle and high altitudes near thunderstorms or the jet stream, and near the ground in the vicinity of thunderstorms or temperature inversions. The shear can be either vertical or horizontal.

6-64   PA.I.C.K3b, K3g

When may hazardous wind shear be expected?

A – When stable air crosses a mountain barrier where it tends to flow in layers forming lenticular clouds.

B – In areas of low-level temperature inversion, frontal zones, and clear air turbulence.

C – Following frontal passage when stratocumulus clouds form indicating mechanical mixing.

6-64. Answer B. GFDPP 6C, AW

Wind shear can be found above a temperature inversion when the surface air is cold and calm, and the warmer layer above it is moving at 25 knots or more. Because frontal zones are identified by a shift in the wind, wind shear can be expected. Clear air turbulence can be associated with either vertical or horizontal wind shear.

6-65   PA.I.C.K3b

A pilot can expect a wind-shear zone in a temperature inversion whenever the wind speed at 2,000 to 4,000 feet above the surface is at least

A – 10 knots.

B – 15 knots.

C – 25 knots.

6-65. Answer C. GFDPP 6C, AW

A temperature inversion with light surface winds may form near the surface on a clear night. If the winds at 2,000–4,000 feet are 25 knots or more, you can expect a shear zone in the inversion.

6-66   PA.I.C.K3i

One in-flight condition necessary for structural icing to form is

A – small temperature/dewpoint spread.

B – stratiform clouds.

C – visible moisture.

6-66. Answer C. GFDPP 6C, AW

Structural icing requires two conditions to form: (1) visible moisture, such as rain or cloud droplets, and (2) temperature of the aircraft surface must be at or below freezing. A small temperature/dewpoint spread may be present without visible moisture. Stratiform clouds are not the only cloud types in which icing can occur.

6-67   PA.I.C.K3i

In which environment is aircraft structural ice most likely to have the highest accumulation rate?

A – Cumulus clouds with below freezing temperatures.

B – Freezing drizzle.

C – Freezing rain.

6-67. Answer C. GFDPP 6C, AW

The rate of structural ice accumulation is usually the highest in freezing rain below a frontal surface. As the rain falls through air with temperatures below freezing, it becomes supercooled. The supercooled drops freeze on impact with the large water droplets, and heavy rain accelerates the buildup.

6-68   PA.I.G.K1j

Considering the significant hazard that freezing rain presents to aircraft, how does it affect the operation of deicing and anti-icing equipment?

A – Deicing and anti-icing equipment should not be used in freezing rain conditions.

B – Freezing rain has no impact on the operation of deicing and anti-icing equipment.

C – Freezing rain can overwhelm deicing and anti-icing equipment, leading to ice accumulation.

6-68. Answer C. GFDPP 6C, PHB, AW

According to the FAA Pilot's Handbook of Aeronautical Knowledge, operation of aircraft anti-icing and deicing systems should be checked prior to encountering icing conditions. Encounters with structural ice require immediate action. Anti-icing and deicing equipment are not intended to sustain long-term flight in icing conditions.

According to Aviation Weather Services, AC 00-45, Freezing rain is particularly dangerous because it can lead to rapid ice accumulation on an aircraft's critical surfaces, even when deicing and anti-icing systems are in use. These systems are designed to reduce the risk of ice buildup, but freezing rain can exceed their capacity to remove or prevent ice, potentially compromising the aircraft's performance and safety.

6-69   PA.I.C.K3l

If you inadvertently enter a volcanic ash cloud, the recommended course of action is to

A – reverse course.

B – attempt to climb up and out of the ash cloud.

C – continue straight ahead and exit on the opposite side.

6-69. Answer A. GFDPP 6C, AIM

If you suspect that you are in the vicinity of a volcanic ash cloud, you should attempt to stay upwind. If you inadvertently enter a volcanic ash cloud, you should not attempt to fly straight through or climb out of the cloud because the ash cloud can be hundreds of miles wide and extend to great heights. You should reduce power to a minimum, altitude permitting, and reverse course to escape the ash cloud.

SECTION C ■ **Weather Hazards**

SECTION C ■ **Weather Hazards**

CHAPTER 7

Interpreting Weather Data

NOTE: No FAA questions apply to GFD Private Pilot textbook Chapter 7, Section A — The Forecasting Process.

SECTION B
Aviation Weather Reports and Forecasts

OBSERVATIONS

AVIATION ROUTINE WEATHER REPORT

- An aviation routine weather report (METAR) is an observation of surface weather taken every hour that is reported in a standard format.

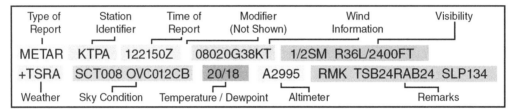

- Winds on an aviation routine weather report are referenced to true north.

- Peak gusts on an aviation routine weather report are denoted by a number following a "G" after the wind direction and base speed.

- Cloud heights or visibility into an obscuration are reported with three digits in hundreds of feet. Visibility is reported in statute miles and is indicated by the abbreviation "SM."

- For aviation purposes, ceiling is defined as the height above the surface of the earth of the lowest broken or overcast layer or vertical visibility into an obscuration.

- The definition of VFR is visibility of at least 3 statute miles and a ceiling of at least 1,000 feet.

- The remarks section of a METAR is used to report weather considered significant to aircraft operations. The contraction "RMK" precedes the remarks.

PILOT WEATHER REPORT

- In a pilot weather report (PIREP), identified by the letters "UA," sky condition is designated by the letters "SK," followed by the base and top of each cloud layer.

- The wind direction and velocity in a PIREP are shown as "WV" followed by the direction and speed, with the last digit of the wind direction dropped.

- The ceiling is the lowest cloud layer reported as broken, overcast, or obscured.

- Turbulence is reported in a PIREP as "TB" followed by an intensity designation, such as "SVR," "MDT," or "LGT." The altitude of the turbulence layer is also reported.

- Icing is reported in a PIREP after the letters "IC"—followed by the intensity of the icing, and the altitude of the layers in which it was encountered.

FORECASTS

TERMINAL AERODROME FORECAST

- The terminal aerodrome forecast (TAF) is a concise statement of the expected meteorological conditions within a 5 statute mile radius from the center of an airport's runway complex.

- TAFs are usually valid for a 24-hour period and are scheduled four times a day (0000Z, 0600Z, 1200Z, and 1800Z). The six-digit issuance date-time group is followed by the valid date-time group.

- The abbreviation "BECMG" precedes a gradual change in the weather with the Zulu time period over which the weather is forecast to change. For example, "BECMG 1012" means that the weather change is expected to occur between 1000Z and 1200Z. The time frame is followed by the change expected, such as "3 SM" would mean the visibility is forecast to change to three statute miles.

- Wind blocks read as follows: wind direction comes first, followed by speed, and then any gust factor expected. Ceilings are written with the amount of coverage, followed by the cloud base height, in hundreds of feet.

- When the abbreviation "VRB" appears before the wind speed, the wind is expected to be variable at that speed.

- In a TAF, the abbreviation "SHRA" stands for rain showers.

- Cumulonimbus clouds are the only cloud type included in the TAFs.

- A change group is used when a significant, lasting change to the weather conditions is forecast during the valid time.

- If a rapid change, usually within one hour, is expected, the code for from (FM) is used with the time of change.

- The code "NSW" means that no significant weather change is forecast to occur.

WIND AND TEMPERATURE ALOFT FORECAST

- The first two digits represent the wind direction in relation to true north. The next two digits are the speed. Temperatures follow the wind block. Temperatures are assumed to be negative above 24,000 feet MSL.

- In some formats, winds of 100 to 199 knots have 50 added to the direction. For example, when you observe a wind direction above 360, subtract 50 to get wind direction, and add 100 to the listed wind speed.

- "Light and variable" in a winds aloft forecast is coded as 9900 for forecast winds less than five knots.

AIRMETS AND SIGMETS

- An AIRMET is a warning of weather conditions that are especially hazardous to light aircraft or aircraft having limited capability because of lack of equipment, instrumentation, or pilot qualifications. The three types of AIRMETS are:

  - AIRMET Sierra—IFR conditions and/or extensive mountain obscurations.

  - AIRMET Tango—moderate turbulence, sustained surface winds of 30 knots or greater, and/or nonconvective low-level wind shear.

  - AIRMET Zulu—moderate icing and provides freezing level heights.

- A SIGMET is issued for hazardous weather (other than convective activity) that is considered significant to all aircraft. SIGMETs are issued for:

  - Severe icing.

  - Severe or extreme turbulence.

  - Clear air turbulence (CAT).

  - Dust storms and sandstorms lowering visibility to less than three miles.

  - Volcanic ash.

SECTION B ■ Aviation Weather Reports and Forecasts

CONVECTIVE SIGMETS

- Convective SIGMETs are issued for hazardous convective weather that is significant to the safety of all aircraft.

- Because convective SIGMETs always imply severe or greater turbulence, severe icing, and low-level wind shear, these items are not specified in the advisory. Convective SIGMETS are issued for:

  - Severe thunderstorms due to surface winds greater than or equal to 50 knots; hail at the surface greater than or equal to 3/4 inches in diameter; tornadoes.

  - Embedded thunderstorms (expected by obscured by massive cloud layers).

  - A line of thunderstorms.

  - Thunderstorms producing precipitation greater than or equal to heavy precipitation affecting 40 percent or more of an area at least 3,000 square miles.

NOTE: An asterisk appearing after an ACS code (i.e. PA.II.F.R3*) indicates that the question subject appears more than one time in the ACS. The code shown corresponds to the first instance of the subject in the ACS.

7-1   PA.I.C.K2, K2a

(Refer to figure 12.) Which of the reporting stations have VFR weather?

A – All.

B – KINK, KBOI, and KJFK.

C – KINK, KBOI, and KLAX.

7-1. Answer C. GFDPP 7B, AWS

To answer this question, you must know that the definition of VFR is visibility of at least three statute miles and ceiling of at least 1,000 feet. KINK has 15 miles visibility with clear skies, KBOI has 30 miles visibility with a scattered layer at 15,000 feet, and KLAX has 6 miles visibility, with scattered layers at 700 feet and 25,000 feet. Remember, a scattered layer does not constitute a ceiling.

7-2   PA.I.C.K2

For aviation purposes, ceiling is defined as the height above the earth's surface of the

A – lowest reported obscuration and the highest layer of clouds reported as overcast.

B – lowest broken or overcast layer or vertical visibility into an obscuration.

C – lowest layer of clouds reported as scattered, broken, or thin.

7-2. Answer B. GFDPP 7B, AWS

According to Aviation Weather Services, AC 00-45, a ceiling is defined as the lowest broken or overcast layer, or vertical visibility into an obscuration.

7-3   PA.I.C.K2, K2a

(Refer to Figure 12.) The wind direction and velocity at KJFK is from

A – 180° true at 4 knots.

B – 180° magnetic at 4 knots.

C – 040° true at 18 knots.

7-3. Answer A. GFDPP 7B, AWS

The wind entry for KJFK is "18004KT," meaning the wind is from 180 degrees at four knots. Winds on an aviation routine weather report are referenced to true north.

7-4   PA.I.C.K2, K2a

(Refer to Figure 12.) What are the wind conditions at Wink, Texas (KINK)?

A – Calm.

B – 110° at 12 knots, peak gusts 18 knots.

C – 111° at 2 knots, peak gusts 18 knots.

7-4. Answer B. GFDPP 7B, AWS

The wind entry for KINK is "11012G18KT," meaning the wind direction is 110 degrees, and the velocity is 12 knots, with peak gusts of 18 knots.

7-5   PA.I.C.K2, K2a

(Refer to Figure 12.) The remarks section for KMDW has RAB35 listed. This entry means

A – blowing mist has reduced the visibility to 1-1/2 SM.

B – rain began at 1835Z.

C – the barometer has risen 0.35 inches Hg.

7-5. Answer B. GFDPP 7B, AWS

The remarks section of a METAR is used to report weather considered significant to aircraft operations. The contraction "RMK" precedes remarks, which include the beginning and ending times of certain weather phenomena. In this case, "RAB35" means that the rain began at 35 minutes past the hour, or 1835Z.

7-6   PA.I.C.K2, K2a

(Refer to Figure 12.) What are the current conditions depicted for Chicago Midway Airport (KMDW)?

A – Sky 700 feet overcast, visibility 1-1/2 SM, rain.

B – Sky 7000 feet overcast, visibility 1-1/2 SM, heavy rain.

C – Sky 700 feet overcast, visibility 11, occasionally 2SM, with rain.

7-6. Answer A. GFDPP 7B, AWS

Cloud heights or the vertical visibility into an obscuration is reported with three digits in hundreds of feet. Visibility is reported in statute miles (SM). In this case, the METAR from KMDW indicates that Midway's visibility is 1-1/2 miles and the sky is overcast at 700 feet. "RA" indicates precipitation in the form of rain.

7-7   PA.I.C.K2, K2a

(Refer to Figure 14.) The base and tops of the overcast layer reported by a pilot are

A – 1,800 feet MSL and 5,500 feet MSL.

B – 5,500 feet AGL and 7,200 feet MSL.

C – 7,200 feet MSL and 8,900 feet MSL.

7-7. Answer C. GFDPP 7B, AWS

In the pilot report (PIREP), which is identified by the letters UA, sky cover is designated by the letters SK, followed by the base and top of each cloud layer. The overcast layer is shown as OVC 072-TOP 089, which means the base is 7,200 feet and the tops are 8,900 feet. Altitudes are MSL unless otherwise noted.

SECTION B ■ Aviation Weather Reports and Forecasts

7-8   PA.I.C.K2, K2a

(Refer to Figure 14.) The wind and temperature at 12,000 feet MSL as reported by a pilot are

A – 080° at 21 knots and -7°C.

B – 090° at 21 MPH and -9°F.

C – 090° at 21 knots and -9°C.

7-8. Answer A. GFDPP 7B, AWS

The ambient temperature and wind velocity appear in the part of the PIREP that indicates "/TA M7/WV 08021/". All temperatures aloft are given in degrees Celsius, and the "M" indicates temperatures below zero. Wind speed is reported in knots.

7-9   PA.I.C.K2, K2a, K3

(Refer to Figure 14.) If the terrain elevation is 1,295 feet MSL, what is the height above ground level of the base of the ceiling?

A – 505 feet AGL.

B – 1,295 feet AGL.

C – 6,586 feet AGL.

7-9. Answer A. GFDPP 7B, AWS

The ceiling is the lowest cloud layer reported as broken, overcast, or obscured. In this PIREP, the lowest layer is 1,800 feet broken (MSL). To find the AGL height, subtract the ground elevation (1,800 – 1,295 = 505 feet AGL).

7-10   PA.I.C.K2, K2a, K3

(Refer to Figure 14.) The intensity of the turbulence reported at a specific altitude is

A – moderate from 5,500 feet to 7,200 feet.

B – moderate at 5,500 feet and at 7,200 feet.

C – light from 5,500 feet to 7,200 feet.

7-10. Answer C. GFDPP 7B, AWS

In this PIREP, turbulence is reported as "/TB LGT 055-072/", meaning light turbulence between 5,500 and 7,200 feet MSL.

7-11   PA.I.C.K2, K2a, K3

(Refer to Figure 14.) The intensity and type of icing reported by a pilot is

A – light to moderate rime.

B – light to moderate.

C – light to moderate clear.

7-11. Answer A. GFDPP 7B, AWS

Icing intensity and type is shown in this PIREP as "/IC LGT-MDT RIME/" or light to moderate rime.

7-12   PA.I.C.K2c, K3

(Refer to Figure 15.) What is the valid period for the TAF for KMEM?

A – 1200Z to 1200Z.

B – 1200Z to 1800Z.

C – 1800Z to 1800Z.

7-12. Answer C. GFDPP 7B, AWS

Terminal Area Forecasts (TAFs) are usually valid for a 24-hour period and are scheduled four times a day (0000Z, 0600Z, 1200Z, and 1800Z). The six-digit issuance date-time group is followed by the valid date-time group. Therefore, "121720Z 1218/1324" indicates that the KMEM TAF was issued on the 12th at 1720 Zulu. This report is valid from 1800 Zulu on the 12th until 1800 Zulu on the 13th.

7-13   PA.I.C.K2c, K3

(Refer to Figure 15.) In the TAF for KMEM, what does "SHRA" stand for?

A – Rain showers are expected.

B – A shift in wind direction is expected.

C – A significant change in precipitation is possible.

7-13. Answer A. GFDPP 7B, AWS
This group of the TAF, "PROB40 2202 3SM SHRA," indicates a 40-percent probability, between 2200 Zulu and 0200 Zulu, that the visibility will be 3 statute miles with showery precipitation or rain showers (SHRA). The next entry, "FM0200 35012KT OVC008" indicates, from 0200 Zulu, the wind is expected to be from 350° at 12 knots.

7-14   PA.I.C.K2c, K3

(Refer to Figure 15.) Between 1000Z and 1200Z, the visibility at KMEM is forecast to be?

A – 1/2 statute mile.

B – 3 statute miles.

C – 6 statute miles.

7-14. Answer B. GFDPP 7B, AWS
During a specified time period, when changes in the weather conditions are forecast, a change group is appended to the forecast. In this TAF, "BECMG 1310/1312" indicates that a change in the weather will occur between 1000Z and 1200Z. The "3SM" indicates that the visibility is forecast to become 3 statute miles.

7-15   PA.I.C.K2c, K3

(Refer to Figure 15.) What is the forecast wind for KMEM from 1600Z until the end of the forecast?

A – Variable in direction at 6 knots.

B – No significant wind.

C – Variable in direction at 4 knots.

7-15. Answer A. GFDPP 7B, AWS
This part of the TAF reads, "131600 VRB06KT P6SM SKC=" From 1600Z until the end of forecast, the wind is variable in direction at 6 knots, with visibility greater than 6 miles.

7-16   PA.I.C.K2c, K3

(Refer to Figure 15.) In the TAF from KOKC, the "FM (FROM) Group" is forecast for the hours from 1600Z to 2200Z with the wind from

A – 180° at 10 knots, becoming 200° at 13 knots.

B – 160° at 10 knots.

C – 180° at 10 knots.

7-16. Answer C. GFDPP 7B, AWS
The TAF code "FM051600 18010KT" indicates a rapid change in the forecast conditions starting at that time. 18010KT indicates wind from 180 degrees at 10 knots. "BECMG 0522/0524" means that the next change will happen gradually from 2200 to 2400Z.

SECTION B ■ **Aviation Weather Reports and Forecasts**

7-17   PA.I.C.K2c, K3

(Refer to Figure 15.) In the TAF from KOKC, the clear sky becomes

A – overcast at 2,000 feet during the forecast period between 2200Z and 2400Z.

B – overcast at 200 feet with a 40% probability of becoming overcast at 600 feet during the forecast period between 2200Z and 2400Z.

C – overcast at 200 feet with the probability of becoming overcast at 400 feet during the forecast period between 2200Z and 2400Z.

7-17. Answer A. GFDPP 7B, AWS

When a gradual change in the forecast weather is expected, the becoming (BECMG) change group is used, followed by the beginning and ending times. The TAF from KOKC "BECMG 0522/0524 20013G20KT 4SM SHRA OVC020" means between 2200Z and 2400Z, the weather will gradually change to winds from 200° at 13 knots, 4 miles visibility in rain showers, and overcast skies at 2,000 feet.

7-18   PA.I.C.K2c, K3

(Refer to Figure 15.) During the time period from 0600Z to 0800Z, what visibility is forecast for KOKC?

A – Greater than six statute miles.

B – Not forecasted.

C – Possibly six statute miles.

7-18. Answer A. GFDPP 7B, AWS

This TAF section reads, "BECMG 0606/0608 21015KT P6SM SCT040=." These abbreviations mean that between 0600-0800Z, the wind will become 210° at 15 knots, visibility is forecast to be more than (not possibly) 6 statute miles, and clouds will become scattered at 4,000 feet.

7-19   PA.I.C.K2c, K3

Terminal aerodrome forecasts (TAF) are issued how many times a day and cover what period of time?

A – Four times daily and are usually valid for a 24-hour period.

B – Six times daily and are usually valid for a 24-hour period including a 4-hour categorical outlook.

C – Four times daily and are valid for 12 hours including a 6-hour categorical outlook.

7-19. Answer A. GFDPP 7B, AWS

The primary source for the forecast weather at a specific airport is the terminal aerodrome forecast (TAF). TAFs normally are valid for a 24-hour period and scheduled four times a day at 0000Z, 0600Z, 1200Z, and 1800Z for the area within a 5-statute mile radius from the center of an airport's runway complex.

7-20   PA.I.C.K2c, K3

What does the contraction VRB in the terminal aerodrome forecast (TAF) mean?

A – Wind speed is variable throughout the period.

B – Cloud base is variable.

C – Wind direction is variable.

7-20. Answer C. GFDPP 7B, AWS

The prevailing wind direction is forecast for any speed greater than or equal to 7 knots. When the prevailing wind direction is variable (variations in wind direction of 30 degrees or more), the wind direction is encoded as VRB. Two conditions where this can occur are very light winds (1 to 6 knots inclusive) and convective activity.

7-21 PA.I.C.K2c, K3

(Refer to Figure 15.) The only cloud type forecast in TAF reports is

A – Nimbostratus.

B – Cumulonimbus.

C – Scattered cumulus.

7-21. Answer B. GFDPP 7B, AWS

If cumulonimbus clouds are expected at the airport, the contraction "CB" in a TAF is appended to the height of the cloud layer to indicate the base of the cumulonimbus cloud.

7-22 PA.I.C.K2e, K3

(Refer to Figure 17.) What wind is forecast for STL at 9,000 feet?

A – 230° true at 32 knots.

B – 230° magnetic at 25 knots.

C – 230° true at 25 knots.

7-22. Answer A. GFDPP 7B, AWS

In the Wind and Temperature Aloft Forecast, directions are relative to true north and rounded to the nearest 10 degrees. The wind information for STL at 9,000 feet is 2332+02. The first two digits represent the wind direction in relation to true north, 230°. The next two digits are the speed, which in this case is 32 knots. The temperature is +2°C.

7-23 PA.I.C.K2e, K3

(Refer to Figure 17.) What wind is forecast for STL at 12,000 feet?

A – 230° true at 56 knots.

B – 230° true at 39 knots.

C – 230° magnetic at 56 knots.

7-23. Answer B. GFDPP 7B, AWS

In the Wind and Temperature Aloft Forecast, directions are relative to true north and rounded to the nearest 10 degrees. The wind information for STL at 12,000 feet is 2339–04. The first two digits represent the wind direction, 230° true. The next two digits are the speed, 39 knots. The temperature is –4°C.

7-24 PA.I.C.K2e, K3

When the term "light and variable" is used in a winds aloft forecast, the coded group and wind speed is

A – 0000 and less than 7 knots.

B – 9900 and less than 5 knots.

C – 9999 and less than 10 knots.

7-24. Answer B. GFDPP 7B, AWS

The direction is shown as 99, indicating that the direction is variable. When the second two digits are listed as 00, the speed is less than 5 knots.

7-25 PA.I.C.K2e, K3

What values are used for wind aloft forecasts?

A – Magnetic direction and knots.

B – Magnetic direction and miles per hour.

C – True direction and knots.

7-25. Answer C. GFDPP 7B, AWS

All forecast and ASOS/AWOS-reported winds are given in true direction, and speed is always in knots. The only time wind direction is given in magnetic is when it is provided by the tower or ATIS.

SECTION B ■ **Aviation Weather Reports and Forecasts**

SECTION B ■ Aviation Weather Reports and Forecasts

7-26   PA.I.C.K2g, K3

AIRMETs are advisories of significant weather phenomena but of lower intensities than SIGMETs and are intended for dissemination to

A – only IFR pilots.

B – all pilots.

C – only VFR pilots.

7-27   PA.I.C.K2g, K3

What type of in-flight weather advisory provides an enroute pilot with information regarding the possibility of moderate icing, moderate turbulence, winds of 30 knots or more at the surface and extensive mountain obscuration?

A – Convective SIGMETs and SIGMETs.

B – Terminal aerodrome forecasts.

C – AIRMETs.

7-28   PA.I.C.K2g, K3

SIGMETs are issued as a warning of weather conditions hazardous to which aircraft?

A – Small aircraft only.

B – Large aircraft only.

C – All aircraft.

7-26. Answer B. GFDPP 7B, AWS

An AIRMET advises of weather that is of operational interest to all aircraft, but may be hazardous to aircraft with limited capabilities, such as light single-engine airplanes.

7-27. Answer C. GFDPP 7B, AWS

AIRMETs, SIGMETs, and convective SIGMETs are textual forecasts that advise enroute aircraft of the development of potentially hazardous weather. In addition, you use these advisories during your preflight weather briefing to learn of the latest adverse conditions affecting your proposed flight.

AIRMETs (WAs) are issued every six hours, with amendments issued as necessary, for weather phenomena that are of operational interest to all aircraft, but hazardous mainly to light aircraft. AIRMETs are issued for:

- IFR conditions (ceilings less than 1,000 feet or visibility less than 3 miles) and/or extensive Mountain obscuration.

- Moderate turbulence.

- Sustained surface winds of 30 knots or greater.

- Nonconvective low-level wind shear.

7-28. Answer C. GFDPP 7B, AWS

SIGMETs are issued for weather potentially hazardous to all aircraft. An AIRMET advises of weather that is of operational interest to all aircraft, but that could be hazardous to aircraft with limited capabilities, such as light single-engine airplanes.

7-29   PA.I.C.K2g, K3

Which in-flight advisory would contain information on severe icing not associated with thunderstorms?

A – Convective SIGMET.

B – SIGMET.

C – AIRMET.

7-29. Answer B. GFDPP 7B, AWS

A SIGMET advises of weather potentially hazardous to all aircraft, which would include severe icing. A convective SIGMET is an advisory of especially hazardous thunderstorm activity.

7-30   PA.I.C.K2g, K3

What is indicated when a Convective SIGMET forecasts thunderstorms?

A – Moderate thunderstorms covering 30 percent of the area.

B – Moderate or severe turbulence.

C – Thunderstorms obscured by massive cloud layers.

7-30. Answer C. GFDPP 7B, AIM

Convective SIGMETs are issued for thunderstorms (very strong, not moderate) covering 40 percent or more (not 30 percent) of an area at least 3,000 sq. miles. They also are issued for embedded thunderstorms; those obscured by massive cloud layers. All convective SIGMETs imply severe or greater turbulence, not moderate turbulence.

7-31   PA.I.C.K2g, K3

What information is contained in a Convective SIGMET?

A – Tornadoes, embedded thunderstorms, and hail 3/4 inch or greater in diameter.

B – Severe icing, severe turbulence, or widespread duststorms lowering visibility to less than 3 miles.

C – Surface winds greater than 40 knots or thunderstorms equal to or greater than video integrator processor (VIP) level 4.

7-31. Answer A. GFDPP 7B, AIM

Convective SIGMETs are issued for any of the following phenomena: tornadoes, lines of thunderstorms, embedded thunderstorms, areas of thunderstorms covering 40 percent or more of the area at least 3,000 sq. miles, and hail of 3/4 inch or greater in diameter.

SECTION B ■ **Aviation Weather Reports and Forecasts**

SECTION C
Graphic Weather Products

Graphic weather products continue to evolve. In the past, pilots referred to specific named weather products to obtain information on certain weather conditions. Many of these products have continue to evolve and you now have the ability to display a wide variety of weather observations and forecast conditions as overlays over a map.

When interpreting graphic weather products and weather overlays, refer to the legend provided by the source for the particular product. However, some standard symbology applies to a variety of analysis and forecast charts to display weather phenomena, precipitation, turbulence, and winds.

OVERVIEW

- Graphic weather products use information gathered from ground observations, weather radar, satellites, and other sources to give you a pictorial view of large scale weather patterns and trends.
- Observations are weather data collected by one or more sensors, and are the basic information upon which forecasts and advisories are made. The most common types of weather observations include radar observations, satellite imagery, and surface observations.
- An analysis is an enhanced depiction of observed data—a map or chart—that can also include interpretation by a weather specialist. Analysis charts show weather conditions that are observed or measured and then plotted graphically for easy interpretation. The most commonly used graphic analysis is the surface analysis chart.
- A forecast is a prediction of the development and movement of weather phenomena based on meteorological observations and mathematical models. Graphic weather products help you visualize the forecast VFR and IFR ceilings and visibilities, surface winds, turbulence, winds and temperatures aloft, hazardous weather, and icing conditions.

SURFACE ANALYSIS CHART

- The surface analysis chart, sometimes referred to as a surface weather map, is an analyzed depiction of surface weather observations. By reviewing this chart, you obtain a picture of atmospheric pressure patterns at the Earth's surface.
- Standard chart symbols depict areas of equal pressure, the positions of highs, lows, ridges, and troughs, the location and type of fronts, and various boundaries, such as drylines, which separate moist and dry air masses.

SURFACE PROGNOSTIC (PROG) CHART

- Use a surface prognostic (prog) chart to obtain an overview of the progression of surface weather—the change in position, size, and intensity of conditions over time.
- These charts depict forecast surface pressure systems, fronts, and precipitation over several days. The charts contain standard symbols for fronts, isobars, and low and high pressure centers.

ISOBARS

- Depicted on surface analysis and surface prog charts, isobars are lines of equal pressure shown as solid lines spaced at intervals of 4 millibars (mb), each labeled with the pressure.
- Isobars that are close together indicate a higher pressure gradient because the pressure changes over a shorter distance, resulting in stronger winds.

SIGNIFICANT WEATHER (SIGWX) CHART

- Low-level significant weather (SIGWX) charts forecast aviation weather hazards from the surface up to flight level 240, and can be used as guidance for a pre-flight briefing.

- SIGWX charts provide forecast depictions of potential aviation hazards such as:

  ° Low visibilities and ceilings—marginal VFR (MVFR) and IFR conditions.

  ° Freezing levels (which can indicate icing).

  ° Turbulence.

- SIGWX charts provide a snapshot of weather expected at the valid forecast time, 12 or 24 hours after chart issuance.

CONVECTIVE OUTLOOK

- The SPC convective outlook is issued by the NWS Storm Prediction Center can be overlaid on a thunderstorms forecast chart.

- Convective outlooks depict predicted general and severe thunderstorm activity and is issued 5 times daily, with the first issuance valid beginning at 1200Z and all 5 issuances valid until 1200Z the following day.

NOTE: An asterisk appearing after an ACS code (i.e. PA.II.F.R3*) indicates that the question subject appears more than one time in the ACS. The code shown corresponds to the first instance of the subject in the ACS.

7-32   PA.I.C.K2b

What information can you obtain from a surface analysis chart?

A – Atmospheric pressure patterns with chart symbols depicting the positions of highs. lows, and fronts.

B – Radar depiction of precipitation.

C – Clouds and freezing levels associated with frontal activity.

7-32. Answer B. GFDPP 7C, AWS

Surface analysis charts show atmospheric pressure patterns at the Earth's surface. You can also see the locations of high and low pressure systems and associated fronts.

7-33   PA.I.C.K2d

Surface prog charts provide depictions of forecast

A – surface pressure systems, fronts, and precipitation.

B – aviation weather hazards, such as MVFR and IFR conditions, turbulence, and freezing levels.

C – areas of probable turbulence, icing, and IFR conditions.

7-33. Answer A. GFDPP 7C, AWS

Use a surface prognostic (prog) chart to obtain an overview of the progression of surface weather—the change in position, size, and intensity of conditions over time. These charts depict forecast surface pressure systems, fronts, and precipitation over several days. The charts contain standard symbols for fronts, isobars, and low and high pressure centers.

SECTION C ■ **Graphic Weather Products**

7-34   PA.I.C.K2d

What information can be derived from the isobars on a surface analysis or surface prog chart?

A – The isobars can be an indicator of freezing levels.

B – When isobars are close together the visibility is low.

C – Isobars can used to determine areas with strong winds.

7-34. Answer C. GFDPP 7C, AWS

Isobars—lines of equal pressure—are depicted as solid lines spaced at intervals of 4 millibars (mb), each labeled with the pressure. Isobars that are close together indicate a higher pressure gradient because the pressure changes over a shorter distance, resulting in stronger winds.

7-35   PA.I.C.K2d, K3

You can refer to a low-level significant weather (SIGWX) chart

A – for overall planning at all altitudes.

B – to determine areas to avoid, such as those with turbulence or IFR conditions.

C – for analyzing current frontal activity and cloud coverage.

7-35. Answer B. GFDPP 7C, AWS

Low-level significant weather (SIGWX) charts forecast aviation weather hazards from the surface up to flight level 240, and can be used as guidance for a pre-flight briefing. SIGWX charts provide forecast depictions of potential aviation hazards such as:

• Low visibilities and ceilings—marginal VFR (MVFR) and IFR conditions.

• Freezing levels (which can indicate icing).

• Turbulence.

SIGWX charts provide a snapshot of weather expected at the valid forecast time, 12 or 24 hours after chart issuance.

7-36   PA.I.C.K2d

Low-level significant weather (SIGWX) charts provide depictions of forecast

A – surface pressure systems and fronts.

B – aviation weather hazards such as MFVR and IFR conditions, turbulence, and freezing levels.

C – areas of probable precipitation, including ice, snow, and thunderstorms.

7-36. Answer B. GFDPP 7C, AWS

Low-level significant weather (SIGWX) charts forecast aviation weather hazards from the surface up to flight level 240, and can be used as guidance for a pre-flight briefing. SIGWX charts provide forecast depictions of potential aviation hazards such as:

• Low visibilities and ceilings—marginal VFR (MVFR) and IFR conditions.

• Freezing levels (which can indicate icing).

• Turbulence.

SIGWX charts provide a snapshot of weather expected at the valid forecast time, 12 or 24 hours after chart issuance.

Surface pressure systems, fronts, and areas of probable precipitation are depicted on the surface prog chart, not the low-level SIGWX chart.

7-37   PA.I.C.K2d, K2f, K3

Which weather product provides a forecast for both severe and general thunderstorms?

A – Convective outlook.

B – Surface prog chart.

C – U.S. low-level significant weather chart.

7-37. Answer A. GFDPP 7C, AWS

The SPC convective outlook is issued by the NWS Storm Prediction Center can be overlaid on a thunderstorms forecast chart. Convective outlooks depict predicted general and severe thunderstorm activity and is issued 5 times daily, with the first issuance valid beginning at 1200Z and all 5 issuances valid until 1200Z the following day.

7-38   PA.I.C.K2d, K3

(Refer to Figure 19.) What is the forecast freezing level over Montana at 06Z?

A – 8,000 feet MSL

B – 10,000 feet MSL.

C – 4,000 feet MSL.

7-38. Answer C. GFDPP 7C, AWS

The 24-hour U.S. low-level significant weather chart on the bottom half of Figure 19 shows a valid time (VT) of 06Z. Freezing levels above the surface are depicted by green dashed lines labeled in hundreds of feet MSL beginning at 4,000 feet, and using 4,000-foot intervals. The freezing level over Montana is 4,000 feet MSL (040).

7-39   PA.I.C.K2d, K3

(Refer to Figure 19.) What type of weather can be expected over Northern Georgia and South Carolina at 18Z?

A – Moderate or greater turbulence.

B – Ceilings between 1,000 and 3,000 feet MSL and/or visibility 3 to 5 miles.

C – Ceilings less than 1,000 feet MSL and/or visibility less than 3 miles.

7-39. Answer C. GFDPP 7C, AWS

The 12-hour U.S. low-level significant weather chart on the top half of Figure 19 shows a valid time (VT) of 18Z. The bold red line encloses an area forecast to have IFR conditions, with ceilings less than 1,000 feet, and/or visibility less than 3 miles. This area extends from central Georgia to North Carolina.

7-40   PA.I.C.K2d, K3

(Refer to figure 19.) What type of condition can be expected over Iowa at 06Z?

A – Severe turbulence at 8,000 feet MSL.

B – Moderate turbulence from the surface to 8,000 feet MSL.

C – Freezing level at the surface.

7-40. Answer B. GFDPP 7C, AWS

The 12-hour U.S. low level significant weather chart on the upper half of Figure 19 shows a valid time (VT) of 18Z. The single line white chevron turbulence indicator over Iowa depicts moderate turbulence. The number below it indicates the top of the turbulence at 8,000 ft MSL. The omission of an altitude after the forward slash (/) indicates that the turbulence begins at the surface.

SECTION C ■ **Graphic Weather Products**

SECTION D
Sources of Weather Information

PREFLIGHT WEATHER SOURCES

- You can obtain an official weather briefing by calling 1-800-WX-BRIEF or by going to 1800wxbrief.com.

- When calling Flight Service for a preflight weather briefing, identify yourself as a pilot and supply the following information: type of flight planned (VFR or IFR), aircraft tail number or your name, aircraft type, departure airport, route of flight, destination, flight altitude(s), estimated time of departure (ETD), and estimated time enroute (ETE).

- To get a complete weather briefing for the planned flight, request a standard briefing, which assumes you do not already have a detailed comprehensive weather picture for your flight.

- Obtain an abbreviated briefing when you need only one or two specific items or to update weather information from a previous briefing or other weather sources.

- An outlook briefing is the weather briefing provided when the information requested is six or more hours before the proposed departure time. When requesting information for the following morning, ask for an outlook briefing.

IN-FLIGHT WEATHER SOURCES

Because forecasting is an inexact science, weather conditions can change rapidly and unexpectedly over a few hours. Often, you need to receive updated information during flight:

- Obtain actual weather information and thunderstorm activity along the route from Flight Service.

- Refer to communication boxes on VFR charts or your navigation database for nearby Flight Service frequencies.

- FIS-B is an example of a data-link weather service you can receive if your airplane has ADS-B capability and the proper receiver.

- When using an FIS-B radar mosaic to avoid thunderstorm activity, know that the echoes you see on your screen can be up to 20 minutes old, and that significant changes can occur during that time period.

CENTER WEATHER ADVISORIES

- A center weather advisory (CWA) is an unscheduled weather advisory issued by an ARTCC to alert pilots of existing or anticipated adverse weather conditions within the next two hours.

- A CWA can be initiated when an AIRMET or SIGMET has not been issued but based on current PIREPs, conditions meet those criteria.

- A CWA can be issued to supplement an existing in-flight advisory as well as any conditions that currently or will soon adversely affect the safe flow of traffic within the ARTCC area of responsibility.

- ARTCCs broadcast CWAs and other hazardous weather information once on all frequencies, except emergency frequencies (121.5), when any part of the area described is within 150 miles of the airspace under the ARTCC jurisdiction.

- In terminal areas, local control and approach control might limit CWA broadcasts to weather occurring within 50 miles of the airspace under their jurisdiction.

AUTOMATED WEATHER REPORTING SYSTEMS

- The automated weather observation system (AWOS) uses various sensors, a voice synthesizer, and a radio transmitter to provide real-time weather data. There are four versions of the AWOS system. AWOS 3 has the ability to report the altimeter setting, wind speed, direction and gusts, temperature, dewpoint, visibility, and cloud and ceiling data.

- The automated surface observing system (ASOS), in addition to broadcasting the same elements as the AWOS, measures and reports variable cloud height, variable visibility, and rapid pressure changes, as well as precipitation type, intensity, accumulation, and beginning and ending times. In addition, ASOS is capable of measuring wind shifts, and peak winds, and some ASOS stations can determine the difference between liquid precipitation and frozen or freezing precipitation.

NOTE: An asterisk appearing after an ACS code (i.e. PA.II.F.R3*) indicates that the question subject appears more than one time in the ACS. The code shown corresponds to the first instance of the subject in the ACS.

7-41   PA.I.C.K1

You planned a route and obtained preliminary weather information to determine the feasibility of your flight. On the morning of your flight, an FAA-approved briefing consists of

A – obtaining a standard briefing at 1800wxbrief.com.

B – obtaining recorded weather by telephone at 1-800-WX-BRIEF (1-800-992-7433).

C – viewing official weather reports, forecasts, and charts at AviationWeather.gov.

7-41. Answer A. GFDPP 7D, PHB

Obtain your FAA-approved weather briefing by telephone or online. You can fill out a preliminary flight plan and select Standard Briefing at 1800wxbrief.com or call a Flight Service briefer at 1-800-WX-BRIEF (1-800-992-7433). While you can obtain official weather service information by listening to recorded weather at 1-800-WX-BRIEF or online at AviationWeather.gov, these methods are not substitutes for an approved briefing, and do not include temporary flight restrictions (TFRs) or other essential NOTAMs.

7-42   PA.I.C.K1

When telephoning a weather briefing facility for preflight weather information, you should state

A – the aircraft identification or your name.

B – true airspeed.

C – fuel on board.

7-42. Answer A. GFDPP 7D, AWS

When calling Flight Service for a preflight weather briefing, identify yourself as a pilot and supply the following information: type of flight planned (VFR or IFR), aircraft tail number or your name, aircraft type, departure airport, route of flight, destination, flight altitude(s), estimated time of departure (ETD), and estimated time enroute (ETE).

7-43   PA.I.C.K2

To get a complete weather briefing for the planned flight, you should request

A – a general briefing.

B – an abbreviated briefing.

C – a standard briefing.

7-43. Answer C. GFDPP 7D, AIM

To get a complete weather briefing for the planned flight, request a standard briefing, which assumes you do not already have a detailed comprehensive weather picture for your flight.

SECTION D ■ Sources of Weather Information

7-44   PA.I.C.K2

Which type of weather briefing should you request, when departing within the hour, if no preliminary weather information has been received?

A – Outlook briefing.

B – Abbreviated briefing.

C – Standard briefing.

7-44. Answer C. GFDPP 7D, AIM

To get a complete weather briefing for the planned flight, request a standard briefing, which assumes you do not already have a detailed comprehensive weather picture for your flight.

7-45   PA.I.C.K2

Which type of weather briefing should you request to supplement mass disseminated data?

A – An outlook briefing.

B – A supplemental briefing.

C – An abbreviated briefing.

7-45. Answer C. GFDPP 7D, AIM

Obtain an abbreviated briefing when you need only one or two specific items or to update weather information from a previous briefing or other weather sources.

7-46   PA.I.C.K2

To update a previous weather briefing, a pilot should request

A – an abbreviated briefing.

B – a standard briefing.

C – an outlook briefing.

7-46. Answer A. GFDPP 7D, AIM

Obtain an abbreviated briefing when you need only one or two specific items or to update weather information from a previous briefing or other weather sources..

7-47   PA.I.C.K2

A weather briefing that is provided when the information requested is six or more hours before the proposed departure time is

A – an outlook briefing.

B – a forecast briefing.

C – a prognostic briefing.

7-47. Answer A. GFDPP 7D, AIM

An outlook briefing is appropriate for flights with proposed departure times that are at least six hours in the future.

7-48   PA.I.C.K2

When requesting weather information for the following morning, you should request

A – an outlook briefing.

B – a standard briefing.

C – an abbreviated briefing.

7-48. Answer A. GFDPP 7D, AIM

Assuming your flight is six or more hours away, you would request an outlook briefing.

7-49　PA.I.C.K1

You plan to phone a weather briefing facility for preflight weather information. You should

A – provide the number of occupants on board.

B – identify yourself as a pilot.

C – begin with your route of flight.

7-49. Answer B. GFDPP 7D, AWS

When calling Flight Service for a preflight weather briefing, identify yourself as a pilot and supply the following information: type of flight planned (VFR or IFR), aircraft tail number or your name, aircraft type, departure airport, route of flight, destination, flight altitude(s), estimated time of departure (ETD), and estimated time enroute (ETE).

7-50　PA.I.C.K1

When speaking to a Flight Service weather briefer, you should state

A – the pilot in command's full name and address.

B – a summary of your qualifications.

C – whether the flight is VFR or IFR.

7-50. Answer C. GFDPP 7D, AWS

Ensure the briefer knows whether you intend to fly VFR or IFR, so that the information can help you make an effective go/no-go decision.

7-51　PA.I.C.K1

Obtain an online Flight Service briefing at

A – Flightservice.gov and enter your pilot certificate number.

B – 1800wxbrief.com and create an account by using your pilot credentials.

C – AviationWeather.gov and select Standard Briefing.

7-51. Answer B. GFDPP 7D, AWS

The site for official Flight Service briefings is 1800wxbrief.com. This website is for pilots only, you must create an account before you can obtain an official briefing or file a flight plan. You can also obtain official NWS briefings at AviationWeather.gov, but that website is not hosted by Flight Service, and flightservice.gov is not a valid website.

7-52　PA.I.C.K2. K2d

What is a good source to obtain an overall snapshot of weather in addition to a weather briefing?

A – Utilizing your phones weather app.

B – Watching your local weather news station on tv.

C – Reviewing the surface prognostic (prog) chart on AviationWeather.gov.

7-52. Answer C. GFDPP 7D, AWS

Among other weather reporting tools, AviationWeather.gov features a surface prognostic (prog) chart that forecasts surface conditions every three hours. .

SECTION D ▪ Sources of Weather Information

7-53   PA.I.C.K4, PA.I.C.R2

What considerations apply when using a flight deck display of radar data obtained from FIS-B?

A – Echoes and terrain that are close to your aircraft can shield echoes that are farther away.

B – The radar echoes that are depicted on the flight deck display can sometimes be more than 20 minutes old.

C – The flight deck radar display lacks the accuracy and integrity of radar mosaics obtained from Flight Service.

7-53. Answer B. GFDPP 7D, AWS
FIS-B is an example of a data-link weather service you can receive if an airplane has ADS-B capability and the proper receiver. The main concern when using a FIS-B flight deck radar display is that the data can be up to 20 minutes older than the age shown on the display and significant changes can occur over that time.

7-54   PA.I.C.R2c

Weather advisory broadcasts, including center weather advisories and SIGMETS, are provided by

A – ARTCCs on all frequencies, except emergency, when any part of the area described is within 150 miles of the airspace under their jurisdiction.

B – Flight service on 122.2 MHz and adjacent VORs, when any part of the area described is within 200 miles of the airspace under their jurisdiction.

C – ARTCCs on all frequencies, except emergency, only when any part of the area described is within 50 miles of the airspace under their jurisdiction.

7-54. Answer A. GFDPP 7D, AWS
ARTCCs broadcast CWAs and other hazardous weather information once on all frequencies, except emergency, when any part of the area described is within the area described is within when any part of the area described is within 150 miles of the airspace under the ARTCC jurisdiction. In terminal areas, local control and approach control might limit these broadcasts to weather occurring within 50 miles of the airspace under their jurisdiction.

7-55   PA.I.C.R2c, PA.III.B.K4

Which information is broadcast by an ASOS?

A – AIRMETs and SIGMETs that affect an area within 50 miles of the transmission site.

B – variable cloud height, variable visibility, precipitation type, intensity, and beginning and ending times.

C – visibility, cloud and ceiling data, and instrument approach in use.

7-55. Answer B. GFDPP 7D, AWS
An automated surface observing system (ASOS) broadcasts the same elements as an AWOS, such as the altimeter setting, wind speed, direction and gusts, temperature, dew point, visibility, and cloud and ceiling data. In addition, ASOS measures and reports variable cloud height, variable visibility, and rapid pressure changes, as well as precipitation type, intensity, accumulation, and beginning and ending times.

7-56   PA.I.C.R2c

Which is true regarding flight deck weather information?

A – Data link weather is instantaneous up-to-date weather information

B – Airborne weather radar can detect water vapor, lightning, or wind shear.

C – NEXRAD radar images on a flight deck display are should not be used to navigate through a line of thunderstorms.

7-56. Answer C. GFDPP 7D, AWS

Data link weather services are often included with GPS and electronic flight bag (EFB) flight deck or tablet display systems. The weather information displayed using a data link is near real-time but should not be thought of as instantaneous, up-to-date information.

The NEXRAD radar image that you view on your cockpit display can be up to 5 minutes old. Even small-time differences between the age indicator and actual conditions can be critical to flight safety. At no time should you use the images as storm penetrating radar to navigate through a line of thunderstorms.

Airborne weather radar is prone to many of the same limitations as ground-based systems—it cannot detect water vapor, lightning, or wind shear.

SECTION D ■ **Sources of Weather Information**

SECTION D ■ **Sources of Weather Information**

CHAPTER 8

Airplane Performance

SECTION A
Predicting Performance

FACTORS AFFECTING AIRCRAFT PERFORMANCE

- When altitude is corrected for nonstandard pressure, the result is pressure altitude.

- Density altitude is the term for pressure altitude that has been corrected for nonstandard temperature.

- Density altitude and pressure altitude are the same value at standard temperature.

- If the outside air temperature at an airport is warmer than standard, the density altitude is higher than pressure altitude and as a result, your airplane will perform as though the airport were at the higher elevation.

- When the air is less dense, aircraft performance decreases. For example:

  ◦ The wing must move through the air faster to develop enough lift for takeoff, resulting in a longer takeoff roll.

  ◦ Engine power is decreased because the engine must take in a larger volume of air to get enough air molecules for combustion.

  ◦ Propeller efficiency is decreased because a smaller mass of air is accelerated by the propeller, which reduces the force exerted.

- High temperature decreases the density of the air, which increases density altitude and reduces aircraft takeoff, climb, and landing performance.

- High relative humidity decreases the density of the air, which reduces aircraft takeoff, climb, and landing performance.

- (Refer to Figure 8.) To find the density altitude for given conditions, first find the pressure altitude, using the pressure altitude conversion factor scale and interpolating for the current pressure. Then, find the temperature on the OAT scale at the bottom of the graph and follow its line vertically to where it intersects the pressure altitude line. From this point, follow the horizontal density altitude line to the left scale to find an approximate density altitude.

TAKEOFF AND LANDING PERFORMANCE

- (Refer to Figure 36.) To find the headwind and crosswind components:

  1. Determine the difference between the runway heading and the wind direction.

  2. Find the intersection of the wind/runway angle line and the wind velocity arc.

- (Refer to Figure 36.) To find the velocity of the maximum crosswind component for an aircraft:

  1. Begin with the crosswind component at the bottom of the graph.

  2. Follow the line up to where it intersects the wind/runway angle line. Then, read the wind velocity.

- (Refer to Figure 37.) To determine landing distance:

  1. Start at the bottom left side of the graph. Find the OAT and follow the line up to the corresponding pressure altitude.

  2. Move right to the reference line and parallel the diagonal guide line downward to intersect the weight line.

  3. Move straight across to the next reference line, and parallel the diagonal headwind or tailwind guide line to intersect the wind component line.

  4. Move straight across to the next reference line and parallel the diagonal obstacle height guide line up to the obstacle height given.

  5. Read the landing distance on the right side of the graph. If no wind or obstacle exists, move straight across the corresponding section to the next reference line.

- (Refer to Figure 38.) To determine the landing distance,:
    1. Find the table that corresponds to the temperature and pressure altitudes that most closely resemble the given conditions.
    2. To consider an obstacle, select the distance to clear an obstacle.
    3. Use the information in the Notes section to adjust the distance for a headwind, nonstandard temperature, and/or runway surface conditions.

- (Refer to Figure 40.) To determine takeoff distance:
    1. Start at the bottom left side of the graph. Find the OAT and follow the line up to the corresponding pressure altitude.
    2. Move right to the reference line and parallel the diagonal guide line downward to intersect the weight line.
    3. Move straight across to the next reference line, and parallel the diagonal headwind or tailwind guide line to intersect the wind component line.
    4. Move straight across to the next reference line and parallel the diagonal obstacle height guide line up to the obstacle height given.
    5. Read the takeoff distance on the right side of the graph. If no wind or obstacle exists, move straight across the corresponding section to the next reference line.

CLIMB PERFORMANCE

- Best rate-of-climb speed (V_Y) provides the greatest gain in altitude in the least amount of time. Use this speed during a normal takeoff to reach a safe maneuvering altitude quickly.

- Best angle-of-climb speed (V_X) provides the greatest altitude gain in the shortest horizontal distance. Use this speed to clear obstacles at the departure end of the runway.

- You can find the operating limitations for an aircraft in the current, FAA-approved flight manual, approved manual material, markings, placards, or a combination of these references.

CRUISE PERFORMANCE

- The International Standard Atmosphere (ISA) provides a common reference for temperature and pressure—15°C (59°F), and a barometric pressure of 29.92 in. Hg. (1013.2 mb) at sea level. You can calculate ISA values for various altitudes using standard lapse rates.

- (Refer to Figure 35.) To determine the true airspeed (TAS) under given conditions, find the pressure altitude on the left side of the table and then read the TAS in the correct column under the appropriate temperature (ISA -20°C/-36°F, standard temperature, or ISA +20°C/+36°F). If necessary, interpolate to determine the TAS if the pressure altitude is not listed.

- (Refer to Figure 35.) To determine fuel consumption under given conditions:
    1. Determine the fuel flow in gallons per hour—find the pressure altitude on the left side of the table and follow the row across to read fuel flow in the correct column under the appropriate temperature (ISA -20°C/-36°F, standard temperature, or ISA +20°C/+36°F).
    2. Find the time enroute by dividing the distance by the TAS. Multiply the time by the fuel flow.

- (Refer to Figure 35.) To determine the manifold pressure (MP) setting, find the pressure altitude on the left side of the table and then read the MP in the correct column under the appropriate temperature (ISA -20°C/-36°F, standard temperature, or ISA +20°C/+36°F). Note: all RPM values are the same.

SECTION A ■ Predicting Performance

NOTE: An asterisk appearing after an ACS code (i.e. PA.II.F.R3*) indicates that the question subject appears more than one time in the ACS. The code shown corresponds to the first instance of the subject in the ACS.

8-1   PA.I.F.K2a, PA.IV.A.K1*

What effect does high density altitude, as compared to low density altitude, have on propeller efficiency and why?

A – Efficiency is increased due to less friction on the propeller blades.

B – Efficiency is reduced because the propeller exerts less force at high density altitudes than at low density altitudes.

C – Efficiency is reduced due to the increased force of the propeller in the thinner air.

8-1. Answer B. GFDPP 8A, PHB
High temperature decreases the density of the air, which increases density altitude and reduces propeller efficiency—a smaller mass of air is accelerated by the propeller, which reduces the force exerted.

8-2   PA.I.F.K2a, PA.IV.A.K1*

Which combination of atmospheric conditions reduces aircraft takeoff and climb performance?

A – Low temperature, low relative humidity, and low density altitude.

B – High temperature, low relative humidity, and low density altitude.

C – High temperature, high relative humidity, and high density altitude.

8-2. Answer C. GFDPP 8A, PHB
High temperature increases density altitude with a resulting decrease in aircraft performance. In addition, high relative humidity decreases the density of the air, which reduces aircraft takeoff, climb, and landing performance.

8-3   PA.I.F.K2a, PA.IV.A.K1*

What effect does high density altitude have on aircraft performance?

A – It increases engine performance.

B – It reduces climb performance.

C – It increases takeoff performance.

8-3. Answer B. GFDPP 8A, PHB
High temperature decreases the density of the air, which increases density altitude and reduces aircraft takeoff, climb, and landing performance. A high density altitude *decreases* engine performance.

8-4   PA.I.F.K2a, PA.IV.A.K1*, PA.IV.B.K2*

What effect, if any, does high humidity have on aircraft performance?

A – It increases performance.

B – It decreases performance.

C – It has no effect on performance.

8-4. Answer B. GFDPP 8A, PHB
High relative humidity decreases the density of the air, which reduces aircraft performance. For example, the wing must move through the air faster to develop enough lift for takeoff, resulting in a longer takeoff roll. Engine power is decreased because the engine must take in a larger volume of air to get enough air molecules for combustion; and propeller efficiency is decreased.

8-5   PA.I.F.K1, K2a

(Refer to Figure 8.) What is the effect of a temperature increase from 25°F to 50°F on the density altitude if the pressure altitude remains at 5,000 feet?

A – 1,200-foot increase.

B – 1,400-foot increase.

C – 1,650-foot increase.

8-5. Answer C. GFDPP 8A, PHB

Follow the line above 25°F up to where it intersects 5,000 feet pressure altitude and identify 3,750 feet density altitude on the left scale. Do the same with 50°F, up to 5,000 feet, and then left to read 5,400 feet. The difference is an increase of 1,650 feet.

8-6   PA.I.F.K1, K2a

(Refer to Figure 8.) Determine the pressure altitude with an indicated altitude of 1,380 feet MSL with an altimeter setting of 28.22 at standard temperature.

A – 3,010 feet.

B – 2,991 feet.

C – 2,913 feet.

8-6. Answer B. GFDPP 8A, PHB

Using the table on the right side of the chart, interpolate between 28.2 and 28.3, to find the conversion factor for the altimeter setting of 28.22, using the pressure altitude conversion factors of 1,630 and 1,533. The difference between 1,630 and 1,533 is 97. Multiply 97 times 20 percent or 0.20 (20 percent of the way between 1,630 and 1,533), which provides a product of 19.4. Subtract 19 from 1,630. The difference is 1611. Add 1,611 to the indicated altitude of 1,380 for a pressure altitude of 2,991 feet.

8-7   PA.I.F.K1, K2a

(Refer to Figure 8.) Determine the density altitude for these conditions:

Altimeter setting...29.25

Runway temperature...+81°F

Airport elevation...5,250 ft MSL

A – 4,600 feet MSL.

B – 5,877 feet MSL.

C – 8,500 feet MSL.

8-7. Answer C. GFDPP 8A, PHB

First find the pressure altitude by using the pressure altitude conversion factor scale and interpolate for 29.25. The conversion factor is 626 feet; (673 − 579) ÷ 2 + 579 = 626. Add 626 feet to 5,250 feet to find a pressure altitude of 5,876 feet. Next, find 81°F on the OAT scale at the bottom of the graph and follow its line vertically to where it intersects with the 5,876-ft pressure altitude line. From this point, follow the horizontal density altitude line to the left scale to find an approximate density altitude of 8,500 feet.

8-8   PA.I.F.K1, K2a

(Refer to Figure 8.) Determine the pressure altitude at an airport that is 3,563 feet MSL with an altimeter setting of 29.96.

A – 3,527 feet MSL.

B – 3,556 feet MSL.

C – 3,639 feet MSL.

8-8. Answer A. GFDPP 8A, PHB

First, interpolate between the two altimeter settings of 29.92 and 30.00, to find the pressure altitude conversion factor for the airport elevation. -73 - 0 = -73; -73 ÷ 2 = -36.5; subtract the conversion factor of -36.5 from the field elevation of 3,563 feet, to find the pressure altitude of 3,526.5 feet, and round up to 3,527 feet.

SECTION A ■ **Predicting Performance**

8-9   PA.I.F.K1, K2a

(Refer to Figure 8.) What is the effect of a temperature increase from 35°F to 50°F on the density altitude if the pressure altitude remains at 3,000 feet MSL?

A – 1,000-foot increase.

B – 1,100-foot decrease.

C – 1,300-foot increase.

8-9. Answer A. GFDPP 8C, PHB

An increase in temperature increases density altitude (DA). Find the DA for 35°F—about 1,900 feet. At 50°F, the DA is about 2,900 feet, an increase of 1,000 feet.

8-10   PA.I.F.K1, K2a

(Refer to Figure 8.) Determine the pressure altitude at an airport that has a field elevation of 1,386 feet MSL, and an altimeter setting of 29.97.

A – 1,341 feet.

B – 1,451 feet.

C – 1,562 feet.

8-10. Answer A. GFDPP 8C, PHB

First you must interpolate to find the conversion factor for 29.97 (−73 − 0 = −73 ÷ 8 increments = −9 × 5 increments = −45). Subtract 45 from 1,386 to find the pressure altitude of 1,341 feet.

8-11   PA.I.F.K1, K2a

(Refer to Figure 8.) What is the effect of a temperature decrease and a pressure altitude increase on the density altitude from 90°F and 1,250 feet pressure altitude to 55°F and 1,750 feet pressure altitude?

A – 1,750-foot increase.

B – 1,350-foot decrease.

C – 1,750-foot decrease.

8-11. Answer C. GFDPP 8C, PHB

1. Enter the graph at 90°F on the bottom scale. Draw a line straight up to meet the upsloping 1,250-ft pressure altitude line (visualize this line or draw it in between the 1,000-ft and 2,000-ft lines), then go left to 3,600 feet on the density altitude scale.

2. Repeat for 55°F and 1,750-ft pressure altitude to get a density altitude of 1,850 feet.

3. The difference is (1,850 − 3,600) feet = −1,750 feet.

8-12   PA.I.F.K2c, PA.IV.A.K3*

Which is true about takeoff performance?

A – The use of flaps during takeoff typically increases takeoff distance.

B – The takeoff distance determined on performance charts is based on a specific airplane configuration.

C – Airplane weight is not a factor in determining takeoff performance.

8-12. Answer B. GFDPP 8A, PHB

Takeoff performance charts, graphs, and tables typically indicate a specific airplane configuration. For example, takeoff flaps might be indicated. Partially extended flaps on takeoff—based on the manufacturer's recommendation—provide greater lift at low speeds and can decrease the takeoff distance.

Takeoff distance increases with an increase in weight. You can readily see the effect of weight when using takeoff performance charts.

8-13   PA.IV.M.K4

In a forward slip, the power is normally

A – reduced to idle and the airplane is pitched down with down elevator.

B – reduced to idle and the airplane is pitched up with up elevator.

C – increased and airplane is pitched up with up elevator.

8-13. Answer A. GFDPP 8A, AFH

A forward slip is used to dissipate altitude and increase descent rate without increasing airspeed. A forward slip is a maneuver that is used to lose altitude. Power is normally reduced to idle, and you control airspeed using elevator control.

8-14   PA.I.F.K1, K2a

(Refer to Figure 36.) What is the headwind component for a landing on Runway 18 if the tower reports the wind as 220° at 30 knots?

A – 19 knots.

B – 23 knots.

C – 26 knots.

8-14. Answer B. GFDPP 8A, PHB

First, compute the difference between the runway (180°) and the wind (220°). The result is an angle of 40°. Find the intersection of the 40° angle line and the 30 knot wind velocity arc, then read across to the left side to find the headwind component of 23 knots.

8-15   PA.I.F.K1, K2a

(Refer to Figure 36.) Determine the maximum wind velocity for a 45° crosswind if the maximum crosswind component for the airplane is 25 knots.

A – 25 knots.

B – 29 knots.

C – 35 knots.

8-15. Answer C. GFDPP 8A, PHB

Start with the crosswind component of 25 knots at the bottom of the graph and follow the line straight up to where it intersects the 45° angle line. This intersection is midway between the 30 and 40 knot wind velocity lines, or 35 knots.

8-16   PA.I.F.K1, K2a

(Refer to Figure 36.) What is the maximum wind velocity for a 30° crosswind if the maximum crosswind component for the airplane is 12 knots?

A – 16 knots.

B – 20 knots.

C – 24 knots.

8-16. Answer C. GFDPP 8A, PHB

Start with the crosswind component of 12 knots at the bottom of the graph and follow the line straight up to where it intersects the 30° angle line. This intersection is approximately 24 knots on the wind velocity scale.

SECTION A ■ **Predicting Performance**

8-17   PA.I.F.K1, K2a

(Refer to Figure 36.) With a reported wind of north at 20 knots, which runway (6, 29, or 32) is acceptable for use for an airplane with a 13-knot maximum crosswind component?

A – Runway 6.

B – Runway 29.

C – Runway 32.

8-17. Answer C. GFDPP 8A, PHB
At first glance, Runway 32 is most closely aligned with north (360°). To verify, find the crosswind component for each runway. Runway 32 is 40° from the wind, and because the wind speed is 20 knots, the crosswind component is slightly less than 13 knots, so Runway 32 is acceptable. Runway 6 is 60 degrees from the wind, and the crosswind component is about 17.5 knots. Runway 29 is 70 degrees from the wind, and the crosswind component is about 19 knots. Both Runways 6 and 29 exceed the 13 knot maximum crosswind component.

8-18   PA.I.F.K1, K2a

(Refer to Figure 36.) With a reported wind of south at 20 knots, which runway (10, 14, or 24) is appropriate for an airplane with a 13-knot maximum crosswind component?

A – Runway 10.

B – Runway 14.

C – Runway 24.

8-18. Answer B. GFDPP 8A, PHB
Runway 14 is most closely aligned with the wind and would have the least crosswind. The crosswind angle and component for each runway is: Runway 14, 40° at 12.5 knots; Runway 10, 80° at 19.7 knots; Runway 24, 60° at 17.5 knots. Runway 14 is the only appropriate runway because the crosswind component is less than 13 knots.

8-19   PA.I.F.K1, K2a

(Refer to Figure 36.) What is the crosswind component for a landing on Runway 18 if the tower reports the wind as 220° at 30 knots?

A – 19 knots.

B – 23 knots.

C – 30 knots.

8-19. Answer A. GFDPP 8A, PHB
The crosswind angle is 40° (220° − 180° = 40°). Find the intersection of 40° and 30 knots. Then read down to find the crosswind component of approximately 19 knots.

8-20   PA.I.F.K1, K2a, PA.IV.K1*

(Refer to Figure 40.) Determine the approximate ground roll distance required for takeoff.

OAT...32°C

Pressure altitude...2,000 ft

Takeoff weight...2,500 lb

Headwind component...20 knots

A – 650 feet.

B – 800 feet.

C – 1,000 feet.

8-20. Answer A. GFDPP 8A, PHB
Start at 32°C and move up to the 2,000 foot pressure altitude line, then right to the reference line. Follow the guide line diagonally down to the 2,500 pound line. Move across to the next reference line, and follow the headwind guide line diagonally down to 20 knots. Because no obstacle exists, move straight across to the right-hand border. The ground roll is approximately 650 feet.

8-21   PA.I.F.K1, K2a, PA.IV.K1

(Refer to Figure 40.) Determine the total distance required for takeoff to clear a 50-foot obstacle.

OAT...Std

Pressure altitude...4,000 ft

Takeoff weight...2,800 lb

Headwind component...Calm

A – 1,500 feet.

B – 1,750 feet.

C – 2,000 feet.

8-21. Answer B. GFDPP 8A, PHB

Because the temperature is standard, start at the intersection of the ISA and 4,000 foot pressure altitude line. Move right to the reference line and follow the guide line diagonally downward to the 2,800 pound line. Because the winds are calm, move straight across to the obstacle height reference line. Follow the guide line upward to the 50 foot line, which is on the right-hand border. The takeoff distance is approximately 1,700 feet.

8-22   PA.I.F.K1, K2a, PA.IV.K1

(Refer to Figure 40.) Determine the total distance required for takeoff to clear a 50-foot obstacle.

OAT...Std

Pressure altitude...Sea level

Takeoff weight...2,700 lb

Headwind component...Calm

A – 1,000 feet.

B – 1,400 feet.

C – 1,700 feet.

8-22. Answer B. GFDPP 8A, PHB

Because the temperature is standard, start at the intersection of the ISA and the seal level (S.L.) pressure altitude line. Move right to the reference line and follow the guide line diagonally downward to the 2,700 pound line. Because the winds are calm, move straight across to the obstacle height reference line. Follow the guide line upward to the 50 foot line, which is on the right-hand border. The takeoff distance is approximately 1,400 feet.

8-23   PA.I.F.K1, K2a, PA.IV.K1

(Refer to Figure 40.) Determine the approximate ground roll distance required for takeoff.

OAT...38°C

Pressure altitude...2,000 ft

Takeoff weight...2,750 lb

Headwind component...Calm

A – 1,150 feet.

B – 1,300 feet.

C – 1,800 feet.

8-23. Answer A. GFDPP 8A, PHB

Start at 38°C and move up to the 2,000 foot pressure altitude line, then right to the reference line. Follow the guide line diagonally downward to the 2,750 pound line. Because the winds are calm, and no obstacle exists, move straight across to the right-hand border. The ground roll is approximately 1,150 feet.

SECTION A ■ **Predicting Performance**

8-24   PA.I.F.K1, K2a, PA.IV.B.K2*

(Refer to Figure 37.) Determine the approximate total distance required to land over a 50-foot obstacle.

   OAT...90°F

   Pressure altitude...4,000 ft

   Weight...2,800 lb

   Headwind component...10 knots

A – 1,525 feet.

B – 1,775 feet.

C – 1,950 feet.

8-24. Answer B. GFDPP 8A, PHB

Start at the lower left at the 90°F (32°C) line and move up to where it intersects the 4,000-foot pressure altitude line. Go right to the weight reference line and then down and to the right to 2,800 lb. Go straight right to the wind component reference line, and then down and to the right to the 10-knot headwind line, and then straight right to the obstacle height reference line. Move up and to the right through the obstacle height and read the landing distance over a 50-foot obstacle on the right-hand scale, which is 1,775 feet.

8-25   PA.I.F.K1, K2a, PA.IV.B.K2*

(Refer to Figure 38.) Determine the approximate landing ground roll distance.

   Pressure altitude...Sea level

   Headwind...4 knots

   Temperature...Std

A – 356 feet.

B – 401 feet.

C – 490 feet.

8-25. Answer B. GFDPP 8A, PHB

Use the table listed under sea level and 59°F, which is the standard temperature. Because you need the landing ground roll distance, do not include obstacle clearance. The ground roll is given as 445, but according to Note 1, you must correct for headwind by decreasing the distance shown by 10 percent for each four knots of headwind. In this case, subtract 10 percent of 445 (44.5) from 445 = 401 feet.

8-26   PA.I.F.K1, K2a, K2d, PA.IV.B.K2*

(Refer to Figure 38.) Determine the total distance required to land over a 50-foot obstacle.

   Pressure altitude...7,500 ft

   Headwind...8 knots

   Temperature...32°F

   Runway...Hard surface

A – 1,004 feet.

B – 1,205 feet.

C – 1,506 feet.

8-26. Answer A. GFDPP 8A, PHB

According to the table, the landing distance over a 50-foot obstacle is 1,255 feet at 7,500 feet MSL and 32°F. Note 1 states, decrease the distances shown by 10 percent for each four knots of headwind; with 8 knots headwind, subtract 20 percent. 1,255 ft × 0.80 = 1,004 feet.

SECTION A ■ Predicting Performance

8-27　PA.I.F.K1, K2a, 2d, PA.IV.B.K2*

(Refer to Figure 38.) Determine the total distance required to land over a 50-foot obstacle.

　Pressure altitude...5,000 ft

　Headwind...8 knots

　Temperature...41°F

　Runway...Hard surface

A – 837 feet.

B – 956 feet.

C – 1,076 feet.

8-27. Answer B. GFDPP 8A, PHB
Use the table at 5,000 feet and 41°F. The distance to land over a 50-foot obstacle is 1,195 feet. According to Note 1, decrease the distances shown by 10 percent for each 4 knots of headwind. For the 8 knots of headwind, decrease the distance by 20 percent, which is 239 feet. Subtract 239 feet from 1,195 feet, which results in a total landing distance of 954 feet.

8-28　PA.I.F.K1, K2a, PA.IV.B.K2

(Refer to Figure 38.) Determine the approximate landing ground roll distance.

　Pressure altitude...5,000 ft

　Headwind...Calm

　Temperature...101°F

A – 445 feet.

B – 545 feet.

C – 495 feet.

8-28. Answer B. GFDPP 8A, PHB
At 5,000 feet and 41°F (ISA Standard Temperature), the ground roll distance is 495 feet. According to Note 2, this distance is increased 10% for every 60°F above standard. 495 x 0.10 = 49.5, 495 + 49.5 = 544.5, and round this up to 545, feet.

8-29　PA.I.F.K1, K2a, PA.IV.B.K2

(Refer to Figure 38.) Determine the total distance required to land over a 50-foot obstacle.

　Pressure altitude...3,750 ft

　Headwind...12 knots

　Temperature...Std

A – 794 feet.

B – 836 feet.

C – 816 feet.

8-29. Answer C. GFDPP 8A, PHB
1. At 2,500 feet MSL and standard ISA temperature, the landing distance over a 50-ft obstacle with zero wind is 1,135 feet. At 5,000 feet MSL, this distance is 1,195 feet. At 3,750 feet MSL, assume that the landing distance is half way between 1,135 and 1,195 feet. (1,135 + 1,195) feet ÷ 2 = 1,165 feet.

2. Note 1 says to decrease the distance 10% for every four knots of headwind. The headwind is 12 knots. 12 knots × 0.10 decrease/(4 knots) = 0.30 decrease. 1,165 feet × (1.0 – 0.30) = 816 feet.

SECTION A ■ **Predicting Performance**

8-30   PA.I.F.K1, K2a, PA.IV.B.K2

(Refer to Figure 38.) Determine the approximate landing ground roll distance.

Pressure altitude...1,250 ft

Headwind...8 knots

Temperature...Std

A – 275 feet.

B – 366 feet.

C – 470 feet.

8-30. Answer B. GFDPP 8A, PHB

This question requires that you interpolate between the ground roll distances at sea level and 2,500 feet pressure altitude. Because 1,250 feet is midway between the two values, the ground roll is 458-ft—(470 – 445) ÷ 2 + 445). To correct for headwind, subtract 20% of the distance (10% for every 4 knots). 20% of 457.5 is 91.5. The landing distance is 457.5 – 91.5 = 366 feet.

8-31   PA.IX.B.K1, K2c, PA.I.F.K2b, PA.IV.B.K2*

If an emergency situation requires a downwind landing, pilots should expect a faster

A – airspeed at touchdown, a longer ground roll, and better control throughout the landing roll.

B – groundspeed at touchdown, a longer ground roll, and the likelihood of overshooting the desired touchdown point.

C – groundspeed at touchdown, a shorter ground roll, and the likelihood of undershooting the desired touchdown point.

8-31. Answer B. GFDPP 8A, PHB

Consider the direction of the wind for all landings. A headwind lowers the groundspeed at touchdown, resulting in a shorter ground roll than without any wind. Landing with a tailwind increases the groundspeed at touchdown, resulting in a longer ground roll. In addition, the higher groundspeed produced by a tailwind results in the aircraft traveling further in the roundout and flare, which can result in overshooting the desired touchdown point.

8-32   PA.IV.A.K2*

Which would provide the greatest gain in altitude in the shortest distance during climb after takeoff?

A – V_Y.

B – V_A.

C – V_X.

8-32. Answer C. GFDPP 8A, AFH

V_X is the best angle-of-climb speed, which provides the greatest gain in altitude in the shortest horizontal distance. V_Y is the best rate-of-climb speed, which provides the greatest altitude gain in the shortest time. V_A is maneuvering speed, which is the maximum speed at which you may apply full and abrupt control movement without overstressing the airframe and causing structural damage.

8-33   PA.IV.A.K2*

After takeoff, which airspeed would the pilot use to gain the most altitude in a given time?

A – V_Y.

B – V_X.

C – V_A.

8-33. Answer A. GFDPP 8A, AFH, PHB

V_Y is the best rate-of-climb speed, which provides the greatest gain in altitude in the shortest period of time. V_X is the best angle-of-climb speed, which provides the greatest gain in altitude in the shortest horizontal distance. V_A is maneuvering speed, which is the maximum speed at which you may apply full and abrupt control movement without overstressing the airframe and causing structural damage.

8-34   PA.I.F.K1, K2a

(Refer to Figure 35.) Approximately what true airspeed should a pilot expect with 65 percent maximum continuous power at 9,500 feet with a temperature of 36°F below standard?

A – 158 knots.

B – 161 knots.

C – 163 knots.

8-34. Answer A. GFDPP 8A, PHB

Use the left-hand portion of the table, under ISA –36°F. Interpolate between the TAS values for 8,000 feet (157 knots) and 10,000 feet (160 knots). The closest answer is 158 knots.

8-35   PA.I.F.K1, K2a

(Refer to Figure 35.) What is the expected fuel consumption for a 1,000-nautical mile flight under the following conditions?

Pressure altitude...8,000 ft

Temperature...22°C

Manifold pressure...20.8 inches Hg.

Wind...Calm

A – 60.2 gallons.

B – 70.1 gallons.

C – 73.2 gallons.

8-35. Answer B. GFDPP 8A, PHB

Find the pressure altitude (8,000 feet) on the left side of the table and then follow this row to the right-hand column (ISA + 20°C/+36°F) to read the fuel flow of 11.5 GPH, and TAS of 164 knots (use knots because the distance is in nautical miles). Now, find the time enroute by dividing 1,000 NM by 164 knots (normally you would use groundspeed, but with a calm wind, TAS equals groundspeed.) The time enroute is approximately 6.1 hours. Multiply 6.1 hours by the fuel flow rate of 11.5 GPH. The total fuel consumption is 70.15 gallons, or 70.2 gallons.

SECTION A ■ **Predicting Performance**

8-36   PA.I.F.K1, K2a

(Refer to Figure 35.) What fuel flow should a pilot expect at 11,000 feet on a standard day with 65 percent maximum continuous power?

A – 10.6 gallons per hour.

B – 11.2 gallons per hour.

C – 11.8 gallons per hour.

8-36. Answer B. GFDPP 8A, PHB

Use the center portion of the table for a standard day. You can interpolate to find the fuel flow for 11,000 feet, which is halfway between 12,000 feet and 10,000 feet.

$(11.5 - 10.9)/2 + 10.9$

$= (0.6 \div 2) + 10.9$

$= 0.3 + 10.9$

$= 11.2$ gallons per hour

8-37   PA.I.F.K1, K2a

(Refer to Figure 35.) Determine the approximate manifold pressure setting with 2,450 RPM to achieve 65 percent maximum continuous power at 6,500 feet with a temperature of 36°F higher than standard.

A – 19.8 inches Hg.

B – 20.8 inches Hg.

C – 21.0 inches Hg.

8-37. Answer C. GFDPP 8A, PHB

The RPM is the same for all altitudes. Therefore, to determine what manifold pressure (MP) is required to achieve 65% maximum continuous power. Under ISA + 36°F, the MP for 6,000 feet is 21.0", and for 8,000 feet it is 20.8". To interpolate the MP for 6,500 feet, 21.0 - 20.8 = 0.2; half of 0.2 is 0.1. Subtract 0.1 from 21.0 = 20.9, and round this up to 21.0 in Hg for the approximate manifold pressure.

NOTE: *The following questions refer to the effect of wind when performing approaches and landings and ground reference maneuvers. Although Chapter 8, Section A of the GFD Private Pilot textbook addresses content regarding the effects of wind on the airplane, Private Pilot Maneuvers, Chapter 3 — Airport Operations covers crosswind approaches and landings and Chapter 6 — Ground Reference Maneuvers covers how to perform the rectangular course, S-turns, and turns around a point.*

8-38   PA.I.F.K2c, PA.IV.E.K3

When performing a short-field takeoff, you should operate the airplane to obtain maximum takeoff performance by

A – using full power, full flaps and accelerating to V_Y.

B – consulting and following the performance section of the AFM/POH to obtain the power setting, flap setting, airspeed, and procedures prescribed by the manufacturer.

C – rotating at V_X and then climbing to clear the 50-foot obstacle at V_Y.

8-38. Answer B. GFDPP 8A, AFH

When performing takeoffs and climbs from fields where the takeoff area is short or the available takeoff area is restricted by obstructions, you should consult and follow the performance section of the AFM/POH to obtain the power setting, flap setting, airspeed, and procedures prescribed by the airplane's manufacturer. You should have adequate knowledge in the use and effectiveness of the best angle-of-climb speed (V_X) and the best rate-of climb speed (V_Y) for the specific make and model of airplane being flown to safely accomplish a takeoff at maximum performance.

8-39   PA.I.F.K2c, PA.IV.B.K3*

Which is a proper technique to perform an approach and landing in a crosswind?

A – After correcting for drift during final approach, neutralize the ailerons and rudder in the landing flare and during rollout.

B – Correct for drift by lowering the upwind wing and apply opposite rudder to keep the longitudinal axis aligned with the runway.

C – Correct for drift by lowering the downwind wing and apply opposite rudder to keep the longitudinal axis aligned with the runway.

8-39. Answer B. GFDPP 8A, AFH

To perform a crosswind approach and landing, on final approach, align the airplane's heading with the runway centerline and then correct sideways drift by lowering the upwind wing. Apply opposite rudder to prevent the airplane from turning and to keep the longitudinal axis aligned with the runway.

As airspeed decreases in the flare, the flight controls become less effective and you must gradually increase the rudder and aileron deflection to maintain the proper amount of wind correction. Rollout maintaining the crosswind correction.

8-40   PA.IV.F.K2*, PA.IV.F.K3, PA.IV.N.K3

For a crosswind from the right during a short-field landing,

A – a sideslip with the right wing low and left rudder as needed will help to maintain the desired ground track for the airplane and keep the longitudinal axis aligned with the runway centerline.

B – a sideslip with the left wing low and right rudder as needed will help to maintain the desired ground track for the airplane and keep the longitudinal axis aligned with the runway centerline.

C – a forward slip with left wing low and right rudder as needed will help to maintain the desired ground track for the airplane and keep the longitudinal axis aligned with the runway centerline.

8-40. Answer A. GFDPP 8A, AFH

While the wing-low (sideslip) method also compensates for a crosswind from any angle, it keeps the airplane's ground track and longitudinal axis aligned with the runway centerline throughout the final approach, round out, touchdown, and after-landing roll. This prevents the airplane from touching down in a sideward motion and imposing damaging side loads on the landing gear.

SECTION A ■ **Predicting Performance**

8-41 PA.V.B.K1, K2, K4

(Refer to Figure 62.) In flying the rectangular course, when would the aircraft be turned less than 90°?

A – Corners 1 and 4.

B – Corners 1 and 2.

C – Corners 2 and 4.

8-41. Answer A. GFDPP 8A, GFDPPM, AFH

To maintain the rectangular ground track at turn 1, the aircraft is turned less than 90° to establish a right crab into the wind. Approaching turn 4, the aircraft is crabbed left into the wind and is therefore turned less than 90° into the direct headwind. Performing a rectangular course helps you develop the skill to compensate for effects of the wind and prepare for flying the airport traffic pattern.

8-42 PA.V.B.K1, K2, K3

(Refer to Figure 66.) While practicing S-turns, a consistently smaller half-circle is made on one side of the road than on the other, and this turn is not completed before crossing the road or reference line. This would most likely occur in turn

A – 1-2-3 because the bank is decreased too rapidly during the latter part of the turn.

B – 4-5-6 because the bank is increased too rapidly during the early part of the turn.

C – 4-5-6 because the bank is increased too slowly during the latter part of the turn.

8-42. Answer B. GFDPP 8A, GFDPPM, AFH

Turn 4-5-6 begins with an upwind leg (traveling into a headwind), resulting in a slower groundspeed. The slower groundspeed requires you to increase the bank more slowly than on the downwind leg, so the aircraft has more time to travel equidistant from the "road" as it did on the other side, helping to create semicircles with the same radii. Performing S-turns helps you improve your ability to compensate for wind drift during turns.

SECTION B
Weight and Balance

WEIGHT AND BALANCE TERMS

- The standard empty weight of an airplane includes hydraulic fluid, unusable fuel, and full oil.

- The basic empty weight is the standard empty weight plus the weight of optional equipment installed. This weight typically serves as the starting for weight and balance calculations.

- The standard weight of gasoline is 6 pounds per gallon. To determine the amount of fuel to drain, if necessary, divide the excess weight by six.

PRINCIPLES OF WEIGHT AND BALANCE

- The center of gravity (CG) is the total moment divided by the total weight.

- Datum is a vertical plane in the aircraft from which weight and balance distances are measured.

- Arm is the distance from datum of a particular station, or place in the aircraft.

- To calculate aircraft moment, multiply the weight at a station by the arm. Positive CG values are aft of datum, negative CG values are forward of the datum.

DETERMINING TOTAL WEIGHT AND CENTER OF GRAVITY

Several methods can aid you in calculating the center of gravity and total weight of your aircraft, and the change in CG with a shift in weight.

TABLE METHOD

- The best way to determine aircraft weight and balance is to construct a table, or spreadsheet, that lists the stations of the aircraft, the weight at each station, and the arm of each station. From here, you can find the moment at each station, and add up the total weight and moments. The CG is the total moment divided by the total weight.

- (Refer to Figures 32 and 33.) To find the arm at each station, look for the station, such as usable fuel, on the table and read the arm listed at the top. Many graphs calculate moments for a specific weight range, so that you can simply read these values off the table as well.

- (Refer to Figure 34.) Other graphs provide arm and moment information graphically. To read the moment from the graph, find the line that corresponds to the station, and follow it to the weight at that station. To find the moment, move down to the bottom of the graph.

WEIGHT SHIFT FORMULA

Use the weight shift formula to determine how far the center of gravity shifts when weight is added to or removed from the aircraft:

- Weight Moved ÷ Weight of Airplane = Distance CG Moves ÷ Distance between Arms

- Some weight shift questions require you to construct a table of weights and moments first.

NOTE: *An asterisk appearing after an ACS code (i.e. PA.VII.B.K1*) indicates that the question subject appears more than one time in the ACS. The code shown corresponds to the first instance of the subject in the ACS.*

SECTION B ■ **Weight and Balance**

8-43   PA.I.F.K2f

Which items are included in the basic empty weight of an aircraft?

A – Hydraulic fluid, unusable fuel, full engine oil, and optional equipment.

B – Only the airframe, powerplant, and optional equipment.

C – Full fuel tanks and engine oil to capacity.

8-43. Answer A. GFDPP 8B, PHB

The standard empty weight is the weight of an empty airplane, which includes unusable fuel, full operating fluids (like hydraulic fluids), and full engine oil. The basic empty weight is the standard empty weight plus the weight of optional equipment installed. It serves as the starting point for weight and balance calculations.

8-44   PA.I.F.K2e, K2f

An aircraft is loaded 110 pounds over maximum certificated gross weight. If fuel (gasoline) is drained to bring the aircraft weight within limits, how much fuel should be drained?

A – 15.7 gallons.

B – 16.2 gallons.

C – 18.4 gallons.

8-44. Answer C. GFDPP 8B, PHB

This question requires converting the weight of fuel to gallons. Divide 110 pounds by 6 pounds per gallon, the standard weight of gasoline = 18.33. Therefore, the closest answer is 18.4 gallons.

8-45   PA.I.F.K2e, K2f, PA.I.F.S1

GIVEN:

	WEIGHT (lb)	ARM (in)	MOMENT (lb-in)
Empty weight	1,495	101.4	151,593
Pilot and Pass	380	64.0	-----
Fuel (30 gal usable no reserve)	-----	96.0	-----

The center of gravity (CG) is located how far aft of datum?

A – 92.44 inches.

B – 94.01 inches.

C – 119.8 inches.

8-45. Answer B. GFDPP 8B, PHB

First, fill in the table by entering the fuel weight (30 gallons × 6 pounds/gallon = 180 pounds). Then, multiply each weight by the arm to find the moment. The CG is the total moment divided by the total weight = 193,193 pound-inches ÷ 2,055 lb = 94.01 inches.

8-46   PA.I.F.K2e, K2f, PA.I.F.S1

(Refer to Figures 32 and 33.) What is the maximum amount of baggage that can be carried when the airplane is loaded as follows?

Front seat occupants...387 lb

Rear seat occupants...293 lb

Fuel...35 gal

A – 45 pounds.

B – 63 pounds.

C – 220 pounds.

8-46. Answer A. GFDPP 8B, PHB

Add up all the weights to determine that the airplane is 45 pounds underweight. When adding the 45 pounds of baggage, verify that the resulting center of gravity (CG) is within limits.

8-47   PA.I.F.K2e, K2f, PA.I.F.S1

(Refer to Figures 32 and 33.) Determine if the airplane weight and balance is within limits.

Front seat occupants...415 lb

Rear seat occupants...110 lb

Fuel, main tanks...44 gal

Fuel, aux. tanks...19 gal

Baggage...32 lb

A – 19-pounds overweight, CG within limits.

B – 19-pounds overweight, CG out of limits forward.

C – Weight within limits, CG out of limits.

8-47. Answer C. GFDPP 8B, PHB

First, construct a weight and moment table.

	WEIGHT (lb)	ARM (in)	MOMENT (lb-in/100)
Empty weight	2,015		1,554.0
Front Seat	415	85	352.8
Rear Seat	110	121	133.1
Fuel 44 gal	264	75	198.0
Aux 19 gal	114	94	107.2
Baggage	32	140	44.8
Totals	**2,950**		**2,389.9**

The total weight is at the maximum limit. To find the center of gravity, divide the total moment by the total weight.

Center of Gravity = (2,389 lb-in × 100) ÷ 2,950 lb

The result is 81.0 inches, which is outside the limits.

SECTION B ■ **Weight and Balance**

8-48   PA.I.F.K2e, K2f, PA.I.F.S1

(Refer to Figure 34.) What is the maximum amount of baggage that may be loaded aboard the airplane for the CG to remain within the moment envelope?

	WEIGHT (lb)	MOM/1000
Empty weight	1,350	51.5
Pilot and Front seat passenger	250	-----
Rear seat passengers	400	-----
Baggage	-----	-----
Fuel, 30 gal	-----	-----
Oil, 8 qt	-----	-0.2

A – 105 pounds.

B – 110 pounds.

C – 120 pounds.

8-48. Answer A. GFDPP 8B, PHB

Use Figure 34 to convert oil and fuel to pounds. Add up the known weights, for a total of 2,195 pounds. Subtract 2,195 pounds from 2,300 max weight to find the maximum baggage weight of 105 pounds. This eliminates all answers but choice A, but you still should check the CG limits. Use the LOADING GRAPH and find the moment for each weight.

	WEIGHT (lb)	MOMENT (1,000 lb-in)
Empty Weight	1,350	51.5
Pilot and Front Seat passenger	250	9.4
Rear Seat passengers	400	29.3
Fuel 30 gal	180	8.7
Oil 8 qt	15	–0.2
Subtotal	2,195	98.7
Baggage	105	10.0
Totals	**2,300**	**108.7**

Total the moments and locate the maximum weight on the CENTER OF GRAVITY MOMENT ENVELOPE graph. The intersection of the loaded weight and moment is at the upper right-hand corner of the normal category envelope, and is barely within limits.

8-49   PA.I.F.K2e, K2f, PA.I.F.S1

(Refer to Figure 34.) Calculate the moment of the airplane and determine which category is applicable.

	WEIGHT (lb)	MOM/1000
Empty weight	1,350	51.5
Pilot and Front seat passenger	310	-----
Rear seat passengers	96	-----
Fuel, 38 gal	-----	-----
Oil, 8 qt	-----	-0.2

A – 79.2, utility category.

B – 80.8, utility category.

C – 81.2, normal category.

8-49. Answer B. GFDPP 8B, PHB

Complete the table of weights and moments, using the LOADING GRAPH:

	WEIGHT (lb)	MOMENT (lb-in/1000)
Empty Weight	1,350	51.5
Pilot and Front Seat passenger	310	11.6
Rear Seat passenger	96	7.0
Fuel 38 gal	228	11.0
Oil 8 qt	15	–0.2
Totals	**1,999**	**80.9**

The total moment/1,000 is 80.9 pound-inches. To find the total weight and total moment, use the CENTER OF GRAVITY MOMENT ENVELOPE graph. The intersection falls within the upper right-hand corner of the utility category envelope.

8-50   PA.I.F.K2e, K2f, PA.I.F.S1

(Refer to Figure 34.) If an airplane loaded as follows, what is the maximum amount of fuel that may be aboard on takeoff?

	WEIGHT (lb)	MOM/1000
Empty weight	1,350	51.5
Pilot and Front seat passenger	340	-----
Rear seat passengers	310	-----
Baggage	45	-----
Oil, 8 qt	-----	-----

A – 24 gallons.

B – 32 gallons.

C – 40 gallons.

8-50. Answer C. GFDPP 8B, PHB

Complete the table of weights and moments, using the LOADING GRAPH:

	WEIGHT (lb)	MOMENT (lb-in/1000)
Empty Weight	1,350	51.5
Front Seat passenger	340	12.7
Rear Seat passengers	310	22.6
Baggage	45	4.3
Oil 8 qt	15	–.2
Subtotal	2060	90.9
Fuel 40 gal	240	11.5
Totals	**2300**	**102.4**

The total weight without fuel is 2,060 pounds. This weight is 240 pounds below the maximum of 2,300 pounds. Dividing by 6 pounds/gallon, the maximum fuel load is 40 gallons. Check the moments as well. The total moment of 102.4 is within the center of gravity envelope, therefore 40 gallons is acceptable.

8-51   PA.I.F.K2e, K2f, PA.I.F.S1

(Refer to Figure 34.) Determine the moment with the following data:

	WEIGHT (lb)	MOM/1000
Empty weight	1,350	51.5
Pilot and front seat passenger	340	-----
Fuel (std. tanks)	Capacity	-----
Oil, 8 qt	-----	-----

A – 69.9 pound-inches.

B – 74.9 pound-inches.

C – 77.6 pound-inches.

8-51. Answer B. GFDPP 8B, PHB

To determine the moment for each item, use the LOADING GRAPH. To find the total moment of 74.9, add up the individual moments.

	WEIGHT (lb)	MOMENT (lb-in/1000)
Empty Weight	1,350	51.5
Pilot and front seat passenger	340	12.6
Fuel 38 gal	228	11.0
Oil 8 qt.	15	–0.2
Totals	**1,933**	**74.9**

SECTION B ■ Weight and Balance

8-52   PA.I.F.K2e, K2f, PA.I.F.S1

(Refer to Figure 34.) Determine the aircraft loaded moment and the aircraft category.

	WEIGHT (lb)	MOM/1000
Empty weight	1,350	51.5
Pilot and front seat passenger	380	-----
Fuel, 48 gal	288	-----
Oil, 8 qt	-----	-----

A – 78.2, normal category.

B – 79.2, normal category.

C – 80.4, utility category.

8-53   PA.I.F.K2e, K2f, PA.I.F.S1

(Refer to Figures 32 and 33.) Upon landing, the front passenger (180 pounds) departs the airplane. A rear passenger (204 pounds) moves to the front passenger position. What effect does this weight shift have on the CG if the airplane weighed 2,690 pounds and the MOM/100 was 2,260 before the passenger transfer?

A – The CG moves forward approximately 3 inches.

B – The weight changes, but the CG is not affected.

C – The CG moves forward approximately 0.1 inches.

8-52. Answer B. GFDPP 8B, PHB

Use the LOADING GRAPH to determine the moments:

	WEIGHT (lb)	MOMENT (lb-in/1000)
Empty Weight	1,350	51.5
Pilot and front seat passenger	380	14.2
Fuel 48 gal	288	13.7
Oil, 8 qt	15	–0.2
Totals	**2,033**	**79.2**

The total weight is 2,033 pounds, and the total moment is 79.2. Use the CENTER OF GRAVITY MOMENT ENVELOPE graph with the total weight and total moment. The intersection falls within the normal category, and outside the utility category.

8-53. Answer A. GFDPP 8B, PHB

First, calculate the effect of the front seat passenger disembarking. Use the table in Figure 32 to determine the moment of the departing 180-lb passenger.

- Weight: 2,690 – 180 = 2,510 lb; CG = 84.01 in.

- Moment: 2260 – 153 = 2107 (100 lb-in); CG = 83.94 in.

- The CG from the initial deplaning decreases 0.07 in.

- Then, determine the effect of the rear seat passenger moving to the front seat. Find the arms for the front and rear passenger seats in Figure 32; the difference is 121 – 85 = 36. Use the weight-shift formula to determine how far the CG shifts when the 204-lb passenger moves to the front seat.

- Weight Moved ÷ Weight of Airplane = Distance CG Moves ÷ Distance Between Arms.

- 204 ÷ 2,510 = Distance CG Moves ÷ 36 in.

- Distance CG Moves = 204 × 36 ÷ 2,510 = 2.93 in.

The additional change in CG is 2.93 inches, therefore the total change in CG is 3.0 inches forward.

8-54   PA.I.F.K2e, K2f, PA.I.F.S1

(Refer to Figures 32 and 33.) Which action can adjust the airplane's weight to maximum gross weight and the CG to within limits for takeoff?

    Front seat occupants...425 lb

    Rear seat occupants...300 lb

    Fuel, main tanks...44 gal

A – Drain 12 gallons of fuel.

B – Drain 9 gallons of fuel.

C – Transfer 12 gallons of fuel from the main tanks to the auxiliary tanks.

8-54. Answer B. GFDPP 8B, PHB

Complete the weight and moment table as shown in the following table:

	WEIGHT (lb)	ARM (in)	MOMENT (lb-in/100)
Empty Weight	2,015		1,554.0
Pilot and front seat passenger	425	85	361.3
Rear seat passengers	300	121	363.0
Fuel 44 gal	264	75	198.0
Total	3,004		2,476.3
Max Weight	−2,950		
	54		

The total weight of 3,004 pounds is 54 pounds over maximum weight and 54 pounds of fuel is 9 gallons (54 lb ÷ 6 lb/gal = 9 gallons). Now use Figure 32 to look up the new fuel moment for the 35 gallons that remain, and recalculate the total moment—2,436 (× 100) pounds-inches.

	WEIGHT (lb)	ARM (in)	MOMENT (lb-in/100)
Empty Weight	2,015		1,554.0
Front Seat	425	85	361.3
Rear Seat	300	121	363.0
Fuel 35 gal	210	75	157.5
Total	2,950		2,435.8

Using the table in Figure 33, determine that the total weight and total moment are within the limits.

8-55   PA.I.F.K2e, K2f, PA.I.F.S1

(Refer to Figures 32 and 33.) What effect does a 35-gallon fuel burn (main tanks) have on the weight and balance if the airplane weighed 2,890 pounds and the MOM/100 was 2,452 at takeoff?

A – Weight is reduced by 210 pounds and the CG is aft of limits.

B – Weight is reduced by 210 pounds and the CG is unaffected.

C – Weight is reduced to 2,680 pounds and the CG moves forward.

8-55. Answer A. GFDPP 8B, PHB

Use the chart in Figure 32 to find the weight and moment for 35 gallons of fuel in the main tanks, and subtract these values from the total weight and moment. The result is the total weight and moment after the fuel burn.

	WEIGHT (lb)	MOMENT (lb-in/100)
Total	2,890	2,452
Fuel 35 gal	−210	−158
Adjusted	2,680	2,294

Refer to the chart in Figure 33 for the weight of 2,680 pounds. The moment of 2,294 pounds-inches exceeds the maximum (aft) limit.

SECTION B ■ **Weight and Balance**

8-56　　PA.I.F.K2e, K2f, PA.I.F.S1

(Refer to Figures 33 and 34.) With the airplane loaded as follows, what action can be taken to balance the airplane?

　　Front seat occupants..411 lb

　　Rear seat occupants...100 lb

　　Main wing tanks...44 gal

A – Fill the auxiliary wing tanks.

B – Add a 100-pound weight to the baggage compartment.

C – Transfer 10 gallons of fuel from the main tanks to the auxiliary tanks.

8-56. Answer B. GFDPP 8B, PHB

Construct a table like the following. Find the subtotal weight and moment, and use the chart in Figure 34. The subtotal moment (2,222.4) at the original weight is less than the minimum (forward) limit.

	WEIGHT (lb)	ARM (in)	MOMENT (lb-in/100)
Empty Weight	2,015		1,554.0
Pilot and front seat passenger	411	85	349.4
Rear seat passenger	100	121	121.0
Fuel 44 gal	264	75	198.0
Subtotal	2,790		2,222.4
Baggage	100		140.0
Total	**2,890**		**2,362.4**

Because the baggage compartment is in an aft location, adding weight to this part of the airplane shifts the CG aft. To find the adjusted totals, add the baggage weight and moment to the subtotals. Check the chart in Figure 33 to ensure that the moment is within limits. Answer (A) is wrong because if the auxiliary wing tanks are filled and the total weight and moment are adjusted, the moment is less than the minimum. To check answer (C), find the original CG using the subtotals:

- CG = Total Moments ÷ Total Weight = 2,222.4 lb-in ÷ 2,790 lb = 79.7 in.

- Then use the weight shift formula:

- Weight Moved ÷ Weight of Airplane = Distance CG Moves ÷ Distance between Arms.

- The weight of fuel is 10 gal × 6 lb/gal = 60 lb. The distance between arms is 94 inches – 75 inches = 19 inches. Because the fuel is transferred from an arm of 75 inches to an arm of 94 inches, the CG moves aft 0.4 inches.

- Distance CG Moves = 60 lb × 19 in ÷ 2,790 lb = 0.4 in.

The new CG is 80.1 (79.7 + 0.4). Next, find the new moment on the chart in Figure 33. The new moment is less than the minimum.

8-57   PA.I.F.K2e, K2f, PA.I.F.S1

(Refer to Figure 61.) If 50 pounds of weight is located at point X and 100 pounds at point Z, how much weight must be located at point Y to balance the plank?

A – 30 pounds.

B – 50 pounds.

C – 300 pounds.

8-57. Answer C. GFDPP 8B, PHB

You must equalize the moments generated by the weights on each side of the fulcrum, then solve for the unknown weight as shown in the following calculations:

- (50 lb × 50 in) + (Y lb × 25 in) = (100 lb × 100 in).

- 2,500 lb-in + 25Y lb-in = 10,000 lb-in.

- 25Y lb-in = 7,500 lb-in.

- Y = 300 lb.

8-58   PA.I.F.K2e, K2f, PA.I.F.S1

(Refer to Figure 60.) How should the 500-pound weight be shifted to balance the plank on the fulcrum?

A – 1 inch to the left.

B – 1 inch to the right.

C – 4.5 inches to the right.

8-58. Answer A. GFDPP 8B, PHB

The moment of the 500-lb weight on the left side of the fulcrum must equal the sum of the moments of the 250-lb weight and the unequal weight of the plank on the right side.

- 500 lb × L in = (250 lb × 20 in) + (200 lb × 15 in).

- 500 lb × L in = 5,000 lb-in + 3,000 lb-in.

- 500 lb × L in = 8,000 lb-in.

- L = −16 inches.

Because the 500-lb weight is now sitting at −15 inches from the fulcrum, it must be moved 1 inch to the left.

SECTION B ■ **Weight and Balance**

SECTION C
Flight Computers

ELECTRONIC FLIGHT COMPUTERS

If you are using an electronic flight computer, you must thoroughly read the instructions and understand the features of your specific app or EFB. When you use a flight planning app or EFB, many of these individual calculations are performed for you after you create a route.

MECHANICAL FLIGHT COMPUTERS

The computer has two sides: the computer side, with several scales and small cutout windows, and the wind side, with a large transparent window and a sliding grid. The computer side is used for solving ratio-type problems such as time-speed-distance, fuel consumption, and various conversions. The wind side takes most of the guesswork out of otherwise difficult wind-drift calculations.

FLIGHT COMPUTER CALCULATIONS

With some questions, you will need to calculate the groundspeed and then the estimated time of arrival.

- Determine the groundspeed of the aircraft:

    ◦ Measure the distance between the departure point and the destination.

    ◦ Determine the elapsed time.

    ◦ Divide the distance by the time to derive the groundspeed.

- Determine the estimated time of arrival (ETA) at the destination by adding the elapsed time to the departure time.

Some other questions will require you to calculate groundspeed and then complete a time, speed, and distance problem.

- Measure the distance between the departure point and the destination.

- Determine the true course (TC).

- Determine the groundspeed using your flight computer: Enter the wind direction and velocity. Enter the TC. Enter the true airspeed (TAS) and find the groundspeed.

- Determine the time enroute—distance ÷ groundspeed = time.

- Add departure and climbout time (if any is given). Round as needed to arrive at one of the answer selections.

NOTE: *An asterisk appearing after an ACS code (i.e. PA.VII.B.K1*) indicates that the question subject appears more than one time in the ACS. The code shown corresponds to the first instance of the subject in the ACS.*

8-59   PA.I.D.K3a, K3b

(Refer to Figure 20.) Enroute to First Flight Airport (area 5), your flight passes over Hampton Roads Airport (area 2) at 1456 and then over Chesapeake Municipal at 1501. At what time should your flight arrive at First Flight Airport?

A – 1516.

B – 1521.

C – 1526.

8-59. Answer C. GFDPP 8C, PHB

NOTE: Use the scale at the top of the chart for distance because the chart provided might not be properly scaled to enable accurate measurements with a plotter.

This question requires you to calculate groundspeed and then estimated time of arrival.

1. Determine the actual groundspeed (GS) of the aircraft.

   a. Measure the distance between Hampton Roads Airport (PVG) and Chesapeake Municipal (CPK)—10 NM.

   b. Determine the elapsed time (15:01 – 14:56 = 5 min).

   c. Determine the GS (10 NM in 5 min = 2 NM/min × 60 min = 120 knots groundspeed).

2. Determine the estimated time of arrival (ETA) at First Flight Airport.

   a. Measure the distance between Chesapeake Municipal and First Flight Airport (50 NM).

   b. Determine the time enroute between the two points (50 NM at 120 knots = approximately 25 minutes).

   c. If the aircraft was over Chesapeake Municipal at 15:01, the ETA at First Flight Airport is about 15:26. (15:01 + 25 min = 15:26).

SECTION C ■ Flight Computers

8-60   PA.I.D.K3a, K3b

(Refer to Figure 21.) What is the estimated time enroute from Mercer County Regional Airport (area 3) to Minot International (area 1)? The wind is from 330° at 25 knots and the true airspeed is 100 knots. Add 3-1/2 minutes for departure and climbout.

A – 44 minutes.

B – 48 minutes.

C – 52 minutes.

8-60. Answer B. GFDPP 8C, PHB

NOTE: Use the scale at the top of the chart for distance because the chart provided might not be properly scaled to enable accurate measurements with a plotter.

This question requires you to calculate groundspeed and then complete a time-speed-distance problem.

1. Measure the distance between Mercer County Regional Airport and Minot International—59 NM.

2. Determine the True Course (TC)—012°.

3. Determine the groundspeed using your flight computer:

   a. Enter the wind direction and speed (330° True at 25 knots).

   b. Enter the TC—012°.

   c. Enter the true airspeed (TAS)—100 knots.

   d. GS = 80 knots.

4. Determine the time enroute using your flight computer: (59 NM at 80 NM/hr = 44 min 15 sec).

5. Add departure and climbout time: (3 min 30 sec + 44 min 15 sec = 47 min 45 sec).

Round up to 48 minutes.

8-61   PA.I.D.K3a

If a true heading of 135° results in a ground track of 130° and a true airspeed of 135 knots results in a groundspeed of 140 knots, the wind would be from

A – 019° and 12 knots.

B – 200° and 13 knots.

C – 246° and 13 knots.

8-61. Answer C. GFDPP 8C, PHB

Use a flight computer to solve for wind direction and velocity. This calculation is essentially the reverse of predicting groundspeed from forecast winds aloft.

8-62   PA.I.D.K3a

How far will an aircraft travel in 2-1/2 minutes with a groundspeed of 98 knots?

A – 2.45 NM.

B – 3.35 NM.

C – 4.08 NM.

8-62. Answer C. GFDPP 8C, PHB

You can solve this problem using a flight computer or mathematically. The basic formula for calculating time, speed, and distance is:

Distance = groundspeed × time.

Distance = 98 NM/hr × 2.5 min ÷ 60 min/hr.

8-63   PA.I.D.K3a

How far will an aircraft travel in 7.5 minutes with a groundspeed of 114 knots?

A – 14.25 NM.

B – 15.00 NM.

C – 14.50 NM.

8-63. Answer A. GFDPP 8C, PHB

You can solve this problem using a flight computer or mathematically. The basic formula for calculating time, speed, and distance is:

Distance = groundspeed × time.

Distance = 114 NM/hr × 7.5 min ÷ 60 min/hr.

8-64   PA.I.D.K3a, K3b

On a cross-country flight, point A is crossed at 1500 hours and the plan is to reach point B at 1530 hours. Use the following information to determine the indicated airspeed required to reach point B on schedule.

Distance between A and B: 70 NM

Forecast wind: 310° at 15 knots

Pressure altitude: 8,000 ft

Ambient temperature: -10°C

True course: 270°

The required indicated airspeed would be approximately

A – 126 knots.

B – 137 knots.

C – 152 knots.

8-64. Answer B. GFDPP 8C, PHB

1. Determine the groundspeed required to reach point B by 1530. Use a flight computer or calculate mathematically:
   70 NM ÷ 30 min × 60 min/hr = 140 knots.

2. Use the Winds function of your flight computer to determine the required true airspeed (TAS) with the given right quartering headwind: 152 knots.

3. Use the airspeed function of your flight computer to determine the indicated airspeed or calibrated airspeed: 137 knots.

8-65   PA.I.D.K3c

If an airplane is consuming 12.5 gallons of fuel per hour at a cruising altitude of 8,500 feet and the groundspeed is 145 knots, how much fuel is required to travel 435 NM?

A – 27 gallons.

B – 34 gallons.

C – 38 gallons.

8-65. Answer C. GFDPP 8C, PHB

Use your flight computer or calculate the fuel requirement mathematically.

1. Time enroute = 435 NM ÷ 145 NM/hr = 3 hours

2. Fuel required = 3 hr × 12.5 gallons = 37.5 gallons

SECTION C ■ **Flight Computers**

8-66   PA.I.D.K3c

If an aircraft is consuming 9.5 gallons of fuel per hour at a cruising altitude of 6,000 feet and the groundspeed is 135 knots, how much fuel is required to travel 420 NM?

A – 27 gallons.

B – 30 gallons.

C – 35 gallons.

8-66. Answer B. GFDPP 8C, PHB

Use your flight computer or calculate the fuel requirement mathematically.

1. Time enroute = 420 NM ÷ 135 NM/hr = 3.11 hr (3 hours and 7 minutes)

2. Fuel required = 3.11 hr × 9.5 gal/hr = 29.6 gallons

8-67   PA.I.D.K3c

If an airplane is consuming 14.8 gallons of fuel per hour at a cruising altitude of 7,500 feet and the groundspeed is 167 knots, how much fuel is required to travel 560 NM?

A – 50 gallons.

B – 53 gallons.

C – 57 gallons.

8-67. Answer A. GFDPP 8C, PHB

Use your flight computer or calculate the fuel requirement mathematically.

1. Time enroute = 560 NM ÷ 167 NM/hr = 3.35 hr (3 hours and 21 minutes)

2. Fuel required = 3.35 hr × 14.8 gal/hr = 49.6 gallons

SECTION C ■ Flight Computers

CHAPTER 9

Navigation

SECTION A
Pilotage and Dead Reckoning

COURSE

- When determining a course to fly, you must first select a route based on a variety of elements such as terrain, landmarks, and airspace.

- The methods of pilotage and dead reckoning normally are used together, each acting as a cross-check of the other in order to fly a desired course.

- Determine the true course, using a plotter.

- To change true course to magnetic course, add or subtract the isogonic line value shown on the aeronautical chart. Use the memory aid—east is least; west is best—to remember whether to add or subtract magnetic variation.

- Use a flight computer to determine the wind correction angle.

- The corrections for errors caused by the magnetic fields of items within the airplane are shown on the compass deviation card near the compass.

- You might choose to correct true course for wind before correcting for variation or to correct for variation before wind. The order in which you perform the calculation has no effect on the outcome, but different terms are used to describe some of the intermediate sums, and the wind direction must be changed from true to magnetic if the variation is applied first. Use these formulas to determine true and magnetic values for flight planning:

  TC ± WCA = TH ± VAR = MH ± DEV = CH

  True Course (TC) ± Wind Correction Angle (WCA) = True Heading (TH)

  True Heading (TH) ± Magnetic Variation (VAR) = Magnetic Heading (MH)

  Magnetic Heading ± Compass Deviation (DEV) = Compass Heading (CH)

  TC ± VAR = MC ± WCA = MH ± DEV = CH

  True Course (TC) ± Magnetic Variation (VAR) = Magnetic Course (MC)

  Magnetic Course (MC) ± Wind Correction Angle (WCA) = Magnetic Heading (MH) *[Note: Wind direction must be converted from true to magnetic.]*

  Magnetic Heading (MH) ± Compass Deviation (DEV) = Compass Heading (CH)

VFR CRUISING ALTITUDES

Whenever you are in level cruising flight more than 3,000 feet above ground level (AGL), you must comply with the VFR cruising altitude rule based on your magnetic course.

- Eastbound (0° to 179°)—odd thousands plus 500 feet.

- Westbound (180° to 359°)—even thousands plus 500 feet.

FUEL REQUIREMENTS

The FARs require that you must carry enough fuel (considering wind and forecast weather conditions) to fly to the first point of intended landing, and then, assuming normal cruising speed, for the following time period thereafter:

- Day—30 minutes.

- Night—45 minutes.

FLIGHT PLAN

- Although not required by the FARs, filing a VFR flight plan provides Flight Service with the details of your flight so they can start a search if you are overdue.

- After you arrive at your destination, close your flight plan by contacting Flight Service by phone or use the EasyClose™ service. You can register to receive an EasyClose™ text or email at 1800wxbrief.com.

LOST PROCEDURES

- If you get lost, the five Cs are guidelines to help you take positive action to establish your location: climb, communicate, confess, comply, and conserve.

- In addition to the basic five Cs, check the heading indicator against the magnetic compass and compare the landmarks you see outside to your chart, looking especially for any large, clear landmarks.

NOTE: An asterisk appearing after an ACS code (i.e. PA.VII.B.K1*) indicates that the question subject appears more than one time in the ACS. The code shown corresponds to the first instance of the subject in the ACS.

9-1   PA.VI.A.K1, PA.VI.A.K4c

What is the process of navigating by reference to landmarks called?

A – Dead reckoning.

B – Pilotage.

C – Ground-based navigation.

9-1. Answer B. GFDPP 9A
Although it typically utilizes visual references on the ground, the process of navigating by reference to landmarks is called pilotage. Dead reckoning is process of using time, speed, distance, and direction to navigate without reference to visual landmarks.

9-2   PA.II.A.K4, PA.VI.A.K3, K4a

When selecting a route in cross country planning, you should be careful to

A – avoid terrain, restricted airspace, and large expanses of water.

B – always follow roads in the event of an in-flight emergency.

C – avoid any military operations or training areas.

9-2. Answer A. GFDPP 9A
During route planning, review the general route and then decide whether to avoid certain areas such as special-use airspace, restricted airspace, high terrain, obstacles, or large expanses of open water. Although following roads may be ideal in certain scenarios, it may not always be feasible. While it is permitted to fly through military operations areas, you can also check with the controlling agency for when a military operations area is in use to avoid flight through military activity.

9-3   PA.I.D.K2, PA.VI.A.K4b

Which cruising altitude is appropriate for a VFR flight on a magnetic course of 135°?

A – Even thousands.

B – Even thousands plus 500 feet.

C – Odd thousands plus 500 feet.

9-3. Answer C. GFDPP 9A, FAR 91.159
On an easterly magnetic course (0° to 179°) above 3,000 feet AGL, VFR cruising altitudes are odd thousands plus 500 feet.

SECTION A ■ **Pilotage and Dead Reckoning**

SECTION A ■ **Pilotage and Dead Reckoning**

9-4   PA.I.D.K2, PA.VI.A.K4b

Which VFR cruising altitude is acceptable for a flight on a Victor Airway with a magnetic course of 175°? The terrain is less than 1,000 feet.

A – 4,500 feet.

B – 5,000 feet.

C – 5,500 feet.

9-4. Answer C. GFDPP 9A, FAR 91.159

On an easterly magnetic course (0° to 179°) above 3,000 feet AGL, VFR cruising altitudes are odd thousands plus 500 feet.

9-5   PA.I.D.K2, PA.VI.A.K4b

Which VFR cruising altitude is appropriate when flying above 3,000 feet AGL on a magnetic course of 185°?

A – 4,000 feet.

B – 4,500 feet.

C – 5,000 feet.

9-5. Answer B. GFDPP 9A, FAR 91.159

On a westerly magnetic course (180° to 359°) above 3,000 feet AGL, VFR cruising altitudes are even thousands plus 500 feet.

9-6   PA.I.D.K2, PA.VI.A.K4b

Each person operating an aircraft at a VFR cruising altitude shall maintain an odd-thousand plus 500-foot altitude while on a

A – magnetic heading of 0° through 179°.

B – magnetic course of 0° through 179°.

C – true course of 0° through 179°.

9-6. Answer B. GFDPP 9A, FAR 91.159

On an easterly magnetic course (0° to 179°) above 3,000 feet AGL, VFR cruising altitudes are even thousands plus 500 feet.

9-7   PA.VI.A.K5a, K5b

When converting from true course to magnetic heading, you should

A – subtract easterly variation and right wind correction angle.

B – add westerly variation and subtract left wind correction angle.

C – subtract westerly variation and add right wind correction angle.

9-7. Answer B. GFDPP 9A, PHB

Remember "east is least, west is best" to recall that easterly variation is subtracted and westerly variation is added. This calculation is often performed in conjunction with wind correction calculations using the formula:

TC ± WCA = TH ± VAR = MH

If you are using the wind side of a mechanical flight computer, add the wind correction if your wind dot is to the right of the centerline, and subtract the wind correction if the wind dot is to the left of the centerline. You can easily visualize this by remembering that compass headings decrease as you turn left, so a correction to the left requires that you subtract the correction angle.

9-8   PA.I.D.K3a

(Refer to Figure 20.) Determine the magnetic course from First Flight Airport (area 5) to Hampton Roads Airport (area 2).

A – 141°.

B – 321°.

C – 332°.

9-8. Answer C. GFDPP 9A, PHB

This question requires you to find the magnetic course. Determine true course, then correct for magnetic variation.

TC ± VAR = MC

1. Determine the True Course with a plotter (321°).

2. Locate the nearest isogonic line (11° West).

3. Convert TC to MC by adding west variation (321° + 11° = 332°).

9-9   PA.I.D.K3a

(Refer to Figure 21.) Determine the magnetic heading for a flight from Mercer County Regional Airport (area 3) to Minot International (area 1). The wind is from 330° at 25 knots, the true airspeed is 100 knots, and the magnetic variation is 10° east.

A – 002°.

B – 012°.

C – 352°.

9-9. Answer C. GFDPP 9A, PHB

This question requires you to find magnetic heading. First, determine the true heading by correcting true course for wind. Then, correct true heading for magnetic variation.

TC ± WCA = TH ± VAR = MH

1. Use the plotter to measure true course (012°).

2. Use the flight computer to determine true heading.

   a. Enter wind direction and speed (330° at 25 knots).

   b. Enter the true course (012°).

   c. Enter the TAS (100 knots).

   d. TH = 002°.

3. Convert TH to MH by correcting for magnetic variation (10°E). Because this is an east variation, subtract it from the true heading. (002° − 10° = 352°).

SECTION A ■ **Pilotage and Dead Reckoning**

9-10   PA.I.D.K3a

(Refer to Figure 22.) Determine the magnetic heading for a flight from Sandpoint Airport (area 1) to St. Maries Airport (area 4). The wind is from 215° at 25 knots, and the true airspeed is 125 knots.

A – 172°.

B – 187°.

C – 351°.

9-10. Answer A. GFDPP 9A, PHB

This question requires you to find magnetic heading. First, determine the true heading by correcting true course for wind. Then, correct true heading for magnetic variation.

TC ± WCA = TH ± VAR = MH

1. Use the plotter to measure true course (181°).

2. Use the flight computer to determine true heading.

    a. Enter the wind direction and speed (215° at 25 knots).

    b. Enter the true course (181°).

    c. Enter the TAS (125 knots).

    d. TH = 187°.

3. Convert TH to MH by correcting for magnetic variation (round to 15°E). Because this is an east variation, subtract it from the true heading. (187° − 15° = 172°).

9-11   PA.I.D.K3a

(Refer to Figure 22.) What is the magnetic heading for a flight from Priest River Airport (area 1) to Shoshone County Airport (area 3)? The wind is from 030° at 12 knots and the true airspeed is 95 knots.

A – 121°.

B – 136°.

C – 143°.

9-11. Answer A. GFDPP 9A, PHB

This question requires you to find magnetic heading. First, determine the true heading by correcting true course for wind. Then, correct true heading for magnetic variation.

TC ± WCA = TH ± VAR = MH

1. Use the plotter to measure true course (143°).

2. Use the flight computer to calculate true heading.

    a. Enter the wind direction and speed (030° at 12 knots).

    b. Enter the true course (143°).

    c. Enter the TAS (95 knots).

    d. TH = 136°.

3. Convert TH to MH by correcting for magnetic variation (round to 15°E). Because this is an east variation, subtract it from the true heading. (136° − 15° = 121°).

9-12   PA.I.D.K3a

(Refer to Figure 22.) Determine the magnetic heading for a flight from St. Maries Airport (area 4) to Priest River Airport (area 1). The wind is from 340° at 10 knots and the true airspeed is 90 knots.

A – 320°.

B – 330°.

C – 345°.

9-12. Answer B. GFDPP 9A, PHB

This question requires you to find magnetic heading. First, determine the true heading by correcting true course for wind. Then, correct true heading for magnetic variation.

TC ± WCA = TH ± VAR = MH

1. Use the plotter to measure true course (345°).

2. Use the flight computer to calculate true heading.

    a. Enter the wind direction and speed (340° at 10 knots).

    b. Enter the true course (345°).

    c. Enter the TAS (90 knots).

    d. TH = 345°.

3. Convert TH to MH by correcting for magnetic variation (round to 15°E). Because this is an east variation, subtract it from the true heading. (345° – 15° = 330°).

9-13   PA.I.D.K3a

(Refer to Figure 23.) Determine the magnetic heading for a flight from Allendale County Airport (area 1) to Claxton-Evans County Airport (area 2). The wind is from 090° at 16 knots and the true airspeed is 90 knots and the magnetic variation is 6°W.

A – 209°.

B – 215°.

C – 230°.

9-13. Answer A. GFDPP 9A, PHB

This question requires you to find magnetic heading. First, determine the true heading by correcting true course for winds. Then, correct true heading for magnetic variation.

TC ± WCA = TH ± VAR = MH

1. Use the plotter to measure true course (212°).

2. Use the flight computer to calculate true heading.

    a. Enter the wind direction and speed (090° at 16 knots).

    b. Enter the true course (212°).

    c. Enter the TAS (90 knots).

    d. TH = 203°.

3. Convert TH to MH by correcting for magnetic variation (6°W). Because this is a west variation, add it from the true heading. (203° + 6° = 209°).

SECTION A ■ **Pilotage and Dead Reckoning**

9-14   PA.I.D.K3a

(Refer to Figures 23 and 58.) Determine the compass heading for a flight from Claxton-Evans County Airport (area 2) to Hampton Varnville Airport (area 1). The wind is from 280° at 08 knots, and the true airspeed is 85 knots, and the magnetic variation is 6°W.

A – 033°.

B – 038°.

C – 044°.

9-14. Answer C. GFDPP 9A, PHB

This question requires you to find compass heading. First, determine the true heading by correcting true course for wind. Then, correct true heading for magnetic variation. Finally, correct for deviation.

TC ± WCA = TH ± VAR = MH ± DEV = CH

1. Use the plotter to measure true course (045°).

2. Use the flight computer to calculate true heading.

   a. Enter the wind direction and speed (280° at 8 knots).

   b. Enter the true course (045°).

   c. Enter the TAS (85 knots).

   d. TH = 041°.

3. Convert TH to MH by correcting for magnetic variation (6°W). Because this is an west variation, add it to the true heading. (041° + 6° = 047°).

4. Adjust per compass card (–3°) to determine compass heading (044°).

9-15   PA.I.D.K3a

(Refer to Figure 24.) Determine the magnetic course from Airpark East Airport (area 1) to Winnsboro Airport (area 2). Magnetic variation is 3°E.

A – 079°.

B – 082°.

C – 091°.

9-15. Answer B. GFDPP 9A, PHB

This question requires you find the magnetic course. First, determine the true course and then correct it for magnetic variation.

TC ± VAR = MC

1. Use the plotter to measure true course (085°).

2. Convert TC to MC by correcting for magnetic variation. Because this is an east variation, subtract it from the true course. (085° – 3° = 082°).

9-16   PA.I.D.K3a

(Refer to Figure 25.) Determine the magnetic heading for a flight from Fort Worth Meacham (area 4) to Denton Muni (area 1). The wind is from 330° at 25 knots, the true airspeed is 110 knots, and the magnetic variation is 4° east.

A – 007°.

B – 017°.

C – 023°.

9-16. Answer A. GFDPP 9A, PHB

This question requires you to find magnetic heading. First, determine the true heading by correcting true course for wind. Then, correct true heading for magnetic variation.

TC ± WCA = TH ± VAR = MH

1. Use your plotter to determine true course (021°).

2. Use your flight computer to calculate true heading.

    a. Enter the wind direction and speed (330° at 25 knots).

    b. Enter the true course (021°).

    c. Enter the TAS (110 knots).

    d. TH = 011°.

3. Convert TH to MH by correcting for magnetic variation (4°E). Because this is an east variation, subtract it from the true heading.) (011° – 4° = 007°).

9-17   PA.I.D.K3a

(Refer to Figure 26.) Determine the magnetic course from BRYN (Pvt) Airport (area 2) to Jamestown Airport (area 4).

A – 228°.

B – 233°.

C – 360°.

9-17. Answer A. GFDPP 9A, PHB

This question requires you find the magnetic course. First, determine the true course and then correct it for magnetic variation.

TC ± VAR = MC

1. Use the plotter to measure true course (233°).

2. Convert TC to MC by correcting for magnetic variation—the isogonic line down the middle of the figure indicates a 5° east magnetic variation. Because this is an east variation, subtract it from the true course. (233° – 5° = 228°)

9-18   PA.I.D.K3c

What is the specific fuel requirement for flight under VFR during daylight hours in an airplane?

A – Enough to complete the flight at normal cruising speed with adverse wind conditions.

B – Enough to fly to the first point of intended landing and to fly after that for 30 minutes at normal cruising speed.

C – Enough to fly to the first point of intended landing and to fly after that for 45 minutes at normal cruising speed.

9-18. Answer B. GFDPP 9A, FAR 91.151

For day VFR flight in an airplane, you must carry enough fuel (considering wind and forecast weather conditions) to fly to the first point of intended landing, and, assuming normal cruising speed, 30 minutes thereafter.

SECTION A ■ **Pilotage and Dead Reckoning**

9-19   PA.I.D.K3c

What is the specific fuel requirement for flight under VFR at night in an airplane?

A – Enough to complete the flight at normal cruising speed with adverse wind conditions.

B – Enough to fly to the first point of intended landing and to fly after that for 30 minutes at normal cruising speed.

C – Enough to fly to the first point of intended landing and to fly after that for 45 minutes at normal cruising speed.

9-19. Answer C. GFDPP 9A, FAR 91.151

For night VFR flight in an airplane, you must carry enough fuel (considering wind and forecast weather conditions) to fly to the first point of intended landing, and assuming normal cruising speed, 45 minutes thereafter.

9-20   PA.I.D.K4, PA.VI.A.K5c, K6

Why is it beneficial to determine a cruise power setting during preflight calculations?

A – Power setting helps determine fuel burn and time enroute.

B – There is no benefit to determining a cruise power setting before flying.

C – To avoid having to look up a proper power setting in-flight.

9-20. Answer A. GFDPP 9A, FAR 91.153

Cruise power settings help estimate the fuel burn and time enroute. This information helps determine the fuel on board and time elapsed to the first point of intended landing, both required when filing a VFR flight plan.

In-flight, it is best to use this pre-determined power setting. This keeps you close to your estimated fuel burn, avoids unnecessary fuel consumption, and helps ensure your time enroute matches your VFR flight plan.

9-21   PA.I.D.K3a, PA.VI.A.K7

While flying enroute, you arrive at a checkpoint 15 minutes later than the time you calculated in your flight plan. What actions should you take next?

A – Recalculate your ETA and plan to use your fuel reserve as required.

B – Determine the reason for the deviation, add a fuel stop if needed, and notify Flight Service of your revised ETA.

C – Reduce power to increase fuel economy and notify Flight Service of your revised ETA.

9-21. Answer B. GFDPP 9A, PHB

After recognizing a change—the deviation from your ETA at the checkpoint—the next steps are to define the problem and choose a course of action.

If you brought a large fuel reserve, you might be able to use part of it, although you should plan a fuel stop any time the fuel reserve drops below your personal minimums. You could also reduce power to extend your range, if you know that will be sufficient. However, neither of these actions is appropriate before determining the cause of the deviation and the actions needed to manage it successfully.

9-22   PA.I.D.K5

How should a VFR flight plan be closed at the completion of the flight at a controlled airport?

A – The tower automatically closes the flight plan when the airplane turns off the runway.

B – The pilot must close the flight plan by contacting Flight Service.

C – The tower relays the instructions to Flight Service when the aircraft contacts the tower for landing.

9-22. Answer B. GFDPP 9A, PHB

After you arrive at your destination, close your flight plan by contacting Flight Service by phone or use the EasyClose™ service. You can register to receive an EasyClose™ text or email at 1800wxbrief.com.

9-23   PA.VI.D.K1, K2

What are the five Cs in relation to lost procedures?

A – Compare, conserve, cruise, communicate, and choose.

B – Confess, circle, cruise, cross-reference, and call.

C – Climb, communicate, confess, comply, and conserve.

9-23. Answer C. GFDPP 9A, PHB

The five Cs are guidelines to help you take positive action to establish your location while taking care to conserve resources: climb, communicate, confess, comply, and conserve.

9-24   PA.VI.C.K1, K2

Your airplane suffers a partial power loss and you are unable to restore full power so you decide to divert to a nearby airport with maintenance services. What are the appropriate techniques to handle the diversion?

A – Perform precise measurements and calculations before turning toward your new destination to ensure that you do not miss it.

B – Prioritize tasks and avoid using the autopilot, so that you can maintain situational awareness enroute to the alternate airport.

C – Immediately turn the airplane in the general direction of the new landing point and then use rule-of-thumb computations, or the Nearest and Direct-To functions of your GPS, to navigate there.

9-24. Answer C. GFDPP 9A, PHB

In an emergency, divert promptly toward your alternate destination. Apply rule-of-thumb computations, estimates, and other appropriate shortcuts to divert to the new course as soon as possible. Use resources effectively, such as ATC vectors or the nearest airport feature of your GPS equipment.

SECTION A ■ **Pilotage and Dead Reckoning**

SECTION B

VOR Navigation

INTERPRETING VOR INDICATIONS

- To find your position relative to a VOR on the sectional chart, draw a line along a radial until it intersects your position, or the position of the object you wish to locate. Determine the radial using the compass rose depicted on the chart. Measure the distance from the VOR using a plotter.

- To determine your course to a VOR, find your position along a radial, and find the reciprocal by adding or subtracting 180° from the radial.

- The magnetic course can be determined by plotting a line from your position or departure point to the VOR. Remember that the radial you read from the compass rose is the reciprocal of what you would select on the OBS to fly TO the station.

- You can find your position by triangulation—using two or more VORs. Determine the radial you are on from one VOR, and draw a line from the VOR through the compass rose on that radial. Repeat the procedure with another VOR facility. The intersection point is your location.

- Your position is reflected on the VOR receiver by the course selected in the OBS and the position of the CDI needle. If the CDI needle is:

    ◦ Centered with a FROM indication, the OBS heading reflects the radial that the aircraft is on.

    ◦ Centered with a TO indication, fly the course on the OBS to track to the station. The reciprocal course is the radial that the aircraft is on.

    ◦ Deflected to the left or right, the aircraft is left or right of course, respectively.

- Ensure you have the correct radial selected on the OBS; if you have tuned in the reciprocal course, reverse sensing occurs—the CDI is deflected away from the course and the normal procedure of correcting toward the needle will actually take you farther off course.

- If the TO-FROM indicator is blank, you are over the cone of confusion, and the aircraft is either over the station, or on a radial offset 90 degrees from the radial selected on the OBS.

CHECKING VOR ACCURACY

- VOR test facilities (VOTs) enable you to make precise VOR accuracy checks regardless of your airplane's position in relation to the facility because VOTs broadcast a signal for only one radial—360°.

- When the CDI needle is centered during an omnireceiver check using a VOT signal, the OBS and the TO-FROM indicator should read 0° FROM or 180° TO, regardless of the position of the aircraft from the VOT.

NOTE: An asterisk appearing after an ACS code (i.e. PA.VII.B.K1*) indicates that the question subject appears more than one time in the ACS. The code shown corresponds to the first instance of the subject in the ACS.

9-25   PA.VI.B.K1, PA.VIII.F.K2

(Refer to Figure 20, area 3 and Figure 28.) The VOR is tuned to Elizabeth City VOR, and the aircraft is positioned over Shawboro, a small town 3 NM west of Currituck County Regional (ONX). Which VOR indication is correct?

A – 2.

B – 5.

C – 8.

9-25. Answer B. GFDPP 9B, PHB

Shawboro is on the 030° radial of Elizabeth City VOR. The CDI needle would be centered on 030° with a FROM indication (VOR indicator #9) or 210° with a TO indication (VOR indicator #5).

9-26   PA.VI.B.K1, PA.VIII.F.K2

(Refer to Figure 21.) What course should be selected on the omnibearing selector (OBS) to make a direct flight from Mercer County Regional Airport (area 3) to the Minot VORTAC (area 1) with a TO indication?

A – 357°.

B – 177°.

C – 001°.

9-26. Answer A. GFDPP 9B, PHB

The magnetic course can be determined by plotting a line from Mercer County Regional Airport to the Minot VORTAC. The line intersects the Minot VORTAC compass rose at 177°. The reciprocal of 177° is 357°, which is what you would set in the OBS.

9-27   PA.VI.B.K1, PA.VIII.F.K2

(Refer to Figure 23.) What is the approximate position of the aircraft if the VOR receivers indicate the 341° radial of Savannah VORTAC (area 3) and the 184° radial of Allendale VOR (area 1)?

A – Town of Guyton.

B – Town of Springfield.

C – 3 miles east of Marlow.

9-27. Answer B. GFDPP 9B, PHB

The intersection of these two radials places the aircraft near the town of Springfield. The town of Guyton is west of both of these radials.

9-28   PA.VI.B.K1, PA.VIII.F.K2

(Refer to Figure 23.) On what course should the VOR receiver (OBS) be set to navigate direct from Hampton Varnville Airport (area 1) to Savannah VORTAC (area 3)?

A – 015°.

B – 195°.

C – 220°.

9-28. Answer B. GFDPP 9B, PHB

If you draw a line from Hampton Varnville Airport to Savannah VORTAC, which is on the field at Savannah Hilton Head International Airport, it crosses the Savannah compass rose at 015°. To navigate inbound with a TO indication would require the reciprocal of 015°, or 195°, to be set in the course selector.

9-29   PA.VI.B.K1, PA.VIII.F.K2

(Refer to Figure 24.) What is the approximate position of the aircraft if the VOR receivers indicate the 245° radial of Sulphur Springs VOR-DME (area 5) and the 144° radial of Bonham VORTAC (area 3)?

A – The town of Lone Oak.

B – Glenmar Airport.

C – Majors Airport.

9-29. Answer B. GFDPP 9B, PHB

Draw the radials from these VORs. The intersection of these two radials puts the aircraft near the Glenmar Airport (private). Majors Airport and the town of Lone Oak are both west of the Bonham 144° radial.

SECTION B ■ **VOR Navigation**

9-30   PA.VI.B.K1, PA.VIII.F.K2

(Refer to Figure 24.) On what course should the VOR receiver (OBS) be set to navigate direct from Majors Airport (area 1) to Quitman VOR (area 2)?

A – 100°.

B – 108°.

C – 280°.

9-30. Answer A. GFDPP 9B, PHB

A direct course from Majors Airport to Quitman VOR crosses the compass rose at 280°. The inbound course to be set in the OBS is the reciprocal of 280°, or 100°.

9-31   PA.VI.B.K1, PA.VIII.F.K2

(Refer to Figure 24 and 28.) The VOR is tuned to Bonham VORTAC (area 3), and the aircraft is positioned over the Sulphur Springs airport (area 5). Which VOR indication is correct?

A – 1.

B – 3 or 7.

C – 8.

9-31. Answer B. GFDPP 9B, PHB

Sulfur Springs airport is on the 120° radial of the Bonham VORTAC. Because all the VOR indicators in the figure have either 030° or 210° set in the OBS, you must determine the position of the aircraft in relation to these settings and the VOR station. Because the 120° radial is perpendicular to the 030°/210° radials, the TO-FROM indicator indicates OFF.

With the OBS set to 030°, the CDI is deflected to the left—VOR indicator #7.

With the OBS set to 210°, the CDI is deflected to the right—VOR indicator #3.

VOR indicators #1 and #8 are incorrect because the TO-FROM indicators indicate TO instead of OFF.

9-32   PA.VI.B.K1, PA.VIII.F.K2

(Refer to Figure 25, area 5.) The VOR is tuned to the Maverick VOR, which is just south of DFW. The omnibearing selector (OBS) is set on 253°, with a TO indication, and a left course deviation indicator (CDI) deflection. What is the aircraft's position from the VORTAC?

A – East-northeast.

B – East-southeast.

C – West-southwest.

9-32. Answer A. GFDPP 9B, PHB

A course of 253° with a TO indication takes the aircraft to the station. This means that the aircraft is generally on the east side of the station. A CDI deflection to the left means that the aircraft is to the right (north) of the 253° inbound course.

9-33   PA.VI.B.K1, PA.VIII.F.K2

(Refer to Figure 26, areas 2 and 4, and Figure 28.) The VOR is tuned to Jamestown VOR, and the aircraft is positioned over Cooperstown Airport. Which VOR indication is correct?

A – 1.

B – 6.

C – 9.

9-33. Answer C. GFDPP 9B, PHB

Jamestown VOR is in area 4 and the aircraft is positioned over Cooperstown Airport in area 2, which is on the JMS 030° radial. With the OBS set to 030°, the TO-FROM indicator reads FROM, and the CDI shows the aircraft centered on course as shown on VOR indicator #9. If the OBS is set to 210°, the TO-FROM indicator would indicate TO, so VOR indicator #5 would also be correct.

9-34   PA.VI.B.K1, PA.VIII.F.K2

(Refer to Figure 28, illustration 1.) The VOR receiver has the indications shown. What is the aircraft's position relative to the station?

A – North.

B – East.

C – South.

9-34. Answer B. GFDPP 9B, PHB

With the 210° course selected and a TO indication, if the needle were centered, the aircraft would be on the 030° radial and a 210° course would take it to the station. With a right CDI deflection, the aircraft is left of course, between the 030° and 120° radials, east of the station.

9-35   PA.VI.B.K1, PA.VIII.F.K2

(Refer to Figure 28, illustration 3.) The VOR receiver has the indications shown. What is the aircraft's position relative to the station?

A – East.

B – Southeast.

C – West.

9-35. Answer B. GFDPP 9B, PHB

The TO-FROM indicator indicates OFF, which means the aircraft is somewhere on the 120° or 300° radial, 90° from the 210° course. The right CDI deflection puts the aircraft left of the 210° course, which would be southeast of the station.

9-36   PA.VI.B.K1, PA.VIII.F.K2

(Refer to Figure 28, illustration 8.) The VOR receiver has the indications shown. What radial is the aircraft on?

A – 030°.

B – 210°.

C – 300°.

9-36. Answer B. GFDPP 9B, PHB

The aircraft is on the centerline of the selected course. A 030° course takes the aircraft to the station, as indicated by a TO in the TO-FROM window. That means the aircraft is on the reciprocal radial of 030°, the 210° radial.

9-37   PA.VI.B.K1, PA.VIII.F.K2

When the course deviation indicator (CDI) needle is centered using a VOR test signal (VOT), the omnibearing selector (OBS) and the TO-FROM indicator should read

A – 180° FROM, only if the pilot is due north of the VOT.

B – 0° TO or 180° FROM, regardless of the pilot's position from the VOT.

C – 0° FROM or 180° TO, regardless of the pilot's position from the VOT.

9-37. Answer C. GFDPP 9B, AIM

No matter where the aircraft is located in relation to the VOT, the VOR should always read 180° with a TO indication or 0° with a FROM.

SECTION C
Satellite Navigation — GPS

GPS OPERATION

- Satellite navigation is based on a network of satellites that transmit radio signals in medium earth orbit. The global navigation satellite system (GNSS) is the standard generic term for satellite navigation systems.

- The United States global positioning system (GPS) is a primary satellite navigation system that is globally available. GPS consists of three segments—space, control, and user.

- The space segment contains a minimum of 24 GPS satellites in orbits that ensure users can view at least four satellites from virtually any point on the earth.

- The control segment is a global network of ground facilities that track the GPS satellites, monitor and analyze their transmissions, and send commands and data to the constellation.

- The user segment consists of the GPS receivers that receive the signals from the GPS satellites and use the transmitted information to calculate the user's three-dimensional position.

- By using trilateration, a GPS receiver calculates its distance from three satellites to determine its general position for latitude, longitude, and altitude. A fourth satellite is necessary to determine an accurate position—small timing errors from all four satellites are adjusted to determine the exact location of the airplane.

- GPS receivers continuously verify the integrity (usability) of the signals received from the GPS constellation through receiver autonomous integrity monitoring (RAIM). With RAIM, a fifth satellite monitors the position provided by the other four satellites and alerts you of any discrepancy.

- You can verify that RAIM will be available by checking NOTAMs, contacting Flight Service, referring to the FAA RAIM prediction website, or by using your GPS receiver's RAIM monitoring and prediction functions.

- The accuracy of GPS is enhanced with the use of a satellite-based augmentation system (SBAS) known as the Wide Area Augmentation System (WAAS)—a series of ground stations generate a corrective message that is transmitted to aircraft by a geostationary satellite.

NAVIGATING WITH GPS

- A flight management system (FMS) is a computer system containing a database that enables programming of routes, approaches, and departures.

- The FMS can supply navigation data to the flight director or autopilot from various sources and can calculate flight data such as fuel consumption and time remaining.

- Before flight, verify that the navigation database and other databases are current. Database information might include the database type, cycle number, and valid operating dates. Database information typically displays when you power up the GPS equipment.

- GPS equipment installed in the airplane normally displays a course deviation indicator (CDI) on an analog indicator or HSI display, and on the GPS unit.

- When the navigation source is GPS, the CDI displays cross-track error in nautical miles. The CDI has three different sensitivities: enroute, terminal, and approach.

- A GPS moving map provides a pictorial view of the present position of the aircraft, the programmed route, the surrounding airspace, and topographical features.

- Two common errors associated with using a moving map include overreliance on the moving map leading to complacency and using the moving map as a primary navigation instrument.
- Direct-To navigation enables you to fly from your present position directly to a waypoint.
- The desired track to the next waypoint is the active leg, which is normally shown in magenta on a moving map.

NOTE: An asterisk appearing after an ACS code (i.e. PA.VII.B.K1*) indicates that the question subject appears more than one time in the ACS. The code shown corresponds to the first instance of the subject in the ACS.

9-38   PA.VI.B.K2, PA.I.G.K1g

If receiver autonomous integrity monitoring (RAIM) capability is lost in-flight,

A – the pilot may still rely on GPS derived altitude for vertical information.

B – the pilot has no assurance of the accuracy of the GPS position.

C – GPS position is reliable provided at least 3 GPS satellites are available.

9-38. Answer B. GFDPP 9C, PHB
GPS receivers continuously verify the integrity (usability) of the signals received from the GPS constellation through receiver autonomous integrity monitoring (RAIM). With RAIM, a fifth satellite monitors the position provided by the other four satellites and alerts you of any discrepancy. RAIM provides the degree of reliability necessary for IFR operations. Without RAIM, the accuracy of the GPS position solution is not assured.

9-39   PA.VI.B.K2, PA.I.G.K1g

What is RAIM?

A – A method by which the GPS receiver uses a fifth satellite to continuously verify the integrity of the signals received from the GPS constellation and then alerts you of a discrepancy.

B – A series of ground stations that generate a corrective message to improve navigational accuracy by accounting for positional drift of the satellites and signal delays caused by the ionosphere and other atmospheric factors.

C – A method by which the GPS receiver computes your position, track, and groundspeed and displays distance and time estimates to the selected course or waypoint.

9-39. Answer A. GFDPP 9C, PHB
GPS receivers continuously verify the integrity (usability) of the signals received from the GPS constellation through receiver autonomous integrity monitoring (RAIM). With RAIM, a fifth satellite monitors the position provided by the other four satellites and alerts you of any discrepancy. RAIM provides the degree of reliability necessary for IFR operations. Without RAIM, the accuracy of the GPS position solution is not assured.

9-40   PA.VI.B.K2, PA.VIII.F.K2

Select the true statement regarding the CDI used for GPS navigation.

A – The CDI displays the lateral distance from the course.

B – The CDI displays the angular deviation from the course.

C – The CDI has three different sensitivities to select based on your phase of flight.

9-40. Answer A. GFDPP 9C, PHB
Unlike the course deviation indicator (CDI) on a VOR, which displays angular deviation from course, an RNAV CDI displays distance (in nautical miles) off course. GPS CDIs have three different sensitivities based on your phase of flight (enroute, terminal, or approach). However, you cannot select the sensitivity.

SECTION C ■ Satellite Navigation — GPS

9-41   PA.VI.B.K2, PA.VIII.F.K2

What is a common error using a GPS moving map?

A – Using the moving map to enhance situational awareness and cross check primary navigation.

B – Overreliance on the moving map leading to complacency.

C – Incorrectly setting the scale and level of detail on the moving map.

9-41. Answer B. GFDPP 9C, PHB

Two common errors associated with using a moving map are: using the moving map as a primary navigation instrument and overreliance on the moving map leading to complacency.

9-42   PA.II.B.K3, PA.VI.B.K2, PA.VIII.F.K2

Select the true statement regarding GPS navigation.

A – Direct-To navigation enables you to create a route using several waypoints.

B – The desired track to the next waypoint is the active leg and is normally shown in magenta on a moving map.

C – To check the currency of the navigation database, you must select the valid operating dates to be displayed with other navigation data, such as track and groundspeed.

9-42. Answer B. GFDPP 9C, PHB

The desired track for your current (active) leg is typically shown in magenta.

Direct-To navigation enables you to fly from your present position directly to a waypoint.

To determine the currency of your navigation database, you normally view a message when turning on the system; it is not displayed with other navigation data.

9-43   PA.VI.B.K2, PA.VIII.F.K2

What is cross-track error?

A – The distance from your present position to the desired track, measured in nautical miles.

B – The difference between your heading and the course caused by a crosswind.

C – The difference in direction between the desired track and the aircraft's actual track in degrees.

9-43. Answer A. GFDPP 9C, PHB

Cross-track error (XTR) is your deviation from course, or desired track (DTK), in nautical miles.

The difference in direction between the desired track and the aircraft's actual track is called track angle error (TKE).

CHAPTER 10

Applying Human Factors Principles

SECTION A
Aviation Physiology

This section contains questions from Chapter 1, Section C – Introduction to Human Factors as well as Chapter 10, Section A – Aviation Psychology. The FAA questions associated with Single-Pilot resource management such as aeronautical decision making, risk, task management, and situational awareness are in Chapter 10, Section B of this test guide.

NIGHT VISION

- The retina contains many photosensitive cells called cones and rods, which are connected to the optic nerve. The rods are your primary receptors for night vision and also are responsible for much of your peripheral vision.

- Your diet and general physical health also affect how well you can see in the dark. Other factors, such as carbon monoxide poisoning, smoking, alcohol, certain drugs, and a lack of oxygen can greatly decrease your night vision.

- The most effective way to look for traffic during night flight is to scan to permit off-center viewing. Look to the side of an object for the clearest focus. Also, in dim light, you might need to move your eyes more slowly to prevent blurring of images than during the day.

- During a night flight, you can interpret the position lights of another aircraft to determine its location and movement in relation to your aircraft. You observe:

  - A steady red light and a flashing red light ahead and at the same altitude —the other aircraft is crossing to the left.

  - A steady white light and a flashing red light ahead and at the same altitude—the other aircraft is flying away from you.

  - Steady red and green lights ahead and at the same altitude—the other aircraft is approaching head-on.

- Your landing approaches at night should be made the same as during the daytime to reduce the effects of landing illusions.

- To adapt the eyes for night flying, avoid bright white lights for at least 30 minutes before the flight.

DISORIENTATION

- Spatial disorientation is a lack of orientation regarding the position, attitude, or movement of the aircraft in space.

- Spatial disorientation can be caused by misleading information being sent to the brain by your body's various sensory organs. Awareness of your body's position is a result of input from three primary sources: vision, the vestibular system located in your inner ear, and your kinesthetic sense.

- When few outside visual references are available, you need to rely heavily on the flight instruments for accurate information. Spatial disorientation can occur when there is a conflict between the information relayed by your central vision scanning the instruments, and your peripheral vision, which has virtually no references with which to establish orientation (as in IFR conditions).

- If you experience spatial disorientation during flight in restricted visibility, rely on the aircraft instrument indications. You are more susceptible to spatial disorientation if you use body signals to interpret flight attitude.

- Motion sickness is produced by vestibular disorientation. Also known as airsickness, common symptoms include general discomfort, paleness, nausea, dizziness, sweating, and vomiting. If you or your passengers experience motion sickness, look outside, try to relax, open fresh air vents, and consider finding smooth air.

HYPOXIA

- Hypoxia is a state of oxygen deficiency in the body—the tissues in the body do not receive enough oxygen, which can be caused by several factors.
- The forms of hypoxia are divided into four major groups based on their causes:
    - Hypoxic hypoxia—an inadequate supply of oxygen due to the decrease of oxygen molecules at sufficient pressure when at high altitudes.
    - Hypemic hypoxia—the inability of the blood to carry oxygen due to factors such as anemia or carbon monoxide poisoning.
    - Stagnant hypoxia—an oxygen deficiency in the body due to the poor circulation of the blood caused by factors such as the heart failing to pump blood effectively or pulling excessive positive Gs.
    - Histotoxic hypoxia—the inability of the cells to effectively use oxygen caused by factors such as alcohol and other drugs.

CARBON MONOXIDE POISONING

- Large accumulations of carbon monoxide in the body result in the loss of muscular power, vomiting, convulsions, and coma.
- Susceptibility to carbon monoxide poisoning increases as altitude increases.
- If you suspect carbon monoxide poisoning, you should turn the heater off immediately, open the fresh air vents or windows, and use supplemental oxygen if it is available.

SUPPLEMENTAL OXYGEN

- Crewmembers must use supplemental oxygen after 30 minutes when operating an aircraft at cabin pressure altitudes above 12,500 feet mean seal level (MSL) up to and including 14,000 feet MSL.
- Crewmembers must use supplemental oxygen continuously beginning at 14,000 feet MSL.
- Each additional occupant must be provided with supplemental oxygen above 15,000 feet MSL.

DECOMPRESSION SICKNESS

- Decompression sickness (DCS) is caused by a rapid reduction in the ambient pressure surrounding the body.
- Nitrogen and other inert gases expand to form bubbles that rise out of solution and produce symptoms that range from pain in the joints to seizures and unconsciousness.
- Scuba diving can cause DCS so the following wait times before ascending are recommended:
  1. To 8,000 feet MSL—12 hours after a dive that did not require a controlled ascent (nondecompression stop diving)
  2. 24 hours after a dive that required a controlled ascent (decompression stop diving)
  3. Above 8,000 feet MSL—24 hours after any dive.

EAR AND SINUS BLOCK

- During a descent, the outside air pressure in the auditory canal will become higher than the pressure in the middle ear.
- Slow descent rates can help prevent or reduce the severity of ear problems and the eustachian tube can sometimes be opened by yawning, swallowing, or chewing.
- Pressure can be equalized by performing the Valsalva maneuver—holding the nose and mouth shut and forcibly exhaling to force air up the eustachian tube into the middle ear.

SECTION A ■ **Aviation Physiology**

HYPERVENTILATION

- Hyperventilation occurs when you are experience emotional stress, fright, or pain, and your breathing rate and depth increase causing an excessive loss of carbon dioxide from your body.

- Hyperventilation can also be caused by rapid or extra deep breathing while using supplemental oxygen.

- You can overcome the symptoms, or avoid future occurrences of hyperventilation, by slowing your breathing rate, breathing into a bag, or talking aloud.

FITNESS FOR FLIGHT

- Because stress is cumulative, you bring into the flight varying degrees of stress left over from the other areas of your life. Excessive stress interferes with your ability to focus and cope with a given situation.

- Fatigue has been a major factor in many fatal accidents involving very experienced and highly qualified crews.

- Dehydration is a critical loss of water from the body. Although, dehydration can cause headaches, cramps, nausea, and dizziness, the first noticeable effect is typically fatigue.

DRUGS AND ALCOHOL

- You may not act as a crewmember of a civil aircraft if you have consumed alcoholic beverages within the preceding eight hours, or with 0.04 percent or more alcohol in the blood by weight.

- You must provide a written report of each drug or alcohol related motor vehicle action to the FAA no later than 60 days after the action.

- Except in an emergency, you may not allow a person who appears to be intoxicated or who demonstrates by manner or physical indications that the individual is under the influence of drugs (except a patient under proper care) to board your aircraft.

HYPOTHERMIA

- Hypothermia occurs when your body loses heat faster than it can produce heat, causing a dangerously low body temperature below 95°F (35°C).

- You might not be aware of your condition if you have hypothermia because the symptoms often begin gradually and the confused thinking associated with hypothermia prevents self-awareness.

NOTE: *An asterisk appearing after an ACS code (i.e. PA.VII.B.K1*) indicates that the question subject appears more than one time in the ACS. The code shown corresponds to the first instance of the subject in the ACS.*

10-1   PA.XI.A.K1

What is the most effective way to use the eyes during night flight?

A – Look only at far away, dim lights.

B – Scan 5° to 10° off-center for off-center viewing.

C – Concentrate directly on each object for a few seconds.

10-1. Answer B. GFDPP 10A, AFH, PHB
The rods are estimated to be 10,000 times more sensitive to light than the cones once fully adapted to darkness, making them the primary receptors for night vision. Because the cones are concentrated near the fovea, the rods are also responsible for much of the peripheral vision. To see an object clearly at night, you must expose the rods to the image. This can be done by looking 5° to 10° off-center of the object to be seen.

10-2　PA.II.D.K5, PA.XI.A.K4

You are preparing for a night flight. Which of the following is the most appropriate use of your aircraft's lighting system?

A – Turn on all lights to their brightest setting for maximum visibility.

B – Use the minimum necessary lighting to preserve your night vision.

C – Keep the lights off to save battery power.

10-2. Answer B. GFDPP 10A, AFH, PHB

When flying at night, preserving your night vision is essential. Use the minimum necessary lighting. For example, limit use of a personal flashlight or make use of a red light because it does not inhibit your night vision. Although turning on all lights to their brightest setting might seem intuitive, this action can be counterproductive because it can cause glare and make it harder to see outside the aircraft.

10-3　PA.XI.A.K1

The best method to use when looking for other traffic at night is to

A – look to the side of the object and scan slowly.

B – scan the visual field very rapidly.

C – look off-center at the object for more than 2 or 3 seconds to ensure the object focuses properly.

10-3. Answer A. GFDPP 10A, AFH

The rods are estimated to be 10,000 times more sensitive to light than the cones once fully adapted to darkness, making them the primary receptors for night vision. Because the cones are concentrated near the fovea, the rods are also responsible for much of the peripheral vision. To see an object clearly at night, you must expose the rods to the image. This can be done by looking 5° to 10° off-center of the object to be seen. In this manner, the peripheral vision can maintain contact with the object.

With off-center vision, the image of an object viewed longer than 2 to 3 seconds will disappear. To overcome this night vision limitation, pilots should avoid viewing an object for longer than 2 or 3 seconds. The peripheral field of vision will continue to pick up the object when the eyes are shifted from one off-center point to another.

10-4　PA.XI.A.K1

The most effective method of scanning for other aircraft for collision avoidance during nighttime hours is to use

A – regularly spaced concentration on the 3-, 9-, and 12-o'clock positions.

B – a series of short, regularly spaced eye movements to search each 30-degree sector.

C – peripheral vision by scanning small sectors and utilizing off-center viewing.

10-4. Answer C. GFDPP 10A, AFH

At night, scan slowly, in small sectors, using off-center (peripheral) vision. The rods in the retina are used for night vision. Because they are not located directly behind the pupil, you must use off-center viewing. Also, in dim light, you might need to move your eyes more slowly to prevent the blurring of images.

SECTION A ■ **Aviation Physiology**

10-5   PA.XI.A.K1

What preparation should you make to adapt the eyes for night flying?

A – Wear sunglasses after sunset until ready for flight.

B – Avoid red lights at least 30 minutes before the flight.

C – Avoid bright white lights at least 30 minutes before the flight.

10-5. Answer C. GFDPP 10A, AIM
The rods in the human eye can take up to 30 minutes to adapt fully to the dark. Avoid bright lights for this amount of time prior to flight.

10-6   PA.XI.A.K5, K7

During a night flight, you observe a steady red light and a flashing red light ahead and at the same altitude. What is the general direction of movement of the other aircraft?

A – The other aircraft is crossing to the left.

B – The other aircraft is crossing to the right.

C – The other aircraft is approaching head-on.

10-6. Answer A. GFDPP 10A, AFH
The steady red light is a position light on the left wing and the flashing red light is an anticollision beacon. If you are looking at the left wingtip, the other aircraft would be crossing to the left.

10-7   PA.XI.A.K5, K7

During a night flight, you observe a steady white light and a flashing red light ahead and at the same altitude. What is the general direction of movement of the other aircraft?

A – The other aircraft is flying away from you.

B – The other aircraft is crossing to the left.

C – The other aircraft is crossing to the right.

10-7. Answer A. GFDPP 10A, AFH
The white position light is on the tail and the flashing red light is the anticollision light. Therefore, you would be looking at the rear of the aircraft, which indicates it is flying away from you.

10-8   PA.XI.A.K5, K7

During a night flight, you observe steady red and green lights ahead and at the same altitude. What is the general direction of movement of the other aircraft?

A – The other aircraft is crossing to the left.

B – The other aircraft is flying away from you.

C – The other aircraft is approaching head-on.

10-8. Answer C. GFDPP 10A, AFH
In this case, you are seeing both wingtip lights. Without a white tail light in between them, you would most likely be looking head-on at the aircraft. In this situation, the green light is on the left and the red light is on the right.

10-9 **PA.I.H.K1k, PA.XI.A.K1, K8**

Visual flight rule (VFR) approaches to land at night should be accomplished

A – at a higher airspeed.

B – with a steeper descent.

C – the same as during daytime.

10-9. Answer C. GFDPP 10A, AFH

Your landing approaches at night should be made the same as during the daytime to reduce the effects of landing illusions.

10-10 **PA.I.H.K1d, PA.XI.A.K1, K8**

The danger of spatial disorientation during flight in poor visual conditions can be reduced by

A – shifting the eyes quickly between the exterior visual field and the instrument panel.

B – having faith in the instruments rather than taking a chance on the sensory organs.

C – leaning the body in the opposite direction of the motion of the aircraft.

10-10. Answer B. GFDPP 10A, AIM

A lack of orientation regarding the position, attitude, or movement of the aircraft in space is defined as spatial disorientation. If you experience spatial disorientation during flight in restricted visibility, rely on the aircraft instrument indications. You are more susceptible to spatial disorientation if you use body signals to interpret flight attitude.

10-11 **PA.XI.A.K1, K8**

What is the illusion called when at night you stare at single point it and it appears to move?

A – False Horizon

B – Autokinesis

C – Landing illusion

10-11. Answer B. GFDPP 10A, AIM

If you stare at a single point of light against a dark background, such as a ground light or star for more than a few seconds it can appear to move.

10-12 **PA.XI.A.K1, K8**

At night you cannot make out the horizon this known as?

A – False Horizon

B – Autokinesis

C – Landing illusion

10-12. Answer A. GFDPP 10A, AIM

When the natural horizon can become obscured or not readily apparent. This is known as a false horizon.

SECTION A ■ **Aviation Physiology**

10-13   PA.I.H.K1d, PA.XI.A.K1, K8

Pilots are more subject to spatial disorientation if

A – they ignore the sensations of muscles and inner ear.

B – visual cues are taken away, as they are in instrument meteorological conditions (IMC).

C – eyes are moved often in the process of cross-checking the flight instruments.

10-13. Answer B. GFDPP 10A, PHB

When few outside visual references are available, you need to rely heavily on the flight instruments for accurate information. Spatial disorientation can occur when there is a conflict between the information relayed by your central vision scanning the instruments, and your peripheral vision, which has virtually no references with which to establish orientation, as in instrument flight rules (IFR) conditions.

If you experience spatial disorientation during flight in restricted visibility, rely on the aircraft instrument indications. You are more susceptible to spatial disorientation if you use body signals to interpret flight attitude.

10-14   PA.I.H.K1d, PA.VIII.E.K1

If you experience spatial disorientation during flight in a restricted visibility condition, the best way to overcome the effect and avoid an unusual attitude is to

A – rely upon the aircraft instrument indications.

B – concentrate on yaw, pitch, and roll sensations.

C – consciously slow the breathing rate until symptoms clear and then resume normal breathing rate.

10-14. Answer A. GFDPP 10A, PHB

If you experience spatial disorientation during flight in restricted visibility, rely on the aircraft instrument indications. You are more susceptible to spatial disorientation if you use body signals to interpret flight attitude.

10-15   PA.I.H.K1d

A lack of orientation regarding the position, attitude, or movement of the aircraft in space is defined as

A – spatial disorientation.

B – hyperventilation.

C – hypoxia.

10-15. Answer A. GFDPP 10A, PHB

Spatial disorientation is a lack of orientation regarding the position, attitude, or movement of the aircraft in space. Spatial disorientation can be caused by misleading information being sent to the brain by your body's various sensory organs. Awareness of your body's position is a result of input from three primary sources: vision, the vestibular system located in your inner ear, and your kinesthetic sense.

10-16   PA.I.H.K1e

Excessive attention inside the aircraft as well as anxiety and stress can cause passengers to experience

A – hypoxic hypoxia.

B – the inversion illusion.

C – motion sickness.

10-16. Answer C. GFDPP 10A, PHB

Motion sickness is produced by vestibular disorientation, especially when accompanied by anxiety and stress. Common symptoms include general discomfort, paleness, nausea, dizziness, sweating, and vomiting. Have passengers focus outside the airplane and open fresh-air vents to help mitigate motion sickness.

10-17   PA.I.H.K1a

Which statement best defines hypoxia?

A – A state of oxygen deficiency in the body.

B – An abnormal increase in the volume of air breathed.

C – A condition of gas bubble formation around the joints or muscles.

10-17. Answer A. GFDPP 10A, AIM
Hypoxia occurs when the body tissues do not receive enough oxygen.

10-18   PA.I.H.K1a

Anemia, carbon monoxide poisoning, and blood loss can all cause

A – stagnant hypoxia.

B – hypemic hypoxia.

C – histotoxic hypoxia.

10-18. Answer B. GFDPP 10A, AIM
Hypemic hypoxia is a result of a deficiency in the blood or number of healthy functioning blood cells. This can be caused through a variety of factors such as anemia, carbon monoxide poisoning, and disease.

10-19   PA.I.H.K1a, K1i, PA.IX.C.K4

How can pilots reduce susceptibility to hypoxia?

A – Maintaining good physical condition, eating a nutritious diet, and by avoiding alcohol and smoking.

B – Periodically removing supplemental oxygen above 14,000' MSL to acclimate to a higher altitude.

C – There is no way to reduce susceptibility to hypoxia.

10-19. Answer A. GFDPP 10A, FAR 91.211
Your susceptibility to hypoxia is related to many factors, many of which you can control. You can increase your tolerance to hypoxia by maintaining good physical condition, eating a nutritious diet, and by avoiding alcohol and smoking. Although acclimating to a higher altitude can provide an increased tolerance to the conditions that would lead to hypoxia, it is never recommended to remove supplemental oxygen when it is required.

10-20   PA.I.H.K1f

Large accumulations of carbon monoxide in the human body result in

A – tightness across the forehead.

B – loss of muscular power.

C – an increased sense of well-being.

10-20. Answer B. GFDPP 10A, AC 20-32
The key word in this question is "large" accumulations. Large accumulations of carbon monoxide in the body result in the loss of muscular power, vomiting, convulsions, and coma.

10-21   PA.I.H.K1f

Susceptibility to carbon monoxide poisoning increases as

A – altitude increases.

B – altitude decreases.

C – air pressure increases.

10-21. Answer A. GFDPP 10A, AC 20-32
Because carbon monoxide poisoning is a form of hypoxia, its effects are increased with altitude, where less oxygen is available.

SECTION A ■ **Aviation Physiology**

10-22   PA.I.H.K1a, PA.I.G.K1l

What are the oxygen requirements when operating at cabin pressure altitudes above 15,000 feet MSL?

A – The flight crew must use supplemental oxygen after 30 minutes.

B – Both the flight crew and passengers must use supplemental oxygen the entire time.

C – The flight crew must use supplemental oxygen the entire time and your passengers must be provided with supplemental oxygen.

10-22. Answer C. GFDPP 10A, FAR 91.211
When operating an unpressurized aircraft above:

- 12,500 feet MSL up to and including 14,000 feet MSL for more than 30 minutes, the flight crew must use supplemental oxygen.

- Above 14,000 feet MSL, the minimum flight crew must be provided with, and use, supplemental oxygen the entire time.

- Above 15,000 feet MSL, passengers must also be provided supplemental oxygen.

10-23   PA.I.H.K1l

What is the recommended waiting time before ascending to 8,000 feet mean sea level (MSL) after scuba diving during which a controlled ascent was performed?

A – 12 hours.

B – 24 hours.

C – 48 hours.

10-23. Answer B. GFDPP 10A, PHB, AIM
Decompression sickness (DCS) is caused by a rapid reduction in the ambient pressure surrounding the body. Nitrogen and other inert gases expand to form bubbles that rise out of solution and produce symptoms that range from pain in the joints to seizures and unconsciousness.

Scuba diving can cause DCS so the following wait times before ascending are recommended:

- To 8,000 feet MSL—12 hours after a dive that did not require a controlled ascent (nondecompression stop diving); 24 hours after a dive that required a controlled ascent (decompression stop diving).

- Above 8,000 feet MSL—24 hours after any dive.

10-24   PA.I.H.K1c

Which is an action that you can take to help overcome ear pain due to higher pressure in the auditory canal in the middle ear?

A – Increase the descent rate of the aircraft.

B – Recommend that your passenger perform the Valsalva maneuver prior to initiating a climb.

C – Decrease the decent rate of the aircraft.

10-24. Answer C. GFDPP 1C, PHB
During a descent, the outside air pressure in the auditory canal will become higher than the pressure in the middle ear. Slow descent rates can help prevent or reduce the severity of ear problems, and the eustachian tube can sometimes be opened by yawning, swallowing, or chewing. Higher pressure can be equalized by performing the Valsalva maneuver—holding the nose and mouth shut and forcibly exhaling to force air up the eustachian tube into the middle ear.

10-25   PA.I.H.K1b, K1c

When a stressful situation is encountered in flight, an abnormal increase in the volume of air breathed in and out can cause a condition known as

A – hyperventilation.

B – aerosinusitis.

C – aerotitis.

10-25. Answer A. GFDPP 10A, AIM

Rapid or extra deep breathing while using oxygen can cause a condition known as hyperventilation. Emotional tension, anxiety, or fear can also lead to hyperventilation.

Aerosinusitis is an inflammation of the sinuses and aerotitis is an inflammation of the middle ear. Both problems are caused by changes in air pressure.

10-26   PA.I.H.K1b

Which would most likely result in hyperventilation?

A – Emotional tension, anxiety, or fear.

B – The excessive consumption of alcohol.

C – An extremely slow rate of breathing and insufficient oxygen.

10-26. Answer A. GFDPP 10A, AIM

Rapid or extra deep breathing while using oxygen can cause a condition known as hyperventilation. Emotional tension, anxiety, or fear can also lead to hyperventilation.

10-27   PA.I.H.K1b

A pilot experiencing the effects of hyperventilation should be able to restore the proper carbon dioxide level in the body by

A – slowing the breathing rate, breathing into a paper bag, or talking aloud.

B – breathing spontaneously and deeply or gaining mental control of the situation.

C – increasing the breathing rate in order to increase lung ventilation.

10-27. Answer A. GFDPP 10A, AIM

You can overcome the symptoms, or avoid future occurrences of hyperventilation, by slowing your breathing rate, breathing into a bag, or talking aloud.

10-28   PA.I.H.K1g, K1h

What is true about the factors that affect fitness for flight?

A – Fatigue is not a factor in aircraft accidents.

B – The first noticeable effect of dehydration is typically fatigue.

C – Stress is not cumulative.

10-28. Answer B. GFDPP 1C, PHB

The following factors affect your fitness for flight:

- Because stress is cumulative, you bring into the flight varying degrees of stress left over from the other areas of your life.

- Excessive stress interferes with your ability to focus and cope with a given situation.

- Fatigue has been a major factor in many fatal accidents involving very experienced and highly qualified crews.

- Dehydration is a critical loss of water from the body. Although, dehydration can cause headaches, cramps, nausea, and dizziness, the first noticeable effect is typically fatigue.

SECTION A ■ **Aviation Physiology**

10-29   PA.I.H.K2

Under what condition, if any, may a pilot allow a person who is obviously under the influence of drugs to be carried aboard an aircraft?

A – Under no condition.

B – In an emergency, or if the person is a medical patient under proper care.

C – Only if the person does not have access to the cockpit or pilot's compartment.

10-29. Answer B. GFDPP 1C, FAR 91.17
Except in an emergency, no pilot of a civil aircraft may allow a person who appears to be intoxicated or who demonstrates by manner or physical indications that the individual is under the influence of drugs (except a patient under proper care) to be carried in that aircraft.

10-30   PA.I.H.K2

How soon after the conviction for driving while intoxicated by alcohol or drugs shall it be reported to the Federal Aviation Administration (FAA), Civil Aviation Security Division?

A – No later than 60 days after the motor vehicle action.

B – No later than 30 working days after the motor vehicle action.

C – Required to be reported upon renewal of medical certificate.

10-30. Answer A. GFDPP 1C, FAR 61.15, 91.17
Certificated pilots must provide a written report of each motor vehicle action to the FAA no later than 60 days after the action.

10-31   PA.I.H.K2, K3

No person may attempt to act as a crewmember of a civil aircraft with

A – 0.008 percent by weight or more alcohol in the blood.

B – 0.004 percent by weight or more alcohol in the blood.

C – 0.04 percent by weight or more alcohol in the blood.

10-31. Answer C. GFDPP 1C, FAR 91.17
You may not act or attempt to act as a crewmember of a civil aircraft while having 0.04 percent by weight or more alcohol in the blood.

10-32   PA.I.H.K2, K3

A person may not act as a crewmember of a civil aircraft if alcoholic beverages have been consumed by that person with the preceding

A – 8 hours.

B – 12 hours.

C – 24 hours.

10-32. Answer A. GFDPP 1C, FAR 61.15, 91.17

A common saying used in aviation for this regulation is "eight hours from bottle to throttle." In other words, you may not act or attempt to act as a crewmember of a civil aircraft within eight hours after the consumption of any alcoholic beverage.

10-33   PA.I.H.K1j

What is true about hypothermia?

A – Recognizing hypothermia is normally easy because symptoms occur rapidly.

B – A body temperature below 98° is considered hypothermia.

C – Hypothermia symptoms often begin gradually and can include confused thinking.

10-33. Answer C. GFDPP 10A

Hypothermia occurs when your body loses heat faster than it can produce heat, causing a dangerously low body temperature below 95°F (35°C). You might not be aware of your condition if you have hypothermia because the symptoms often begin gradually, and the confused thinking associated with hypothermia prevents self-awareness.

SECTION A ■ **Aviation Physiology**

SECTION B
Single-Pilot Resource Management

AERONAUTICAL DECISION MAKING

- The most common factor in preventable accidents is human error. It is estimated that approximately 75% of all aviation accidents are human factors related.

- Continued VFR flight into adverse weather/IMC is the primary cause of a significant number of fatal accidents. Entering instrument conditions as a non-instrument-qualified pilot can result in spatial disorientation and loss of control, or controlled flight into terrain (CFIT).

- Five hazardous attitudes that can interfere with a pilot's ability to make effective decisions are:

  - Anti-authority— you display this attitude if you resent having someone tell you what to do, or you regard rules and procedures as unnecessary.

  - *Antidote—follow the rules. They are usually right.*

  - Impulsivity— if you feel the need to act immediately and do the first thing that comes to mind without considering the best solution to a problem, then you are exhibiting impulsivity.

  - *Antidote—not so fast. Think first.*

  - Invulnerability—you are more likely to take chances and increase risk if you think accidents will not happen to you.

  - *Antidote—it could happen to me.*

  - Macho—if you have this attitude, you might take risks trying to prove that you are better than anyone else.

  - *Antidote—taking chances is foolish.*

  - Resignation—you are experiencing resignation if you feel that no matter what you do it will have little effect on what happens to you.

  - *Antidote—I'm not helpless. I can make a difference.*

- To combat a hazardous attitude, first recognize the attitude as hazardous, label the thought as one of the hazardous attitudes, and then recite that attitude's antidote.

- Through self assessment, you must ensure that you are current and prepared to act as pilot in command. Include personal minimums and limitations on a risk management checklist to help determine if you are prepared for a particular flight.

- On the day of the flight, use the FAA's I'M SAFE checklist to assess your fitness to fly.

- Risk management is the part of the decision-making process that relies on situational awareness, problem recognition, and good judgment to reduce risks associated with each flight.

- In the event a diversion becomes necessary, first, fly the airplane. Stay calm and avoid letting anxiety impair your judgment. Turn the airplane in the general direction of the new landing point and then use rule-of-thumb computations, or the Nearest and Direct-To functions of your GPS, to navigate there.

AUTOMATION MANAGEMENT

- Automation management typically applies to an airplane with an advanced avionics system that includes digital displays, GPS equipment, a moving map, and an integrated autopilot.

- You can experience information overload when transitioning to an electronic flight display. With the vast amount of information available on GPS equipment and multi-function displays, it is sometimes challenging to locate the exact information you need, avoiding the distraction of less-relevant information.

- Understanding the system and how the information is organized helps locate information and solve problems efficiently.

- A common error associated with using a moving map is overreliance on the moving map leading to complacency.

NOTE: An asterisk appearing after an ACS code (i.e. PA.VII.B.K1*) indicates that the question subject appears more than one time in the ACS. The code shown corresponds to the first instance of the subject in the ACS.

10-34   PA.I.H.R2

What is the antidote when a pilot has a hazardous attitude, such as "Macho"?

A – I can do it.

B – Taking chances is foolish.

C – Nothing will happen.

10-34. Answer B. GFDPP 10B, PHB

If you have the hazardous attitude of "Macho," you might take risks trying to prove that you are better than anyone else. The antidote is: "taking chances is foolish."

"I can do it" and "Nothing will happen" are not antidotes to any of the hazardous attitudes.

10-35   PA.I.H.R2

What is the antidote when a pilot has a hazardous attitude, such as "Resignation"?

A – What is the use?

B – Someone else is responsible.

C – I am not helpless.

10-35. Answer C. GFDPP 10B, PHB

If you have the hazardous attitude of "Resignation," you feel that no matter what you do it will have little effect on what happens to you. The antidote is "I'm not helpless. I can make a difference."

"What is the use" and "Someone else is responsible" are not antidotes to any of the hazardous attitudes.

10-36   PA.I.H.K4, PA.II.A.K1

Who is responsible for determining whether a pilot is fit to fly for a particular flight, even with a current medical certificate?

A – The FAA.

B – The medical examiner.

C – The pilot.

10-36. Answer C. GFDPP 10B, PHB

Through self-assessment, you must ensure that you are current and prepared to act as pilot in command. Use your personal minimums and limitations, and the I'M SAFE checklist to assess your fitness to fly. The FAA and medical examiners determine if you are fit to hold a medical certificate.

SECTION B ■ Single-Pilot Resource Management

10-37   PA.I.H.K4

Which is the most common factor in preventable accidents?

A – Structural failure.

B – Mechanical malfunction

C – Human error.

10-37. Answer C. GFDPP 10B, PHB

The most common factor in preventable accidents is human error. It is estimated that approximately 75% of all aviation accidents are human-factors related. Even when a mechanical malfunction or structural failure is determined to be the primary accident cause, human error is often cited as a contributing or aggravating factor.

10-38   PA.I.H.K4

What often leads to spatial disorientation or collision with ground or obstacles when flying under visual flight rules (VFR)?

A – Continued flight into instrument conditions.

B – Getting behind the aircraft.

C – Duck-under syndrome.

10-38. Answer A. GFDPP 10B, PHB

Continued VFR flight into adverse weather/IMC is the primary cause of a significant number of fatal accidents. Entering instrument conditions as a non-instrument-qualified pilot can result in spatial disorientation and loss of control or controlled flight into terrain (CFIT).

Although the FAA identifies "getting behind the aircraft" and "duck-under syndrome" as common operational pitfalls, these behaviors are not normally associated with spatial disorientation.

10-39   PA.II.B.K2, PA.II.F.K1a

What is one of the neglected items when a pilot relies on short and long-term memory for repetitive tasks?

A – Checklists.

B – Situational awareness.

C – Flying outside the envelope.

10-39. Answer A. GFDPP 10B, PHB

Pilots who rely on their memory for repetitive tasks often neglect the use of checklists.

10-40   PA.I.H.R2

Hazardous attitudes occur to every pilot to some degree at some time. What are some of these hazardous attitudes?

A – Poor risk management and lack of stress management.

B – Anti-authority, impulsivity, macho, resignation, and invulnerability.

C – Poor situational awareness, snap judgments, and lack of a decision-making process.

10-40. Answer B. GFDPP 10B, PHB

Five hazardous attitudes that can interfere with a pilot's ability to make effective decisions are:

Anti-authority. *Antidote—follow the rules, they are usually right.*

Impulsivity. *Antidote—not so fast. Think first.*

Macho. *Antidote—taking chances is foolish.*

Resignation. *Antidote—I'm not helpless. I can make a difference.*

Invulnerability. *Antidote—it could happen to me.*

10-41   PA.I.H.R2

In the aeronautical decision making (ADM) process, what is the first step in neutralizing a hazardous attitude?

A – Making a rational judgment.

B – Recognizing hazardous thoughts.

C – Recognizing the invulnerability of the situation.

10-41. Answer B. GFDPP 10B, PHB

To combat a hazardous attitude, first recognize the attitude as hazardous, label the thought as one of the hazardous attitudes, and then recite that attitude's antidote.

10-42   PA.IX.D.K4, K5

Which is true about airplane systems and equipment to use in an emergency situation?

A – All light aircraft are required to have the same emergency equipment and systems in the event of pilot incapacitation.

B – Emergency systems, such a ballistic parachute, should only be used if a fire is present that cannot be extinguished.

C – You should brief your passengers on how and when to use emergency equipment, such as the fire extinguisher, a ballistic parachute, or auto-land system.

2-42. Answer C. GFDPP 10B, PHB

You should brief your passengers on any systems or equipment that they might be required to use in an emergency in the event that you become incapacitated, such as the fire extinguisher, a ballistic parachute, or auto-land system. Because not all light aircraft are required to have certain emergency systems, such as a ballistic parachute or auto-land system, you must be thoroughly familiar with the operation of your airplane's specific equipment and follow the manufacturer's guidelines as to when and how to activate the system.

Examples of events that might require the use of a ballistic parachute include an engine failure, a loss of control of the airplane, a midair collision, or pilot incapacitation.

SECTION B ■ Single-Pilot Resource Management

APPENDIX 1

FAA LEGENDS

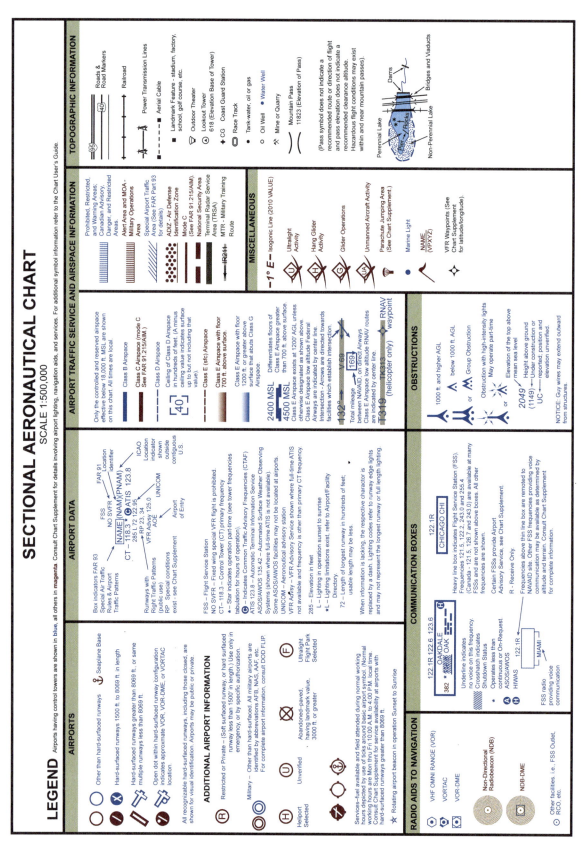

Legend 1. Sectional Aeronautical Chart.

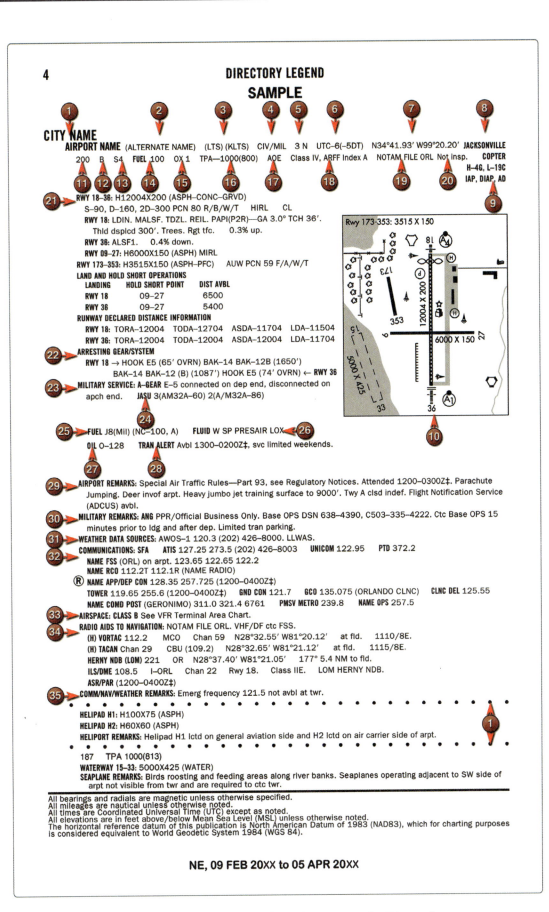

4

DIRECTORY LEGEND
SAMPLE

CITY NAME
AIRPORT NAME (ALTERNATE NAME) (LTS) (KLTS) CIV/MIL 3 N UTC–6(–5DT) N34°41.93' W99°20.20' JACKSONVILLE

200 B S4 FUEL 100 OX 1 TPA—1000(800) AOE Class IV, ARFF Index A NOTAM FILE ORL Not insp. COPTER
H–4G, L–19C
IAP, DIAP, AD

RWY 18–36: H12004X200 (ASPH–CONC–GRVD)
S–90, D–160, 2D–300 PCN 80 R/B/W/T   HIRL   CL
RWY 18: LDIN. MALSF. TDZL. REIL. PAPI(P2R)—GA 3.0° TCH 36'.
Thld dsplcd 300'. Trees. Rgt tfc.   0.3% up.
RWY 36: ALSF1.   0.4% down.
RWY 09–27: H6000X150 (ASPH) MIRL
RWY 173–353: H3515X150 (ASPH–PFC)   AUW PCN 59 F/A/W/T
LAND AND HOLD SHORT OPERATIONS

LANDING	HOLD SHORT POINT	DIST AVBL
RWY 18	09–27	6500
RWY 36	09–27	5400

RUNWAY DECLARED DISTANCE INFORMATION
RWY 18: TORA–12004   TODA–12704   ASDA–11704   LDA–11504
RWY 36: TORA–12004   TODA–12004   ASDA–12004   LDA–11704
ARRESTING GEAR/SYSTEM
RWY 18 → HOOK E5 (65' OVRN) BAK–14 BAK–12B (1650')
BAK–14 BAK–12 (B) (1087') HOOK E5 (74' OVRN) ← RWY 36
MILITARY SERVICE: A–GEAR E–5 connected on dep end, disconnected on
apch end.   JASU 3(AM32A–60) 2(A/M32A–86)

FUEL J8(Mil) (NC–100, A)   FLUID W SP PRESAIR LOX
OIL O–128   TRAN ALERT Avbl 1300–0200Z‡, svc limited weekends.

AIRPORT REMARKS: Special Air Traffic Rules—Part 93, see Regulatory Notices. Attended 1200–0300Z‡. Parachute
Jumping. Deer invof arpt. Heavy jumbo jet training surface to 9000'. Twy A clsd indef. Flight Notification Service
(ADCUS) avbl.
MILITARY REMARKS: ANG PPR/Official Business Only. Base OPS DSN 638–4390, C503–335–4222. Ctc Base OPS 15
minutes prior to ldg and after dep. Limited tran parking.
WEATHER DATA SOURCES: AWOS–1 120.3 (202) 426–8000. LLWAS.
COMMUNICATIONS: SFA   ATIS 127.25 273.5 (202) 426–8003   UNICOM 122.95   PTD 372.2
NAME FSS (ORL) on arpt. 123.65 122.65 122.2
NAME RCO 112.2T 112.1R (NAME RADIO)
Ⓡ NAME APP/DEP CON 128.35 257.725 (1200–0400Z‡)
TOWER 119.65 255.6 (1200–0400Z‡)   GND CON 121.7   GCO 135.075 (ORLANDO CLNC)   CLNC DEL 125.55
NAME COMD POST (GERONIMO) 311.0 321.4 6761   PMSV METRO 239.8   NAME OPS 257.5
AIRSPACE: CLASS B See VFR Terminal Area Chart.
RADIO AIDS TO NAVIGATION: NOTAM FILE ORL. VHF/DF ctc FSS.
(H) VORTAC 112.2   MCO   Chan 59   N28°32.55' W81°20.12'   at fld.   1110/8E.
(H) TACAN Chan 29   CBU (109.2)   N28°32.65' W81°21.12'   at fld.   1115/8E.
HERNY NDB (LOM) 221   OR   N28°37.40' W81°21.05'   177° 5.4 NM to fld.
ILS/DME 108.5   I–ORL   Chan 22   Rwy 18.   Class IIE.   LOM HERNY NDB.
ASR/PAR (1200–0400Z‡)
COMM/NAV/WEATHER REMARKS: Emerg frequency 121.5 not avbl at twr.

· ·
HELIPAD H1: H100X75 (ASPH)
HELIPAD H2: H60X60 (ASPH)
HELIPORT REMARKS: Helipad H1 lctd on general aviation side and H2 lctd on air carrier side of arpt.
· ·
187   TPA 1000(813)
WATERWAY 15–33: 5000X425 (WATER)
SEAPLANE REMARKS: Birds roosting and feeding areas along river banks. Seaplanes operating adjacent to SW side of
arpt not visible from twr and are required to ctc twr.

All bearings and radials are magnetic unless otherwise specified.
All mileages are nautical unless otherwise noted.
All times are Coordinated Universal Time (UTC) except as noted.
All elevations are in feet above/below Mean Sea Level (MSL) unless otherwise noted.
The horizontal reference datum of this publication is North American Datum of 1983 (NAD83), which for charting purposes
is considered equivalent to World Geodetic System 1984 (WGS 84).

NE, 09 FEB 20XX to 05 APR 20XX

(Rwy 173-353: 3515 X 150 / 12004 X 200 / 6000 X 150 / 5000 X 425)

LEGEND 2.—Chart Supplements U.S. (formerly Airport/Facility Directory).

DIRECTORY LEGEND   5

(10) SKETCH LEGEND

RUNWAYS/LANDING AREAS

Hard Surfaced . . . . ▬

Metal Surface . . . . ▨

Sod, Gravel, etc. . . . . . . ▨

Light Plane, ...
Ski Landing Area or Water

Under Construction . . . . .

Closed . . . . ⊠▬⊠

Helicopter Landings Area . . . . Ⓗ

Displaced Threshold . . . . ▬

Taxiway, Apron and Stopways . .

MISCELLANEOUS BASE AND CULTURAL FEATURES

Buildings . . . .

Power Lines . . . . —T—T—

Fence . . . .

Towers . . . .

Tanks . . . .

Oil Well . . . .

Smoke Stack . . . .

Obstruction . . . . Λ 5812

Controlling Obstruction . . . . +5812

Trees . . . .

Populated Places . . . .

Cuts and Fills . . . . . . Cut   Fill

Cliffs and Depressions . .

Ditch . . . .

Hill . . . .

RADIO AIDS TO NAVIGATION

VORTAC . . . ⬡   VOR . . . . ⬡

VOR/DME . . ▢   NDB . . . . ⊙

TACAN . . . . ⬠   NDB/DME . . . . ⊚

MISCELLANEOUS AERONAUTICAL FEATURES

Airport Beacon . . . .   ☆ ✹

Wind Cone . . . .   ➤ ⊥

Landing Tee . . . .   ⊦ ⊣

Tetrahedron . . . .   ▶ ⊣

Control Tower . . . .   🄱 or TWR

When control tower and rotating beacon are co-located beacon symbol will be used and further identified as TWR.

APPROACH LIGHTING SYSTEMS

A dot "•" portrayed with approach lighting letter identifier indicates sequenced flashing lights (F) installed with the approach lighting system e.g. Ⓐ Negative symbology, e.g., Ⓐ₁ Ⓥ Indicates Pilot Controlled Lighting (PCL).

Runway Centerline Lighting . . . . . ▬▬▬

Ⓐ Approach Lighting System ALSF-2 . .

Ⓐ₁ Approach Lighting System ALSF-1 . .

Ⓐ₂ Short Approach Lighting System SALS/SALSF

Ⓐ₃ Simplified Short Approach Lighting System (SSALR) with RAIL . . . . .

Ⓐ₄ Medium Intensity Approach Lighting System (MALS and MALSF)/(SSALS and SSALF)

Ⓐ₅ Medium Intensity Approach Lighting System (MALSR) and RAIL . . . . .

⊕ Omnidirectional Approach Lighting System (ODALS) . . . .

Ⓓ Navy Parallel Row and Cross Bar . . .

⊕ Air Force Overrun . . . .

Ⓥ Visual Approach Slope Indicator with Standard Threshold Clearance provided

Ⓥ₂ Pulsating Visual Approach Slope Indicator (PVASI)

Ⓥ₃ Visual Approach Slope Indicator with a threshold crossing height to accomodate long bodied or jumbo aircraft

Ⓥ₄ Tri-color Visual Approach Slope Indicator (TRCV)

Ⓥ₅ Approach Path Alignment Panel (APAP)

Ⓟ Precision Approach Path Indicator (PAPI)

NE, 09 FEB 20XX to 05 APR 20XX

LEGEND 3.—Chart Supplements U.S. (formerly Airport/Facility Directory).

6   **DIRECTORY LEGEND**

LEGEND

This directory is a listing of data on record with the FAA on all open to the public airports, military facilities and selected private use facilities specifically requested by the Department of Defense (DoD) for which a DoD Instrument Approach Procedure has been published in the U.S. Terminal Procedures Publication. Additionally this listing contains data for associated terminal control facilities, air route traffic control centers, and radio aids to navigation within the conterminous United States, Puerto Rico and the Virgin Islands. Joint civil/military and civil airports are listed alphabetically by state, associated city and airport name and cross-referenced by airport name. Military facilities are listed alphabetically by state and official airport name and cross-referenced by associated city name. Navaids, flight service stations and remote communication outlets that are associated with an airport, but with a different name, are listed alphabetically under their own name, as well as under the airport with which they are associated.

The listing of an open to the public airport in this directory merely indicates the airport operator's willingness to accommodate transient aircraft, and does not represent that the facility conforms with any Federal or local standards, or that it has been approved for use on the part of the general public. Military and private use facilities published in this directory are open to civil pilots only in an emergency or with prior permission. See Special Notice Section, Civil Use of Military Fields.

The information on obstructions is taken from reports submitted to the FAA. Obstruction data has not been verified in all cases. Pilots are cautioned that objects not indicated in this tabulation (or on the airports sketches and/or charts) may exist which can create a hazard to flight operation. Detailed specifics concerning services and facilities tabulated within this directory are contained in the Aeronautical Information Manual, Basic Flight Information and ATC Procedures.

The legend items that follow explain in detail the contents of this Directory and are keyed to the circled numbers on the sample on the preceding pages.

 CITY/AIRPORT NAME

Civil and joint civil/military airports and facilities in this directory are listed alphabetically by state and associated city. Where the city name is different from the airport name the city name will appear on the line above the airport name. Airports with the same associated city name will be listed alphabetically by airport name and will be separated by a dashed rule line. A solid rule line will separate all others. FAA approved helipads and seaplane landing areas associated with a land airport will be separated by a dotted line. Military airports are listed alphabetically by state and official airport name.

 ALTERNATE NAME

Alternate names, if any, will be shown in parentheses.

 LOCATION IDENTIFIER

The location identifier is a three or four character FAA code followed by a four-character ICAO code assigned to airports. ICAO codes will only be published at joint civil/military, and military facilities. If two different military codes are assigned, both codes will be shown with the primary operating agency's code listed first. These identifiers are used by ATC in lieu of the airport name in flight plans, flight strips and other written records and computer operations. Zeros will appear with a slash to differentiate them from the letter ''O''.

 OPERATING AGENCY

Airports within this directory are classified into two categories, Military/Federal Government and Civil airports open to the general public, plus selected private use airports. The operating agency is shown for military, private use and joint civil/military airports. The operating agency is shown by an abbreviation as listed below. When an organization is a tenant, the abbreviation is enclosed in parenthesis. No classification indicates the airport is open to the general public with no military tenant.

A	US Army	MC	Marine Corps
AFRC	Air Force Reserve Command	N	Navy
AF	US Air Force	NAF	Naval Air Facility
ANG	Air National Guard	NAS	Naval Air Station
AR	US Army Reserve	NASA	National Air and Space Administration
ARNG	US Army National Guard	P	US Civil Airport Wherein Permit Covers
CG	US Coast Guard		Use by Transient Military Aircraft
CIV/MIL	Joint Use Civil/Military	PVT	Private Use Only (Closed to the Public)
DND	Department of National Defense Canada		

 AIRPORT LOCATION

Airport location is expressed as distance and direction from the center of the associated city in nautical miles and cardinal points, e.g., 4 NE.

 TIME CONVERSION

Hours of operation of all facilities are expressed in Coordinated Universal Time (UTC) and shown as ''Z'' time. The directory indicates the number of hours to be subtracted from UTC to obtain local standard time and local daylight saving time UTC-5(-4DT). The symbol ‡ indicates that during periods of Daylight Saving Time effective hours will be one hour earlier than shown. In those areas where daylight saving time is not observed the (-4DT) and ‡ will not be shown. Daylight saving time is in effect from 0200 local time the second Sunday in March to 0200 local time the first Sunday in November. Canada and all U.S. Conterminous States observe daylight saving time except Arizona and Puerto Rico, and the Virgin Islands. If the state observes daylight saving time and the operating times are other than daylight saving times, the operating hours will include the dates, times and no ‡ symbol will be shown, i.e., April 15–Aug 31 0630–1700Z, Sep 1–Apr 14 0600–1700Z.

NE, 09 FEB 20XX to 05 APR 20XX

LEGEND 4.—Chart Supplements U.S. (formerly Airport/Facility Directory).

DIRECTORY LEGEND
7

7 GEOGRAPHIC POSITION OF AIRPORT—AIRPORT REFERENCE POINT (ARP)

Positions are shown as hemisphere, degrees, minutes and hundredths of a minute and represent the approximate geometric center of all usable runway surfaces.

8 CHARTS

Charts refer to the Sectional Chart and Low and High Altitude Enroute Chart and panel on which the airport or facility is located. Helicopter Chart locations will be indicated as COPTER. IFR Gulf of Mexico West and IFR Gulf of Mexico Central will be depicted as GOMW and GOMC.

9 INSTRUMENT APPROACH PROCEDURES, AIRPORT DIAGRAMS

IAP indicates an airport for which a prescribed (Public Use) FAA Instrument Approach Procedure has been published. DIAP indicates an airport for which a prescribed DoD Instrument Approach Procedure has been published in the U.S. Terminal Procedures. See the Special Notice Section of this directory, Civil Use of Military Fields and the Aeronautical Information Manual 5–4–5 Instrument Approach Procedure Charts for additional information. AD indicates an airport for which an airport diagram has been published. Airport diagrams are located in the back of each A/FD volume alphabetically by associated city and airport name.

10 AIRPORT SKETCH

The airport sketch, when provided, depicts the airport and related topographical information as seen from the air and should be used in conjunction with the text. It is intended as a guide for pilots in VFR conditions. Symbology that is not self-explanatory will be reflected in the sketch legend. The airport sketch will be oriented with True North at the top. Airport sketches will be added incrementally.

11 ELEVATION

The highest point of an airport's usable runways measured in feet from mean sea level. When elevation is sea level it will be indicated as ''00''. When elevation is below sea level a minus ''−'' sign will precede the figure.

12 ROTATING LIGHT BEACON

B indicates rotating beacon is available. Rotating beacons operate sunset to sunrise unless otherwise indicated in the AIRPORT REMARKS or MILITARY REMARKS segment of the airport entry.

13 SERVICING—CIVIL

S1:	Minor airframe repairs.	S5:	Major airframe repairs.
S2:	Minor airframe and minor powerplant repairs.	S6:	Minor airframe and major powerplant repairs.
S3:	Major airframe and minor powerplant repairs.	S7:	Major powerplant repairs.
S4:	Major airframe and major powerplant repairs.	S8:	Minor powerplant repairs.

14 FUEL

CODE	FUEL	CODE	FUEL
80	Grade 80 gasoline (Red)	B+	Jet B, Wide-cut, turbine fuel with FS–II*, FP** minus 50° C.
100	Grade 100 gasoline (Green)		
100LL	100LL gasoline (low lead) (Blue)	J4 (JP4)	(JP–4 military specification) FP** minus 58° C.
115	Grade 115 gasoline (115/145 military specification) (Purple)		
		J5 (JP5)	(JP–5 military specification) Kerosene with FS–11, FP** minus 46°C.
A	Jet A, Kerosene, without FS–II*, FP** minus 40° C.		
		J8 (JP8)	(JP–8 military specification) Jet A–1, Kerosene with FS–II*, FP** minus 47°C.
A+	Jet A, Kerosene, with FS–II*, FP** minus 40°C.		
A1	Jet A–1, Kerosene, without FS–II*, FP** minus 47°C.	J8+100	(JP–8 military specification) Jet A–1, Kerosene with FS–II*, FP** minus 47°C, with-fuel additive package that improves thermo stability characteristics of JP–8.
A1+	Jet A–1, Kerosene with FS–II*, FP** minus 47° C.		
		J	(Jet Fuel Type Unknown)
B	Jet B, Wide-cut, turbine fuel without FS–II*, FP** minus 50° C.	MOGAS	Automobile gasoline which is to be used as aircraft fuel.

*(Fuel System Icing Inhibitor)

**(Freeze Point)

<u>NOTE:</u> Certain automobile gasoline may be used in specific aircraft engines if a FAA supplemental type certificate has been obtained. Automobile gasoline, which is to be used in aircraft engines, will be identified as ''MOGAS'', however, the grade/type and other octane rating will not be published.

Data shown on fuel availability represents the most recent information the publisher has been able to acquire. Because of a variety of factors, the fuel listed may not always be obtainable by transient civil pilots. Confirmation of availability of fuel should be made directly with fuel suppliers at locations where refueling is planned.

15 OXYGEN—CIVIL

OX 1	High Pressure	OX 3	High Pressure—Replacement Bottles
OX 2	Low Pressure	OX 4	Low Pressure—Replacement Bottles

16 TRAFFIC PATTERN ALTITUDE

Traffic Pattern Altitude (TPA)—The first figure shown is TPA above mean sea level. The second figure in parentheses is TPA above airport elevation. Multiple TPA shall be shown as ''TPA—See Remarks'' and detailed information shall be shown in the Airport or Military Remarks Section. Traffic pattern data for USAF bases, USN facilities, and U.S. Army airports (including those on which ACC or U.S. Army is a tenant) that deviate from standard pattern altitudes shall be shown in Military Remarks.

NE, 09 FEB 20XX to 05 APR 20XX

LEGEND 5.—Chart Supplements U.S. (formerly Airport/Facility Directory).

8   **DIRECTORY LEGEND**

 AIRPORT OF ENTRY, LANDING RIGHTS, AND CUSTOMS USER FEE AIRPORTS

U.S. CUSTOMS USER FEE AIRPORT—Private Aircraft operators are frequently required to pay the costs associated with customs processing.

AOE—Airport of Entry. A customs Airport of Entry where permission from U.S. Customs is not required to land. However, at least one hour advance notice of arrival is required.

LRA—Landing Rights Airport. Application for permission to land must be submitted in advance to U.S. Customs. At least one hour advance notice of arrival is required.

NOTE: Advance notice of arrival at both an AOE and LRA airport may be included in the flight plan when filed in Canada or Mexico. Where Flight Notification Service (ADCUS) is available the airport remark will indicate this service. This notice will also be treated as an application for permission to land in the case of an LRA. Although advance notice of arrival may be relayed to Customs through Mexico, Canada, and U.S. Communications facilities by flight plan, the aircraft operator is solely responsible for ensuring that Customs receives the notification. (See Customs, Immigration and Naturalization, Public Health and Agriculture Department requirements in the International Flight Information Manual for further details.)

US Customs Air and Sea Ports, Inspectors and Agents

Northeast Sector (New England and Atlantic States—ME to MD)	407–975–1740
Southeast Sector (Atlantic States—DC, WV, VA to FL)	407–975–1780
Central Sector (Interior of the US, including Gulf states—MS, AL, LA)	407–975–1760
Southwest East Sector (OK and eastern TX)	407–975–1840
Southwest West Sector (Western TX, NM and AZ)	407–975–1820
Pacific Sector (WA, OR, CA, HI and AK)	407–975–1800

 CERTIFICATED AIRPORT (14 CFR PART 139)

Airports serving Department of Transportation certified carriers and certified under 14 CFR part 139 are indicated by the Class and the ARFF Index; e.g. Class I, ARFF Index A, which relates to the availability of crash, fire, rescue equipment. Class I airports can have an ARFF Index A through E, depending on the aircraft length and scheduled departures. Class II, III, and IV will always carry an Index A.

14 CFR PART 139 CERTIFICATED AIRPORTS
AIRPORT CLASSIFICATIONS

Type of Air Carrier Operation	Class I	Class II	Class III	Class IV
Scheduled Air Carrier Aircraft with 31 or more passenger seats	X			
Unscheduled Air Carrier Aircraft with 31 or more passengers seats	X	X		X
Scheduled Air Carrier Aircraft with 10 to 30 passenger seats	X	X	X	

14 CFR–PART 139 CERTIFICATED AIRPORTS
INDICES AND AIRCRAFT RESCUE AND FIRE FIGHTING EQUIPMENT REQUIREMENTS

Airport Index	Required No. Vehicles	Aircraft Length	Scheduled Departures	Agent + Water for Foam
A	1	<90'	≥1	500#DC or HALON 1211 or 450#DC + 100 gal H$_2$O
B	1 or 2	≥90', <126'	≥5	Index A + 1500 gal H$_2$O
		≥126', <159'	<5	
C	2 or 3	≥126', <159'	≥5	Index A + 3000 gal H$_2$O
		≥159', <200'	<5	
D	3	≥159', <200'		Index A + 4000 gal H$_2$O
		>200'	<5	
E	3	≥200'	≥5	Index A + 6000 gal H$_2$O

> Greater Than; < Less Than; ≥ Equal or Greater Than; ≤ Equal or Less Than;   H$_2$O–Water; DC–Dry Chemical.

NOTE: The listing of ARFF index does not necessarily assure coverage for non-air carrier operations or at other than prescribed times for air carrier. ARFF Index Ltd.—indicates ARFF coverage may or may not be available, for information contact airport manager prior to flight.

 NOTAM SERVICE

All public use landing areas are provided NOTAM service. A NOTAM FILE identifier is shown for individual landing areas, e.g., "NOTAM FILE BNA". See the AIM, Basic Flight Information and ATC Procedures for a detailed description of NOTAMs.

NE, 09 FEB 20XX to 05 APR 20XX

LEGEND 6.—Chart Supplements U.S. (formerly Airport/Facility Directory).

DIRECTORY LEGEND
9

Current NOTAMs are available from flight service stations at 1–800–WX–BRIEF (992–7433) or online through the FAA PilotWeb at https://pilotweb.nas.faa.gov. Military NOTAMs are available using the Defense Internet NOTAM Service (DINS) at https://www.notams.jcs.mil.

Pilots flying to or from airports not available through the FAA PilotWeb or DINS can obtain assistance from Flight Service.

 FAA INSPECTION

All airports not inspected by FAA will be identified by the note: Not insp. This indicates that the airport information has been provided by the owner or operator of the field.

 RUNWAY DATA

Runway information is shown on two lines. That information common to the entire runway is shown on the first line while information concerning the runway ends is shown on the second or following line. Runway direction, surface, length, width, weight bearing capacity, lighting, and slope, when available are shown for each runway. Multiple runways are shown with the longest runway first. Direction, length, width, and lighting are shown for sea-lanes. The full dimensions of helipads are shown, e.g., 50X150. Runway data that requires clarification will be placed in the remarks section.

RUNWAY DESIGNATION

Runways are normally numbered in relation to their magnetic orientation rounded off to the nearest 10 degrees. Parallel runways can be designated L (left)/R (right)/C (center). Runways may be designated as Ultralight or assault strips. Assault strips are shown by magnetic bearing.

RUNWAY DIMENSIONS

Runway length and width are shown in feet. Length shown is runway end to end including displaced thresholds, but excluding those areas designed as overruns.

RUNWAY SURFACE AND LENGTH

Runway lengths prefixed by the letter ''H'' indicate that the runways are hard surfaced (concrete, asphalt, or part asphalt–concrete). If the runway length is not prefixed, the surface is sod, clay, etc. The runway surface composition is indicated in parentheses after runway length as follows:

(AFSC)—Aggregate friction seal coat	(GRVD)—Grooved	(PSP)—Pierced steel plank
(AMS)—Temporary metal planks coated with nonskid material	(GRVL)—Gravel, or cinders	(RFSC)—Rubberized friction seal coat
	(MATS)—Pierced steel planking, landing mats, membranes	(TURF)—Turf
(ASPH)—Asphalt		(TRTD)—Treated
(CONC)—Concrete	(PEM)—Part concrete, part asphalt	(WC)—Wire combed
(DIRT)—Dirt	(PFC)—Porous friction courses	

RUNWAY WEIGHT BEARING CAPACITY

Runway strength data shown in this publication is derived from available information and is a realistic estimate of capability at an average level of activity. It is not intended as a maximum allowable weight or as an operating limitation. Many airport pavements are capable of supporting limited operations with gross weights in excess of the published figures. Permissible operating weights, insofar as runway strengths are concerned, are a matter of agreement between the owner and user. When desiring to operate into any airport at weights in excess of those published in the publication, users should contact the airport management for permission. Runway strength figures are shown in thousand of pounds, with the last three figures being omitted. Add 000 to figure following S, D, 2S, 2T, AUW, SWL, etc., for gross weight capacity. A blank space following the letter designator is used to indicate the runway can sustain aircraft with this type landing gear, although definite runway weight bearing capacity figures are not available, e.g., S, D. Applicable codes for typical gear configurations with S=Single, D=Dual, T=Triple and Q=Quadruple:

CURRENT	NEW	NEW DESCRIPTION
S	S	Single wheel type landing gear (DC3), (C47), (F15), etc.
D	D	Dual wheel type landing gear (BE1900), (B737), (A319), etc.
T	D	Dual wheel type landing gear (P3, C9).
ST	2S	Two single wheels in tandem type landing gear (C130).
TRT	2T	Two triple wheels in tandem type landing gear (C17), etc.
DT	2D	Two dual wheels in tandem type landing gear (B707), etc.
TT	2D	Two dual wheels in tandem type landing gear (B757, KC135).
SBTT	2D/D1	Two dual wheels in tandem/dual wheel body gear type landing gear (KC10).
None	2D/2D1	Two dual wheels in tandem/two dual wheels in tandem body gear type landing gear (A340–600).
DDT	2D/2D2	Two dual wheels in tandem/two dual wheels in double tandem body gear type landing gear (B747, E4).
TTT	3D	Three dual wheels in tandem type landing gear (B777), etc.
TT	D2	Dual wheel gear two struts per side main gear type landing gear (B52).
TDT	C5	Complex dual wheel and quadruple wheel combination landing gear (C5).

NE, 09 FEB 20XX to 05 APR 20XX

Legend 7.—Chart Supplements U.S. (formerly Airport/Facility Directory).

10 ### DIRECTORY LEGEND

AUW—All up weight. Maximum weight bearing capacity for any aircraft irrespective of landing gear configuration.

SWL—Single Wheel Loading. (This includes information submitted in terms of Equivalent Single Wheel Loading (ESWL) and Single Isolated Wheel Loading).

PSI—Pounds per square inch. PSI is the actual figure expressing maximum pounds per square inch runway will support, e.g., (SWL 000/PSI 535).

Omission of weight bearing capacity indicates information unknown.

The ACN/PCN System is the ICAO standard method of reporting pavement strength for pavements with bearing strengths greater than 12,500 pounds. The Pavement Classification Number (PCN) is established by an engineering assessment of the runway. The PCN is for use in conjunction with an Aircraft Classification Number (ACN). Consult the Aircraft Flight Manual, Flight Information Handbook, or other appropriate source for ACN tables or charts. Currently, ACN data may not be available for all aircraft. If an ACN table or chart is available, the ACN can be calculated by taking into account the aircraft weight, the pavement type, and the subgrade category. For runways that have been evaluated under the ACN/PCN system, the PCN will be shown as a five-part code (e.g. PCN 80 R/B/W/T). Details of the coded format are as follows:

(1) The PCN NUMBER—The reported PCN indicates that an aircraft with an ACN equal or less than the reported PCN can operate on the pavement subject to any limitation on the tire pressure.

(2) The type of pavement:
R — Rigid
F — Flexible

(3) The pavement subgrade category:
A — High
B — Medium
C — Low
D — Ultra-low

(4) The maximum tire pressure authorized for the pavement:
W — High, no limit
X — Medium, limited to 217 psi
Y — Low, limited to 145 psi
Z — Very low, limited to 73 psi

(5) Pavement evaluation method:
T — Technical evaluation
U — By experience of aircraft using the pavement

NOTE: Prior permission from the airport controlling authority is required when the ACN of the aircraft exceeds the published PCN or aircraft tire pressure exceeds the published limits.

RUNWAY LIGHTING

Lights are in operation sunset to sunrise. Lighting available by prior arrangement only or operating part of the night and/or pilot controlled lighting with specific operating hours are indicated under airport or military remarks. At USN/USMC facilities lights are available only during airport hours of operation. Since obstructions are usually lighted, obstruction lighting is not included in this code. Unlighted obstructions on or surrounding an airport will be noted in airport or military remarks. Runway lights nonstandard (NSTD) are systems for which the light fixtures are not FAA approved L-800 series: color, intensity, or spacing does not meet FAA standards. Nonstandard runway lights, VASI, or any other system not listed below will be shown in airport remarks or military service. Temporary, emergency or limited runway edge lighting such as flares, smudge pots, lanterns or portable runway lights will also be shown in airport remarks or military service. Types of lighting are shown with the runway or runway end they serve.

NSTD—Light system fails to meet FAA standards.
LIRL—Low Intensity Runway Lights.
MIRL—Medium Intensity Runway Lights.
HIRL—High Intensity Runway Lights.
RAIL—Runway Alignment Indicator Lights.
REIL—Runway End Identifier Lights.
CL—Centerline Lights.
TDZL—Touchdown Zone Lights.
ODALS—Omni Directional Approach Lighting System.
AF OVRN—Air Force Overrun 1000′ Standard Approach Lighting System.
LDIN—Lead-In Lighting System.
MALS—Medium Intensity Approach Lighting System.
MALSF—Medium Intensity Approach Lighting System with Sequenced Flashing Lights.
MALSR—Medium Intensity Approach Lighting System with Runway Alignment Indicator Lights.

SALS—Short Approach Lighting System.
SALSF—Short Approach Lighting System with Sequenced Flashing Lights.
SSALS—Simplified Short Approach Lighting System.
SSALF—Simplified Short Approach Lighting System with Sequenced Flashing Lights.
SSALR—Simplified Short Approach Lighting System with Runway Alignment Indicator Lights.
ALSAF—High Intensity Approach Lighting System with Sequenced Flashing Lights.
ALSF1—High Intensity Approach Lighting System with Sequenced Flashing Lights, Category I, Configuration.
ALSF2—High Intensity Approach Lighting System with Sequenced Flashing Lights, Category II, Configuration.
SF—Sequenced Flashing Lights.
OLS—Optical Landing System.
WAVE–OFF.

NOTE: Civil ALSF2 may be operated as SSALR during favorable weather conditions. When runway edge lights are positioned more than 10 feet from the edge of the usable runway surface a remark will be added in the "Remarks" portion of the airport entry. This is applicable to Air Force, Air National Guard and Air Force Reserve Bases, and those joint civil/military airfields on which they are tenants.

NE, 09 FEB 20XX to 05 APR 20XX

LEGEND 8.—Chart Supplements U.S. (formerly Airport/Facility Directory).

DIRECTORY LEGEND   11

VISUAL GLIDESLOPE INDICATORS

APAP—A system of panels, which may or may not be lighted, used for alignment of approach path.

PNIL	APAP on left side of runway	PNIR	APAP on right side of runway

PAPI—Precision Approach Path Indicator

P2L	2-identical light units placed on left side of runway	P4L	4-identical light units placed on left side of runway
P2R	2-identical light units placed on right side of runway	P4R	4-identical light units placed on right side of runway

PVASI—Pulsating/steady burning visual approach slope indicator, normally a single light unit projecting two colors.

PSIL	PVASI on left side of runway	PSIR	PVASI on right side of runway

SAVASI—Simplified Abbreviated Visual Approach Slope Indicator

S2L	2-box SAVASI on left side of runway	S2R	2-box SAVASI on right side of runway

TRCV—Tri-color visual approach slope indicator, normally a single light unit projecting three colors.

TRIL	TRCV on left side of runway	TRIR	TRCV on right side of runway

VASI—Visual Approach Slope Indicator

V2L	2-box VASI on left side of runway	V6L	6-box VASI on left side of runway
V2R	2-box VASI on right side of runway	V6R	6-box VASI on right side of runway
V4L	4-box VASI on left side of runway	V12	12-box VASI on both sides of runway
V4R	4-box VASI on right side of runway	V16	16-box VASI on both sides of runway

NOTE: Approach slope angle and threshold crossing height will be shown when available; i.e., –GA 3.5° TCH 37'.

PILOT CONTROL OF AIRPORT LIGHTING

Key Mike	Function
7 times within 5 seconds	Highest intensity available
5 times within 5 seconds	Medium or lower intensity (Lower REIL or REIL-Off)
3 times within 5 seconds	Lowest intensity available (Lower REIL or REIL-Off)

Available systems will be indicated in the airport or military remarks, e.g., ACTIVATE HIRL Rwy 07–25, MALSR Rwy 07, and VASI Rwy 07—122.8.

Where the airport is not served by an instrument approach procedure and/or has an independent type system of different specification installed by the airport sponsor, descriptions of the type lights, method of control, and operating frequency will be explained in clear text. See AIM, "Basic Flight Information and ATC Procedures," for detailed description of pilot control of airport lighting.

RUNWAY SLOPE

When available, runway slope data will only be provided for those airports with an approved FAA instrument approach procedure. Runway slope will be shown only when it is 0.3 percent or greater. On runways less than 8000 feet, the direction of the slope up will be indicated, e.g., 0.3% up NW. On runways 8000 feet or greater, the slope will be shown (up or down) on the runway end line, e.g., RWY 13: 0.3% up., RWY 21: Pole. Rgt tfc. 0.4% down.

RUNWAY END DATA

Information pertaining to the runway approach end such as approach lights, touchdown zone lights, runway end identification lights, visual glideslope indicators, displaced thresholds, controlling obstruction, and right hand traffic pattern, will be shown on the specific runway end. "Rgt tfc"—Right traffic indicates right turns should be made on landing and takeoff for specified runway end.

LAND AND HOLD SHORT OPERATIONS (LAHSO)

LAHSO is an acronym for "Land and Hold Short Operations." These operations include landing and holding short of an intersection runway, an intersecting taxiway, or other predetermined points on the runway other than a runway or taxiway. Measured distance represents the available landing distance on the landing runway, in feet.

Specific questions regarding these distances should be referred to the air traffic manager of the facility concerned. The Aeronautical Information Manual contains specific details on hold–short operations and markings.

RUNWAY DECLARED DISTANCE INFORMATION

TORA—Take-off Run Available. The length of runway declared available and suitable for the ground run of an aeroplane take–off.

TODA—Take-off Distance Available. The length of the take–off run available plus the length of the clearway, if provided.

ASDA—Accelerate-Stop Distance Available. The length of the take–off run available plus the length of the stopway, if provided.

LDA—Landing Distance Available. The length of runway which is declared available and suitable for the ground run of an aeroplane landing.

22 ARRESTING GEAR/SYSTEMS

Arresting gear is shown as it is located on the runway. The a–gear distance from the end of the appropriate runway (or into the overrun) is indicated in parentheses. A–Gear which has a bi–direction capability and can be utilized for emergency approach end engagement is indicated by a (B). The direction of engaging device is indicated by an arrow. Up to 15 minutes advance notice may be required for rigging A–Gear for approach and engagement. Airport listing may show availability of other than US Systems. This information is provided for emergency requirements only. Refer to current aircraft operating manuals for specific engagement weight and speed criteria based on aircraft structural restrictions and arresting system limitations.

Following is a list of current systems referenced in this publication identified by both Air Force and Navy terminology:

NE, 09 FEB 20XX to 05 APR 20XX

LEGEND 9.—Chart Supplements U.S. (formerly Airport/Facility Directory).

12   **DIRECTORY LEGEND**

BI–DIRECTIONAL CABLE (B)

TYPE	DESCRIPTION
BAK–9	Rotary friction brake.
BAK–12A	Standard BAK–12 with 950 foot run out, 1–inch cable and 40,000 pound weight setting. Rotary friction brake.
BAK–12B	Extended BAK–12 with 1200 foot run, 1¼ inch Cable and 50,000 pounds weight setting. Rotary friction brake.
E28	Rotary Hydraulic (Water Brake).
M21	Rotary Hydraulic (Water Brake) Mobile.

The following device is used in conjunction with some aircraft arresting systems:

BAK–14	A device that raises a hook cable out of a slot in the runway surface and is remotely positioned for engagement by the tower on request. (In addition to personnel reaction time, the system requires up to five seconds to fully raise the cable.)
H	A device that raises a hook cable out of a slot in the runway surface and is remotely positioned for engagement by the tower on request. (In addition to personnel reaction time, the system requires up to one and one–half seconds to fully raise the cable.)

UNI–DIRECTIONAL CABLE

TYPE	DESCRIPTION
MB60	Textile brake—an emergency one–time use, modular braking system employing the tearing of specially woven textile straps to absorb the kinetic energy.
E5/E5–1/E5–3	Chain Type. At USN/USMC stations E–5 A–GEAR systems are rated, e.g., E–5 RATING–13R–1100 HW (DRY), 31L/R–1200 STD (WET). This rating is a function of the A–GEAR chain weight and length and is used to determine the maximum aircraft engaging speed. A dry rating applies to a stabilized surface (dry or wet) while a wet rating takes into account the amount (if any) of wet overrun that is not capable of withstanding the aircraft weight. These ratings are published under Military Service.

FOREIGN CABLE

TYPE	DESCRIPTION	US EQUIVALENT
44B–3H	Rotary Hydraulic) (Water Brake)	
CHAG	Chain	E–5

UNI–DIRECTIONAL BARRIER

TYPE	DESCRIPTION
MA–1A	Web barrier between stanchions attached to a chain energy absorber.
BAK–15	Web barrier between stanchions attached to an energy absorber (water squeezer, rotary friction, chain). Designed for wing engagement.

NOTE: Landing short of the runway threshold on a runway with a BAK–15 in the underrun is a significant hazard. The barrier in the down position still protrudes several inches above the underrun. Aircraft contact with the barrier short of the runway threshold can cause damage to the barrier and substantial damage to the aircraft.

OTHER

TYPE	DESCRIPTION
EMAS	Engineered Material Arresting System, located beyond the departure end of the runway, consisting of high energy absorbing materials which will crush under the weight of an aircraft.

㉓ MILITARY SERVICE

Specific military services available at the airport are listed under this general heading. Remarks applicable to any military service are shown in the individual service listing.

㉔ JET AIRCRAFT STARTING UNITS (JASU)

The numeral preceding the type of unit indicates the number of units available. The absence of the numeral indicates ten or more units available. If the number of units is unknown, the number one will be shown. Absence of JASU designation indicates non–availability.

The following is a list of current JASU systems referenced in this publication:

USAF JASU (For variations in technical data, refer to T.O. 35–1–7.)

ELECTRICAL STARTING UNITS:

A/M32A–86	AC: 115/200v, 3 phase, 90 kva, 0.8 pf, 4 wire
	DC: 28v, 1500 amp, 72 kw (with TR pack)
MC–1A	AC: 115/208v, 400 cycle, 3 phase, 37.5 kva, 0.8 pf, 108 amp, 4 wire
	DC: 28v, 500 amp, 14 kw
MD–3	AC: 115/208v, 400 cycle, 3 phase, 60 kva, 0.75 pf, 4 wire
	DC: 28v, 1500 amp, 45 kw, split bus
MD–3A	AC: 115/208v, 400 cycle, 3 phase, 60 kva, 0.75 pf, 4 wire
	DC: 28v, 1500 amp, 45 kw, split bus
MD–3M	AC: 115/208v, 400 cycle, 3 phase, 60 kva, 0.75 pf, 4 wire
	DC: 28v, 500 amp, 15 kw

NE, 09 FEB 20XX to 05 APR 20XX

LEGEND 10.—Chart Supplements U.S. (formerly Airport/Facility Directory).

DIRECTORY LEGEND   13

MD–4	AC: 120/208v, 400 cycle, 3 phase, 62.5 kva, 0.8 pf, 175 amp, ''WYE'' neutral ground, 4 wire, 120v, 400 cycle, 3 phase, 62.5 kva, 0.8 pf, 303 amp, ''DELTA'' 3 wire, 120v, 400 cycle, 1 phase, 62.5 kva, 0.8 pf, 520 amp, 2 wire

AIR STARTING UNITS

AM32–95	150 +/– 5 lb/min (2055 +/– 68 cfm) at 51 +/– 2 psia
AM32A–95	150 +/– 5 lb/min @ 49 +/– 2 psia (35 +/– 2 psig)
LASS	150 +/– 5 lb/min @ 49 +/– 2 psia
MA–1A	82 lb/min (1123 cfm) at 130° air inlet temp, 45 psia (min) air outlet press
MC–1	15 cfm, 3500 psia
MC–1A	15 cfm, 3500 psia
MC–2A	15 cfm, 200 psia
MC–11	8,000 cu in cap, 4000 psig, 15 cfm

COMBINED AIR AND ELECTRICAL STARTING UNITS:

AGPU	AC: 115/200v, 400 cycle, 3 phase, 30 kw gen
	DC: 28v, 700 amp
	AIR: 60 lb/min @ 40 psig @ sea level
AM32A–60*	AIR: 120 +/– 4 lb/min (1644 +/– 55 cfm) at 49 +/– 2 psia
	AC: 120/208v, 400 cycle, 3 phase, 75 kva, 0.75 pf, 4 wire, 120v, 1 phase, 25 kva
	DC: 28v, 500 amp, 15 kw
AM32A–60A	AIR: 150 +/– 5 lb/min (2055 +/– 68 cfm at 51 +/– psia
	AC: 120/208v, 400 cycle, 3 phase, 75 kva, 0.75 pf, 4 wire
	DC: 28v, 200 amp, 5.6 kw
AM32A–60B*	AIR: 130 lb/min, 50 psia
	AC: 120/208v, 400 cycle, 3 phase, 75 kva, 0.75 pf, 4 wire
	DC: 28v, 200 amp, 5.6 kw

*NOTE: During combined air and electrical loads, the pneumatic circuitry takes preference and will limit the amount of electrical power available.

USN JASU

ELECTRICAL STARTING UNITS:

NC–8A/A1	DC: 500 amp constant, 750 amp intermittent, 28v;
	AC: 60 kva @ .8 pf, 115/200v, 3 phase, 400 Hz.
NC–10A/A1/B/C	DC: 750 amp constant, 1000 amp intermittent, 28v;
	AC: 90 kva, 115/200v, 3 phase, 400 Hz.

AIR STARTING UNITS:

GTC–85/GTE–85	120 lbs/min @ 45 psi.
MSU–200NAV/A/U47A–5	204 lbs/min @ 56 psia.
WELLS AIR START SYSTEM	180 lbs/min @ 75 psi or 120 lbs/min @ 45 psi. Simultaneous multiple start capability.

COMBINED AIR AND ELECTRICAL STARTING UNITS:

NCPP–105/RCPT	180 lbs/min @ 75 psi or 120 lbs/min @ 45 psi. 700 amp, 28v DC. 120/208v, 400 Hz AC, 30 kva.

JASU (ARMY)

59B2–1B	28v, 7.5 kw, 280 amp.

OTHER JASU

ELECTRICAL STARTING UNITS (DND):

CE12	AC 115/200v, 140 kva, 400 Hz, 3 phase
CE13	AC 115/200v, 60 kva, 400 Hz, 3 phase
CE14	AC/DC 115/200v, 140 kva, 400 Hz, 3 phase, 28vDC, 1500 amp
CE15	DC 22–35v, 500 amp continuous 1100 amp intermittent
CE16	DC 22–35v, 500 amp continuous 1100 amp intermittent soft start

AIR STARTING UNITS (DND):

CA2	ASA 45.5 psig, 116.4 lb/min

COMBINED AIR AND ELECTRICAL STARTING UNITS (DND)

CEA1	AC 120/208v, 60 kva, 400 Hz, 3 phase DC 28v, 75 amp
	AIR 112.5 lb/min, 47 psig

ELECTRICAL STARTING UNITS (OTHER)

C–26	28v 45kw 115–200v 15kw 380–800 Hz 1 phase 2 wire
C–26–B, C–26–C	28v 45kw: Split Bus: 115–200v 15kw 380–800 Hz 1 phase 2 wire
E3	DC 28v/10kw

AIR STARTING UNITS (OTHER):

A4	40 psi/2 lb/sec (LPAS Mk12, Mk12L, Mk12A, Mk1, Mk2B)
MA–1	150 Air HP, 115 lb/min 50 psia
MA–2	250 Air HP, 150 lb/min 75 psia

CARTRIDGE:

MXU–4A	USAF

NE, 09 FEB 20XX to 05 APR 20XX

LEGEND 11.—Chart Supplements U.S. (formerly Airport/Facility Directory).

14   DIRECTORY LEGEND

25 **FUEL—MILITARY**

Fuel available through US Military Base supply, DESC Into–Plane Contracts and/or reciprocal agreement is listed first and is followed by (Mil). At commercial airports where Into–Plane contracts are in place, the name of the refueling agent is shown. Military fuel should be used first if it is available. When military fuel cannot be obtained but Into–Plane contract fuel is available, Government aircraft must refuel with the contract fuel and applicable refueling agent to avoid any breach in contract terms and conditions. Fuel not available through the above is shown preceded by NC (no contract). When fuel is obtained from NC sources, local purchase procedures must be followed. The US Military Aircraft Identaplates DD Form 1896 (Jet Fuel), DD Form 1897 (Avgas) and AF Form 1245 (Avgas) are used at military installations only. The US Government Aviation Into–Plane Reimbursement (AIR) Card (currently issued by AVCARD) is the instrument to be used to obtain fuel under a DESC Into–Plane Contract and for NC purchases if the refueling agent at the commercial airport accepts the AVCARD. A current list of contract fuel locations is available online at www.desc.dla.mil/Static/ProductsAndServices.asp; click on the Commercial Airports button.

See legend item 14 for fuel code and description.

26 **SUPPORTING FLUIDS AND SYSTEMS—MILITARY**

CODE	
ADI	Anti–Detonation Injection Fluid—Reciprocating Engine Aircraft.
W	Water Thrust Augmentation—Jet Aircraft.
WAI	Water–Alcohol Injection Type, Thrust Augmentation—Jet Aircraft.
SP	Single Point Refueling.
PRESAIR	Air Compressors rated 3,000 PSI or more.
De–Ice	Anti–icing/De–icing/Defrosting Fluid (MIL–A–8243).

OXYGEN:

LPOX	Low pressure oxygen servicing.
HPOX	High pressure oxygen servicing.
LHOX	Low and high pressure oxygen servicing.
LOX	Liquid oxygen servicing.
OXRB	Oxygen replacement bottles. (Maintained primarily at Naval stations for use in acft where oxygen can be replenished only by replacement of cylinders.)
OX	Indicates oxygen servicing when type of servicing is unknown.

NOTE: Combinations of above items is used to indicate complete oxygen servicing available;

LHOXRB	Low and high pressure oxygen servicing and replacement bottles;
LPOXRB	Low pressure oxygen replacement bottles only, etc.

NOTE: Aircraft will be serviced with oxygen procured under military specifications only. Aircraft will not be serviced with medical oxygen.

NITROGEN:

LPNIT — Low pressure nitrogen servicing.
HPNIT — High pressure nitrogen servicing.
LHNIT — Low and high pressure nitrogen servicing.

27 **OIL—MILITARY**

US AVIATION OILS (MIL SPECS):

CODE	GRADE, TYPE
O–113	1065, Reciprocating Engine Oil (MIL–L–6082)
O–117	1100, Reciprocating Engine Oil (MIL–L–6082)
O–117+	1100, O–117 plus cyclohexanone (MIL–L–6082)
O–123	1065, (Dispersant), Reciprocating Engine Oil (MIL–L–22851 Type III)
O–128	1100, (Dispersant), Reciprocating Engine Oil (MIL–L–22851 Type II)
O–132	1005, Jet Engine Oil (MIL–L–6081)
O–133	1010, Jet Engine Oil (MIL–L–6081)
O–147	None, MIL–L–6085A Lubricating Oil, Instrument, Synthetic
O–148	None, MIL–L–7808 (Synthetic Base) Turbine Engine Oil
O–149	None, Aircraft Turbine Engine Synthetic, 7.5c St
O–155	None, MIL–L–6086C, Aircraft, Medium Grade
O–156	None, MIL–L–23699 (Synthetic Base), Turboprop and Turboshaft Engines
JOAP/SOAP	Joint Oil Analysis Program. JOAP support is furnished during normal duty hours, other times on request. (JOAP and SOAP programs provide essentially the same service, JOAP is now the standard joint service supported program.)

28 **TRANSIENT ALERT (TRAN ALERT)—MILITARY**

Tran Alert service is considered to include all services required for normal aircraft turn–around, e.g., servicing (fuel, oil, oxygen, etc.), debriefing to determine requirements for maintenance, minor maintenance, inspection and parking assistance of transient aircraft. Drag chute repack, specialized maintenance, or extensive repairs will be provided within the capabilities and priorities of the base. Delays can be anticipated after normal duty hours/holidays/weekends regardless of the hours of transient maintenance operation. Pilots should not expect aircraft to be serviced for TURN–AROUNDS during time periods when servicing or maintenance manpower is not available. In the case of airports not operated exclusively by US military, the servicing indicated by the remarks will not always be available for US military

NE, 09 FEB 20XX to 05 APR 20XX

LEGEND 12.—Chart Supplements U.S. (formerly Airport/Facility Directory).

DIRECTORY LEGEND 15

aircraft. When transient alert services are not shown, facilities are unknown. NO PRIORITY BASIS—means that transient alert services will be provided only after all the requirements for mission/tactical assigned aircraft have been accomplished.

 29 AIRPORT REMARKS

The Attendance Schedule is the months, days and hours the airport is actually attended. Airport attendance does not mean watchman duties or telephone accessibility, but rather an attendant or operator on duty to provide at least minimum services (e.g., repairs, fuel, transportation).

Airport Remarks have been grouped in order of applicability. Airport remarks are limited to those items of information that are determined essential for operational use, i.e., conditions of a permanent or indefinite nature and conditions that will remain in effect for more than 30 days concerning aeronautical facilities, services, maintenance available, procedures or hazards, knowledge of which is essential for safe and efficient operation of aircraft. Information concerning permanent closing of a runway or taxiway will not be shown. A note ''See Special Notices'' shall be applied within this remarks section when a special notice applicable to the entry is contained in the Special Notices section of this publication.

Parachute Jumping indicates parachute jumping areas associated with the airport. See Parachute Jumping Area section of this publication for additional Information.

Landing Fee indicates landing charges for private or non-revenue producing aircraft. In addition, fees may be charged for planes that remain over a couple of hours and buy no services, or at major airline terminals for all aircraft.

Note: Unless otherwise stated, remarks including runway ends refer to the runway's approach end.

30 MILITARY REMARKS

Military Remarks published at a joint Civil/Military facility are remarks that are applicable to the Military. At Military Facilities all remarks will be published under the heading Military Remarks. Remarks contained in this section may not be applicable to civil users. The first group of remarks is applicable to the primary operator of the airport. Remarks applicable to a tenant on the airport are shown preceded by the tenant organization, i.e., (A) (AF) (N) (ANG), etc. Military airports operate 24 hours unless otherwise specified. Airport operating hours are listed first (airport operating hours will only be listed if they are different than the airport attended hours or if the attended hours are unavailable) followed by pertinent remarks in order of applicability. Remarks will include information on restrictions, hazards, traffic pattern, noise abatement, customs/agriculture/immigration, and miscellaneous information applicable to the Military.

Type of restrictions:

CLOSED: When designated closed, the airport is restricted from use by all aircraft unless stated otherwise. Any closure applying to specific type of aircraft or operation will be so stated. USN/USMC/USAF airports are considered closed during non-operating hours. Closed airports may be utilized during an emergency provided there is a safe landing area.

OFFICIAL BUSINESS ONLY: The airfield is closed to all transient military aircraft for obtaining routine services such as fueling, passenger drop off or pickup, practice approaches, parking, etc. The airfield may be used by aircrews and aircraft if official government business (including civilian) must be conducted on or near the airfield and prior permission is received from the airfield manager.

AF OFFICIAL BUSINESS ONLY OR NAVY OFFICIAL BUSINESS ONLY: Indicates that the restriction applies only to service indicated.

PRIOR PERMISSION REQUIRED (PPR): Airport is closed to transient aircraft unless approval for operation is obtained from the appropriate commander through Chief, Airfield Management or Airfield Operations Officer. Official Business or PPR does not preclude the use of US Military airports as an alternate for IFR flights. If a non–US military airport is used as a weather alternate and requires a PPR, the PPR must be requested and confirmed before the flight departs. The purpose of PPR is to control volume and flow of traffic rather than to prohibit it. Prior permission is required for all aircraft requiring transient alert service outside the published transient alert duty hours. All aircraft carrying hazardous materials must obtain prior permission as outlined in AFJI 11–204, AR 95–27, OPNAVINST 3710.7.

Note: OFFICIAL BUSINESS ONLY AND PPR restrictions are not applicable to Special Air Mission (SAM) or Special Air Resource (SPAR) aircraft providing person or persons on aboard are designated Code 6 or higher as explained in AFJMAN 11–213, AR 95–11, OPNAVINST 3722–8J. Official Business Only or PPR do not preclude the use of the airport as an alternate for IFR flights.

31 WEATHER DATA SOURCES

Weather data sources will be listed alphabetically followed by their assigned frequencies and/or telephone number and hours of operation.

ASOS—Automated Surface Observing System. Reports the same as an AWOS–3 plus precipitation identification and intensity, and freezing rain occurrence;

AWOS—Automated Weather Observing System

  AWOS–A—reports altimeter setting (all other information is advisory only).

  AWOS–AV—reports altimeter and visibility.

  AWOS–1—reports altimeter setting, wind data and usually temperature, dew point and density altitude.

  AWOS–2—reports the same as AWOS–1 plus visibility.

  AWOS–3—reports the same as AWOS–1 plus visibility and cloud/ceiling data.

  AWOS–3P reports the same as the AWOS–3 system, plus a precipitation identification sensor.

  AWOS–3PT reports the same as the AWOS–3 system, plus precipitation identification sensor and a thunderstorm/lightning reporting capability.

  AWOS–3T reports the same as AWOS–3 system and includes a thunderstorm/lightning reporting capability.

NE, 09 FEB 20XX to 05 APR 20XX

Legend 13.—Chart Supplements U.S. (formerly Airport/Facility Directory).

16 DIRECTORY LEGEND

See AIM, Basic Flight Information and ATC Procedures for detailed description of Weather Data Sources.

AWOS–4—reports same as AWOS–3 system, plus precipitation occurence, type and accumulation, freezing rain, thunderstorm, and runway surface sensors.

HIWAS—See RADIO AIDS TO NAVIGATION

LAWRS—Limited Aviation Weather Reporting Station where observers report cloud height, weather, obstructions to vision, temperature and dewpoint (in most cases), surface wind, altimeter and pertinent remarks.

LLWAS—Indicates a Low Level Wind Shear Alert System consisting of a center field and several field perimeter anemometers.

SAWRS—identifies airports that have a Supplemental Aviation Weather Reporting Station available to pilots for current weather information.

SWSL—Supplemental Weather Service Location providing current local weather information via radio and telephone.

TDWR—indicates airports that have Terminal Doppler Weather Radar.

WSP—indicates airports that have Weather System Processor.

When the automated weather source is broadcast over an associated airport NAVAID frequency (see NAVAID line), it shall be indicated by a bold ASOS, AWOS, or HIWAS followed by the frequency, identifier and phone number, if available.

32 COMMUNICATIONS

Airport terminal control facilities and radio communications associated with the airport shall be shown. When the call sign is not the same as the airport name the call sign will be shown. Frequencies shall normally be shown in descending order with the primary frequency listed first. Frequencies will be listed, together with sectorization indicated by outbound radials, and hours of operation. Communications will be listed in sequence as follows:

Single Frequency Approach (SFA), Common Traffic Advisory Frequency (CTAF), Automatic Terminal Information Service (ATIS) and Aeronautical Advisory Stations (UNICOM) or (AUNICOM) along with their frequency is shown, where available, on the line following the heading "COMMUNICATIONS." When the CTAF and UNICOM frequencies are the same, the frequency will be shown as CTAF/UNICOM 122.8.

The FSS telephone nationwide is toll free 1–800–WX–BRIEF (1–800–992–7433). When the FSS is located on the field it will be indicated as "on arpt". Frequencies available at the FSS will follow in descending order. Remote Communications Outlet (RCO) providing service to the airport followed by the frequency and FSS RADIO name will be shown when available.

FSS's provide information on airport conditions, radio aids and other facilities, and process flight plans. Airport Advisory Service (AAS) is provided on the CTAF by FSS's for select non-tower airports or airports where the tower is not in operation.

(See AIM, Para 4–1–9 Traffic Advisory Practices at Airports Without Operating Control Towers or AC 90–42C.)

Aviation weather briefing service is provided by FSS specialists. Flight and weather briefing services are also available by calling the telephone numbers listed.

Remote Communications Outlet (RCO)—An unmanned air/ground communications facility that is remotely controlled and provides UHF or VHF communications capability to extend the service range of an FSS.

Civil Communications Frequencies-Civil communications frequencies used in the FSS air/ground system are operated on 122.0, 122.2, 123.6; emergency 121.5; plus receive-only on 122.1.
   a. 122.0 is assigned as the Enroute Flight Advisory Service frequency at selected FSS RADIO outlets.
   b. 122.2 is assigned as a common enroute frequency.
   c. 123.6 is assigned as the airport advisory frequency at select non-tower locations. At airports with a tower, FSS may provide airport advisories on the tower frequency when tower is closed.
   d. 122.1 is the primary receive-only frequency at VOR's.
   e. Some FSS's are assigned 50 kHz frequencies in the 122–126 MHz band (eg. 122.45). Pilots using the FSS A/G system should refer to this directory or appropriate charts to determine frequencies available at the FSS or remoted facility through which they wish to communicate.

Emergency frequency 121.5 and 243.0 are available at all Flight Service Stations, most Towers, Approach Control and RADAR facilities.

Frequencies published followed by the letter "T" or "R", indicate that the facility will only transmit or receive respectively on that frequency. All radio aids to navigation (NAVAID) frequencies are transmit only.

TERMINAL SERVICES

SFA—Single Frequency Approach.

CTAF—A program designed to get all vehicles and aircraft at airports without an operating control tower on a common frequency.

ATIS—A continuous broadcast of recorded non-control information in selected terminal areas.

D–ATIS—Digital ATIS provides ATIS information in text form outside the standard reception range of conventional ATIS via landline & data link communications and voice message within range of existing transmitters.

AUNICOM—Automated UNICOM is a computerized, command response system that provides automated weather, radio check capability and airport advisory information selected from an automated menu by microphone clicks.

UNICOM—A non-government air/ground radio communications facility which may provide airport information.

PTD—Pilot to Dispatcher.

APP CON—Approach Control. The symbol ® indicates radar approach control.

TOWER—Control tower.

GCA—Ground Control Approach System.

GND CON—Ground Control.

GCO—Ground Communication Outlet—An unstaffed, remotely controlled, ground/ground communications facility. Pilots at

NE, 09 FEB 20XX to 05 APR 20XX

Legend 14.—Chart Supplements U.S. (formerly Airport/Facility Directory).

DIRECTORY LEGEND   17

uncontrolled airports may contact ATC and FSS via VHF to a telephone connection to obtain an instrument clearance or close a VFR or IFR flight plan. They may also get an updated weather briefing prior to takeoff. Pilots will use four "key clicks" on the VHF radio to contact the appropriate ATC facility or six "key clicks" to contact the FSS. The GCO system is intended to be used only on the ground.

DEP CON—Departure Control. The symbol ℝ indicates radar departure control.

CLNC DEL—Clearance Delivery.

PRE TAXI CLNC—Pre taxi clearance.

VFR ADVSY SVC—VFR Advisory Service. Service provided by Non-Radar Approach Control.

  Advisory Service for VFR aircraft (upon a workload basis) ctc APP CON.

COMD POST—Command Post followed by the operator call sign in parenthesis.

PMSV—Pilot–to–Metro Service call sign, frequency and hours of operation, when full service is other than continuous. PMSV installations at which weather observation service is available shall be indicated, following the frequency and/or hours of operation as "Wx obsn svc 1900–0000Z‡" or "other times" may be used when no specific time is given. PMSV facilities manned by forecasters are considered "Full Service". PMSV facilities manned by weather observers are listed as "Limited Service".

OPS—Operations followed by the operator call sign in parenthesis.

CON

RANGE

FLT FLW—Flight Following

MEDIVAC

NOTE: Communication frequencies followed by the letter "X" indicate frequency available on request.

 AIRSPACE

Information concerning Class B, C, and part–time D and E surface area airspace shall be published with effective times. Class D and E surface area airspace that is continuous as established by Rulemaking Docket will not be shown.

CLASS B—Radar Sequencing and Separation Service for all aircraft in CLASS B airspace.

CLASS C—Separation between IFR and VFR aircraft and sequencing of VFR arrivals to the primary airport.

TRSA—Radar Sequencing and Separation Service for participating VFR Aircraft within a Terminal Radar Service Area.

Class C, D, and E airspace described in this publication is that airspace usually consisting of a 5 NM radius core surface area that begins at the surface and extends upward to an altitude above the airport elevation (charted in MSL for Class C and Class D). Class E surface airspace normally extends from the surface up to but not including the overlying controlled airspace.

When part–time Class C or Class D airspace defaults to Class E, the core surface area becomes Class E. This will be formatted as:

AIRSPACE: CLASS C svc "times" ctc APP CON other times CLASS E:

or

AIRSPACE: CLASS D svc "times" other times CLASS E.

When a part–time Class C, Class D or Class E surface area defaults to Class G, the core surface area becomes Class G up to, but not including, the overlying controlled airspace. Normally, the overlying controlled airspace is Class E airspace beginning at either 700' or 1200' AGL and may be determined by consulting the relevant VFR Sectional or Terminal Area Charts. This will be formatted as:

AIRSPACE: CLASS C svc "times" ctc APP CON other times CLASS G, with CLASS E 700' (or 1200') AGL & abv:

or

AIRSPACE: CLASS D svc "times" other times CLASS G with CLASS E 700' (or 1200') AGL & abv:

or

AIRSPACE: CLASS E svc "times" other times CLASS G with CLASS E 700' (or 1200') AGL & abv.

NOTE: AIRSPACE SVC "TIMES" INCLUDE ALL ASSOCIATED ARRIVAL EXTENSIONS. Surface area arrival extensions for instrument approach procedures become part of the primary core surface area. These extensions may be either Class D or Class E airspace and are effective concurrent with the times of the primary core surface area. For example, when a part–time Class C, Class D or Class E surface area defaults to Class G, the associated arrival extensions will default to Class G at the same time. When a part–time Class C or Class D surface area defaults to Class E, the arrival extensions will remain in effect as Class E airspace.

NOTE: CLASS E AIRSPACE EXTENDING UPWARD FROM 700 FEET OR MORE ABOVE THE SURFACE, DESIGNATED IN CONJUNCTION WITH AN AIRPORT WITH AN APPROVED INSTRUMENT PROCEDURE.

Class E 700' AGL (shown as magenta vignette on sectional charts) and 1200' AGL (blue vignette) areas are designated when necessary to provide controlled airspace for transitioning to/from the terminal and enroute environments. Unless otherwise specified, these 700'/1200' AGL Class E airspace areas remain in effect continuously, regardless of airport operating hours or surface area status. These transition areas should not be confused with surface areas or arrival extensions.

(See Chapter 3, AIRSPACE, in the Aeronautical Information Manual for further details)

NE, 09 FEB 20XX to 05 APR 20XX

Legend 15.—Chart Supplements U.S. (formerly Airport/Facility Directory).

18　　　　　　　　　　　　**DIRECTORY LEGEND**

 RADIO AIDS TO NAVIGATION

The Airport/Facility Directory lists, by facility name, all Radio Aids to Navigation that appear on FAA, AeroNav Products Visual or IFR Aeronautical Charts and those upon which the FAA has approved an Instrument Approach Procedure, with exception of selected TACANs. Military TACAN information will be published for Military facilities contained in this publication. All VOR, VORTAC, TACAN, ILS and MLS equipment in the National Airspace System has an automatic monitoring and shutdown feature in the event of malfunction. Unmonitored, as used in this publication, for any navigational aid, means that monitoring personnel cannot observe the malfunction or shutdown signal. The NAVAID NOTAM file identifier will be shown as ''NOTAM FILE IAD'' and will be listed on the Radio Aids to Navigation line. When two or more NAVAIDS are listed and the NOTAM file identifier is different from that shown on the Radio Aids to Navigation line, it will be shown with the NAVAID listing. NOTAM file identifiers for ILSs and its components (e.g., NDB (LOM) are the same as the associated airports and are not repeated. Automated Surface Observing System (ASOS), Automated Weather Observing System (AWOS), and Hazardous Inflight Weather Advisory Service (HIWAS) will be shown when this service is broadcast over selected NAVAIDs.

NAVAID information is tabulated as indicated in the following sample:

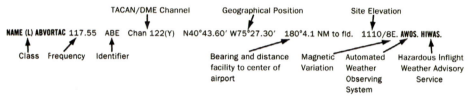

Restriction within the normal altitude/range of the navigational aid (See primary alphabetical listing for restrictions on VORTAC and VOR/DME).

Note: Those DME channel numbers with a (Y) suffix require TACAN to be placed in the ''Y'' mode to receive distance information.

HIWAS—Hazardous Inflight Weather Advisory Service is a continuous broadcast of inflight weather advisories including summarized SIGMETs, convective SIGMETs, AIRMETs and urgent PIREPs. HIWAS is presently broadcast over selected VOR's throughout the U.S.

ASR/PAR—Indicates that Surveillance (ASR) or Precision (PAR) radar instrument approach minimums are published in the U.S. Terminal Procedures. Only part-time hours of operation will be shown.

RADIO CLASS DESIGNATIONS

VOR/DME/TACAN Standard Service Volume (SSV) Classifications

SSV Class	Altitudes	Distance (NM)
(T) Terminal	1000' to 12,000'	25
(L) Low Altitude	1000' to 18,000'	40
(H) High Altitude	1000' to 14,500'	40
	14,500' to 18,000'	100
	18,000' to 45,000'	130
	45,000' to 60,000'	100

NOTE: Additionally, (H) facilities provide (L) and (T) service volume and (L) facilities provide (T) service. Altitudes are with respect to the station's site elevation. Coverage is not available in a cone of airspace directly above the facility.

CONTINUED ON NEXT PAGE

NE, 09 FEB 20XX to 05 APR 20XX

APPENDIX 1 ■ FAA Legends

LEGEND 16.—Chart Supplements U.S. (formerly Airport/Facility Directory).

DIRECTORY LEGEND

CONTINUED FROM PRECEDING PAGE

19

The term VOR is, operationally, a general term covering the VHF omnidirectional bearing type of facility without regard to the fact that the power, the frequency protected service volume, the equipment configuration, and operational requirements may vary between facilities at different locations.

AB	Automatic Weather Broadcast.
DF	Direction Finding Service.
DME	UHF standard (TACAN compatible) distance measuring equipment.
DME(Y)	UHF standard (TACAN compatible) distance measuring equipment that require TACAN to be placed in the ''Y'' mode to receive DME.
GS	Glide slope.
H	Non-directional radio beacon (homing), power 50 watts to less than 2,000 watts (50 NM at all altitudes).
HH	Non-directional radio beacon (homing), power 2,000 watts or more (75 NM at all altitudes).
H-SAB	Non-directional radio beacons providing automatic transcribed weather service.
ILS	Instrument Landing System (voice, where available, on localizer channel).
IM	Inner marker.
ISMLS	Interim Standard Microwave Landing System.
LDA	Localizer Directional Aid.
LMM	Compass locator station when installed at middle marker site (15 NM at all altitudes).
LOM	Compass locator station when installed at outer marker site (15 NM at all altitudes).
MH	Non-directional radio beacon (homing) power less than 50 watts (25 NM at all altitudes).
MLS	Microwave Landing System.
MM	Middle marker.
OM	Outer marker.
S	Simultaneous range homing signal and/or voice.
SABH	Non-directional radio beacon not authorized for IFR or ATC. Provides automatic weather broadcasts.
SDF	Simplified Direction Facility.
TACAN	UHF navigational facility-omnidirectional course and distance information.
VOR	VHF navigational facility-omnidirectional course only.
VOR/DME	Collocated VOR navigational facility and UHF standard distance measuring equipment.
VORTAC	Collocated VOR and TACAN navigational facilities.
W	Without voice on radio facility frequency.
Z	VHF station location marker at a LF radio facility.

NE, 09 FEB 20XX to 05 APR 20XX

Legend 17.—Chart Supplements U.S. (formerly Airport/Facility Directory).

20 DIRECTORY LEGEND

ILS FACILITY PEFORMANCE CLASSIFICATION CODES

Codes define the ability of an ILS to support autoland operations. The two portions of the code represent Official Category and farthest point along a Category I, II, or III approach that the Localizer meets Category III structure tolerances.

Official Category: I, II, or III; the lowest minima on published or unpublished procedures supported by the ILS.

Farthest point of satisfactory Category III Localizer performance for Category I, II, or III approaches: A – 4 NM prior to runway threshold, B – 3500 ft prior to runway threshold, C – glide angle dependent but generally 750–1000 ft prior to threshold, T – runway threshold, D – 3000 ft after runway threshold, and E – 2000 ft prior to stop end of runway.

ILS information is tabulated as indicated in the following sample:

ILS/DME   108.5   I–ORL   Chan 22   Rwy 18.   Class IIE.   LOM HERNY NDB.

ILS Facility Performance
Classification Code

FREQUENCY PAIRING PLAN AND MLS CHANNELING

MLS CHANNEL	VHF FREQUENCY	TACAN CHANNEL	MLS CHANNEL	VHF FREQUENCY	TACAN CHANNEL	MLS CHANNEL	VHF FREQUENCY	TACAN CHANNEL
500	108.10	18X	568	109.45	31Y	636	114.15	88Y
502	108.30	20X	570	109.55	32Y	638	114.25	89Y
504	108.50	22X	572	109.65	33Y	640	114.35	90Y
506	108.70	24X	574	109.75	34Y	642	114.45	91Y
508	108.90	26X	576	109.85	35Y	644	114.55	92Y
510	109.10	28X	578	109.95	36Y	646	114.65	93Y
512	109.30	30X	580	110.05	37Y	648	114.75	94Y
514	109.50	32X	582	110.15	38Y	650	114.85	95Y
516	109.70	34X	584	110.25	39Y	652	114.95	96Y
518	109.90	36X	586	110.35	40Y	654	115.05	97Y
520	110.10	38X	588	110.45	41Y	656	115.15	98Y
522	110.30	40X	590	110.55	42Y	658	115.25	99Y
524	110.50	42X	592	110.65	43Y	660	115.35	100Y
526	110.70	44X	594	110.75	44Y	662	115.45	101Y
528	110.90	46X	596	110.85	45Y	664	115.55	102Y
530	111.10	48X	598	110.95	46Y	666	115.65	103Y
532	111.30	50X	600	111.05	47Y	668	115.75	104Y
534	111.50	52X	602	111.15	48Y	670	115.85	105Y
536	111.70	54X	604	111.25	49Y	672	115.95	106Y
538	111.90	56X	606	111.35	50Y	674	116.05	107Y
540	108.05	17Y	608	111.45	51Y	676	116.15	108Y
542	108.15	18Y	610	111.55	52Y	678	116.25	109Y
544	108.25	19Y	612	111.65	53Y	680	116.35	110Y
546	108.35	20Y	614	111.75	54Y	682	116.45	111Y
548	108.45	21Y	616	111.85	55Y	684	116.55	112Y
550	108.55	22Y	618	111.95	56Y	686	116.65	113Y
552	108.65	23Y	620	113.35	80Y	688	116.75	114Y
554	108.75	24Y	622	113.45	81Y	690	116.85	115Y
556	108.85	25Y	624	113.55	82Y	692	116.95	116Y
558	108.95	26Y	626	113.65	83Y	694	117.05	117Y
560	109.05	27Y	628	113.75	84Y	696	117.15	118Y
562	109.15	28Y	630	113.85	85Y	698	117.25	119Y
564	109.25	29Y	632	113.95	86Y			
566	109.35	30Y	634	114.05	87Y			

FREQUENCY PAIRING PLAN AND MLS CHANNELING

The following is a list of paired VOR/ILS VHF frequencies with TACAN channels and MLS channels.

TACAN CHANNEL	VHF FREQUENCY	MLS CHANNEL	TACAN CHANNEL	VHF FREQUENCY	MLS CHANNEL	TACAN CHANNEL	VHF FREQUENCY	MLS CHANNEL
2X	134.5	-	19Y	108.25	544	25X	108.80	-
2Y	134.55	-	20X	108.30	502	25Y	108.85	556
11X	135.4	-	20Y	108.35	546	26X	108.90	508
11Y	135.45	-	21X	108.40	-	26Y	108.95	558
12X	135.5	-	21Y	108.45	548	27X	109.00	-
12Y	135.55	-	22X	108.50	504	27Y	109.05	560
17X	108.00	-	22Y	108.55	550	28X	109.10	510
17Y	108.05	540	23X	108.60	-	28Y	109.15	562
18X	108.10	500	23Y	108.65	552	29X	109.20	-
18Y	108.15	542	24X	108.70	506	29Y	109.25	564
19X	108.20	-	24Y	108.75	554	30X	109.30	512

NE, 09 FEB 20XX to 05 APR 20XX

LEGEND 18.—Chart Supplements U.S. (formerly Airport/Facility Directory).

DIRECTORY LEGEND   21

TACAN CHANNEL	VHF FREQUENCY	MLS CHANNEL	TACAN CHANNEL	VHF FREQUENCY	MLS CHANNEL	TACAN CHANNEL	VHF FREQUENCY	MLS CHANNEL
30Y	109.35	566	63X	133.60	-	95Y	114.85	650
31X	109.40	-	63Y	133.65	-	96X	114.90	-
31Y	109.45	568	64X	133.70	-	96Y	114.95	652
32X	109.50	514	64Y	133.75	-	97X	115.00	-
32Y	109.55	570	65X	133.80	-	97Y	115.05	654
33X	109.60	-	65Y	133.85	-	98X	115.10	-
33Y	109.65	572	66X	133.90	-	98Y	115.15	656
34X	109.70	516	66Y	133.95	-	99X	115.20	-
34Y	109.75	574	67X	134.00	-	99Y	115.25	658
35X	109.80	-	67Y	134.05	-	100X	115.30	-
35Y	109.85	576	68X	134.10	-	100Y	115.35	660
36X	109.90	518	68Y	134.15	-	101X	115.40	-
36Y	109.95	578	69X	134.20	-	101Y	115.45	662
37X	110.00	-	69Y	134.25	-	102X	115.50	-
37Y	110.05	580	70X	112.30	-	102Y	115.55	664
38X	110.10	520	70Y	112.35	-	103X	115.60	-
38Y	110.15	582	71X	112.40	-	103Y	115.65	666
39X	110.20	-	71Y	112.45	-	104X	115.70	-
39Y	110.25	584	72X	112.50	-	104Y	115.75	668
40X	110.30	522	72Y	112.55	-	105X	115.80	-
40Y	110.35	586	73X	112.60	-	105Y	115.85	670
41X	110.40	-	73Y	112.65	-	106X	115.90	-
41Y	110.45	588	74X	112.70	-	106Y	115.95	672
42X	110.50	524	74Y	112.75	-	107X	116.00	-
42Y	110.55	590	75X	112.80	-	107Y	116.05	674
43X	110.60	-	75Y	112.85	-	108X	116.10	-
43Y	110.65	592	76X	112.90	-	108Y	116.15	676
44X	110.70	526	76Y	112.95	-	109X	116.20	-
44Y	110.75	594	77X	113.00	-	109Y	116.25	678
45X	110.80	-	77Y	113.05	-	110X	116.30	-
45Y	110.85	596	78X	113.10	-	110Y	116.35	680
46X	110.90	528	78Y	113.15	-	111X	116.40	-
46Y	110.95	598	79X	113.20	-	111Y	116.45	682
47X	111.00	-	79Y	113.25	-	112X	116.50	-
47Y	111.05	600	80X	113.30	-	112Y	116.55	684
48X	111.10	530	80Y	113.35	620	113X	116.60	-
48Y	111.15	602	81X	113.40	-	113Y	116.65	686
49X	111.20	-	81Y	113.45	622	114X	116.70	-
49Y	111.25	604	82X	113.50	-	114Y	116.75	688
50X	111.30	532	82Y	113.55	624	115X	116.80	-
50Y	111.35	606	83X	113.60	-	115Y	116.85	690
51X	111.40	-	83Y	113.65	626	116X	116.90	-
51Y	111.45	608	84X	113.70	-	116Y	116.95	692
52X	111.50	534	84Y	113.75	628	117X	117.00	-
52Y	111.55	610	85X	113.80	-	117Y	117.05	694
53X	111.60	-	85Y	113.85	630	118X	117.10	-
53Y	111.65	612	86X	113.90	-	118Y	117.15	696
54X	111.70	536	86Y	113.95	632	119X	117.20	-
54Y	111.75	614	87X	114.00	-	119Y	117.25	698
55X	111.80	-	87Y	114.05	634	120X	117.30	-
55Y	111.85	616	88X	114.10	-	120Y	117.35	-
56X	111.90	538	88Y	114.15	636	121X	117.40	-
56Y	111.95	618	89X	114.20	-	121Y	117.45	-
57X	112.00	-	89Y	114.25	638	122X	117.50	-
57Y	112.05	-	90X	114.30	-	122Y	117.55	-
58X	112.10	-	90Y	114.35	640	123X	117.60	-
58Y	112.15	-	91X	114.40	-	123Y	117.65	-
59X	112.20	-	91Y	114.45	642	124X	117.70	-
59Y	112.25	-	92X	114.50	-	124Y	117.75	-
60X	133.30	-	92Y	114.55	644	125X	117.80	-
60Y	133.35	-	93X	114.60	-	125Y	117.85	-
61X	133.40	-	93Y	114.65	646	126X	117.90	-
61Y	133.45	-	94X	114.70	-	126Y	117.95	-
62X	133.50	-	94Y	114.75	648			
62Y	133.55	-	95X	114.80	-			

(35) COMM/NAV/WEATHER REMARKS:
These remarks consist of pertinent information affecting the current status of communications, NAVAIDs and weather.

NE, 09 FEB 20XX to 05 APR 20XX

LEGEND 19.—Chart Supplements U.S. (formerly Airport/Facility Directory).

APPENDIX 2

FAA FIGURES

Note:
Some figure numbers are skipped in this appendix. Figures that are not used on the Private Pilot Airplane (PAR) test are not included here.

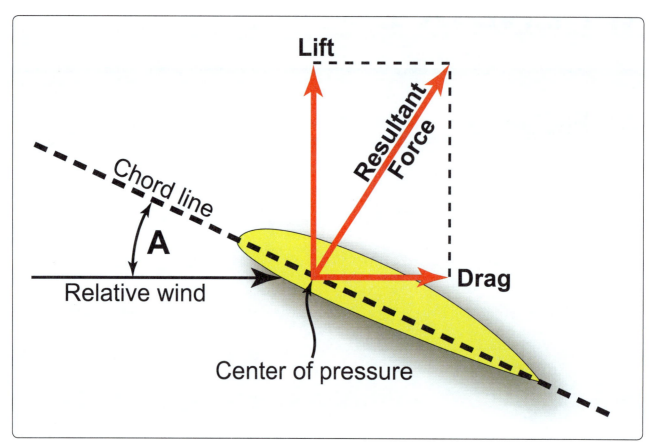

Figure 1. Lift Vector.

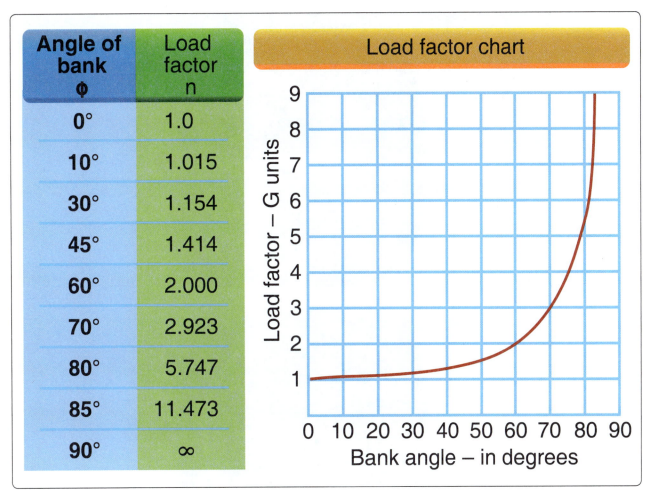

Angle of bank ϕ	Load factor n
0°	1.0
10°	1.015
30°	1.154
45°	1.414
60°	2.000
70°	2.923
80°	5.747
85°	11.473
90°	∞

Figure 2. Load Factor Chart.

Figure 3. Altimeter.

Figure 4. Airspeed Indicator.

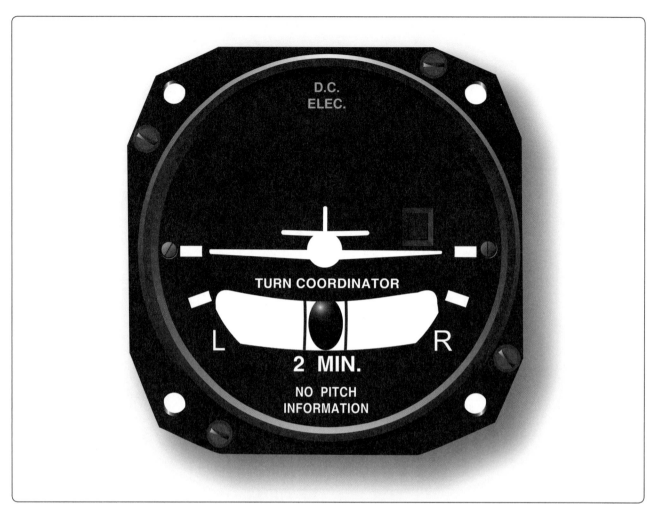

Figure 5. Turn Coordinator.

Figure 6. Heading Indicator.

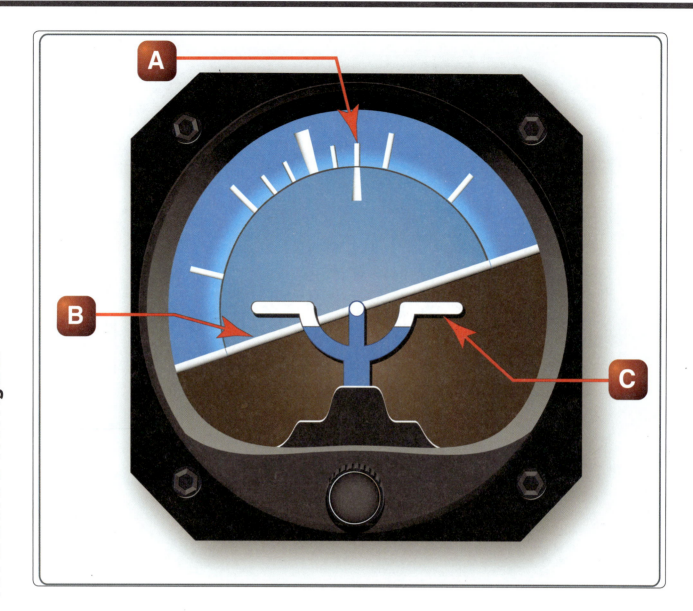

Figure 7. Attitude Indicator.

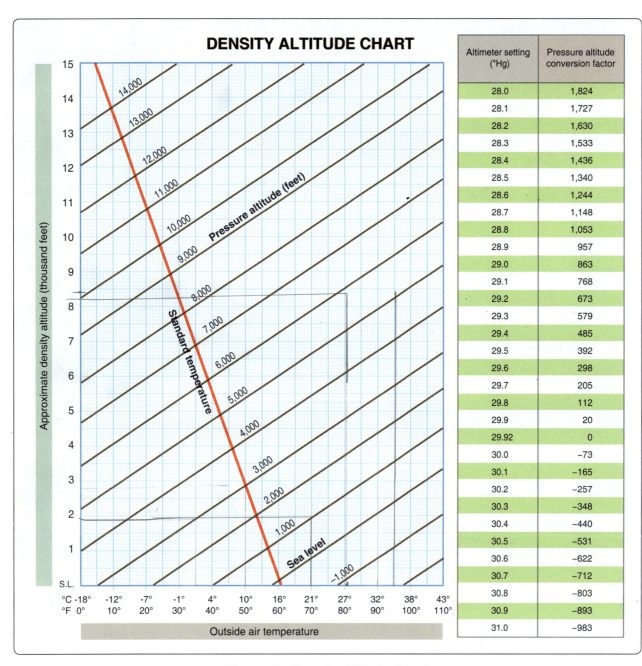

Figure 8. Density Altitude Chart.

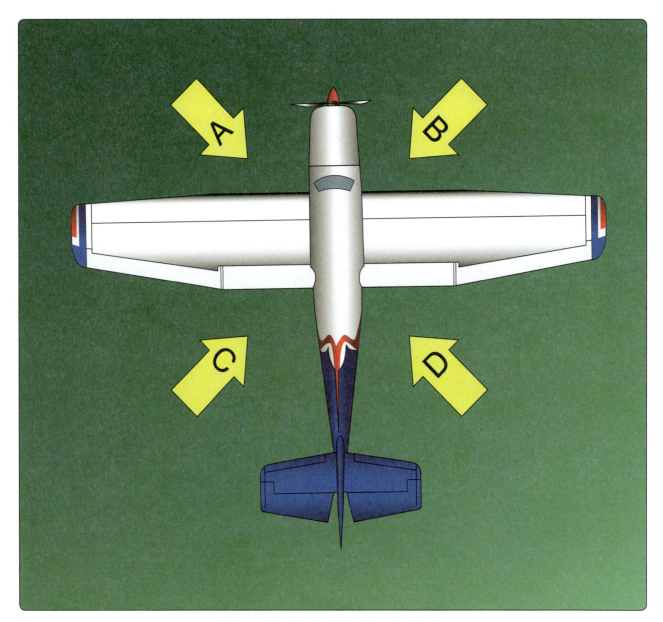

Figure 9. Control Position for Taxi.

METAR KINK 121845Z 11012G18KT 15SM SKC 25/17 A3000

METAR KBOI 121854Z 13004KT 30SM SCT150 17/6 A3015

METAR KLAX 121852Z 25004KT 6SM BR SCT007 SCT250 16/15 A2991

SPECI KMDW 121856Z 32005KT 1 1/2SM RA OVC007 17/16 A2980 RMK RAB35

SPECI KJFK 121853Z 18004KT 1/2SM FG R04/2200 OVC005 20/18 A3006

Figure 12. Aviation Routine Weather Reports (METAR).

This is a telephone weather briefing from the Dallas FSS for local operation of gliders and lighter-than-air at Caddo Mills, Texas (about 30 miles east of Dallas). The briefing is at 13Z.

"There are no adverse conditions reported or forecast for today."

"A weak low pressure over the Texas Panhandle and eastern New Mexico is causing a weak southerly flow over the area."

"Current weather here at Dallas is wind south 5 knots, visibility 12 miles, clear, temperature 21, dewpoint 9, altimeter 29 point 78."

"By 15Z, we should have a few scattered cumuliform clouds at 5 thousand AGL, with higher scattered cirrus at 25 thousand MSL. After 20Z, the wind should pick up to about 15 knots from the south."

"The winds aloft are: 3 thousand 170 at 7, temperature 20; 6 thousand 200 at 18, temperature 14; 9 thousand 210 at 22, temperature 8; 12 thousand 225 at 27, temperature 0; 18 thousand 240 at 30, temperature −7."

Figure 13. Telephone Weather Briefing.

UA/OV KOKC-KTUL/TM 1800/FL120/TP BE90/SK BKN018-TOP055/OVC072-TOP089/CLR ABV/TA M7/WV 08021/TB LGT 055-072/IC LGT-MOD RIME 072-089

Figure 14. Pilot Weather Report.

TAF

KMEM 121720Z 1218/1324 20012KT 5SM HZ BKN030 PROB40 1220/1222 1SM TSRA OVC008CB
      FM122200 33015G20KT P6SM BKN015 OVC025 PROB40 1220/1222 3SM SHRA
      FM120200 35012KT OVC008 PROB40 1202/1205 2SM-RASN BECMG 1306/1308 02008KT BKN012
      BECMG 1310/1312 00000KT 3SM BR SKC TEMPO 1212/1214 1/2SM FG
      FM131600 VRB06KT P6SM SKC=

KOKC 051130Z 0512/0618 14008KT 5SM BR BKN030 TEMPO 0513/0516 1 1/2SM BR
      FM051600 18010KT P6SM SKC BECMG 0522/0524 20013G20KT 4SM SHRA OVC020
      PROB40 0600/0606 2SM TSRA OVC008CB BECMG 0606/0608 21015KT P6SM SCT040=

Figure 15. Terminal Aerodrome Forecasts (TAF).

FB·WBC 151745
DATA BASED ON 151200Z
VALID 1600Z FOR USE 1800-0300Z. TEMPS NEG ABV 24000

FT	3000	6000	9000	12000	18000	24000	30000	34000	39000
ALS			2420	2635-08	2535-18	2444-30	245945	246755	246862
AMA		2714	2725+00	2625-04	2531-15	2542-27	265842	256352	256762
DEN			2321-04	2532-08	2434-19	2441-31	235347	236056	236262
HLC		1707-01	2113-03	2219-07	2330-17	2435-30	244145	244854	245561
MKC	0507	2006+03	2215-01	2322-06	2338-17	2348-29	236143	237252	238160
STL	2113	2325+07	2332+02	2339-04	2356-16	2373-27	239440	730649	731960

Figure 17. Winds and Temperatures Aloft Forecast.

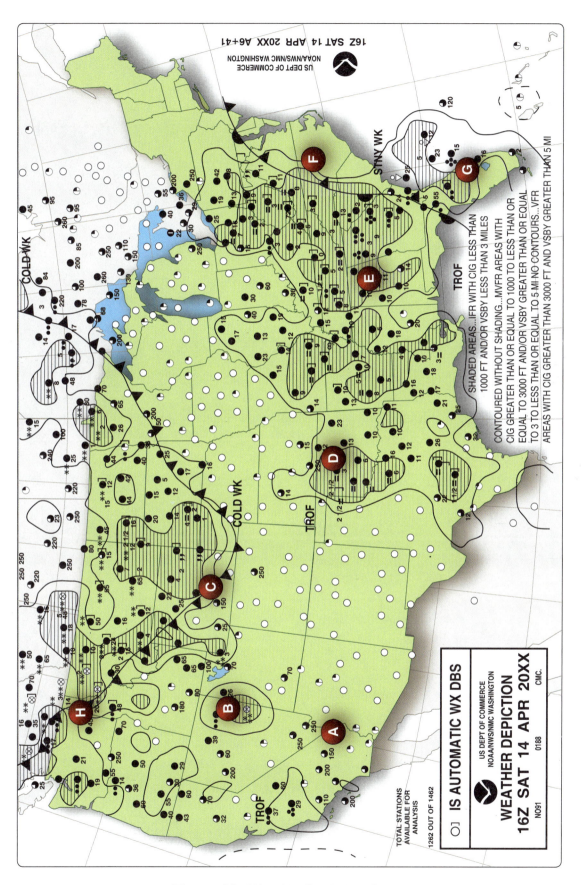

Figure 18. Weather Depiction Chart.

APPENDIX 2 ■ FAA Figures

Figure 19. Low-Level Significant Weather (SIGWX) Prognostic Charts.

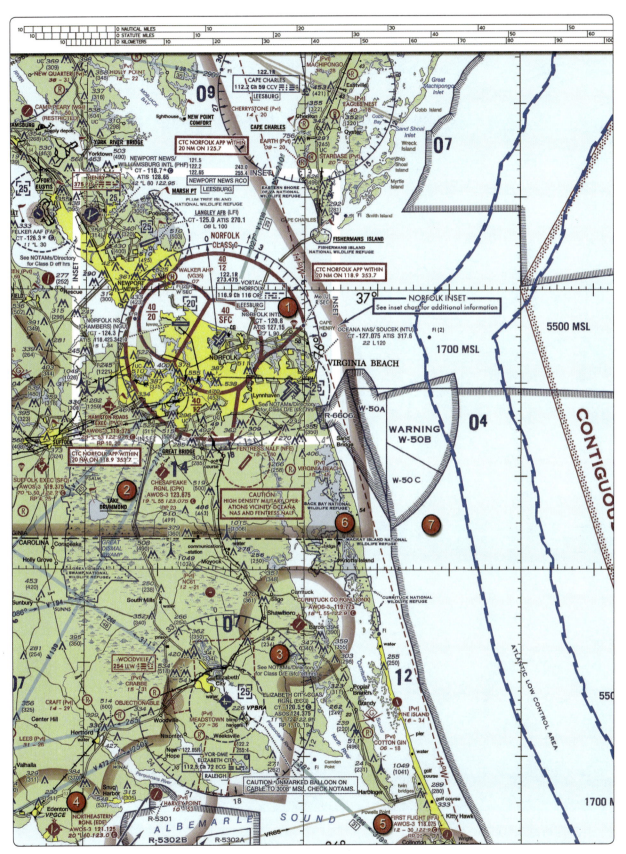

Figure 20. Sectional Chart Excerpt.
NOTE: Chart is not to scale and should not be used for navigation. Use associated scale.

Figure 21. Sectional Chart Excerpt.
NOTE: Chart is not to scale and should not be used for navigation. Use associated scale.

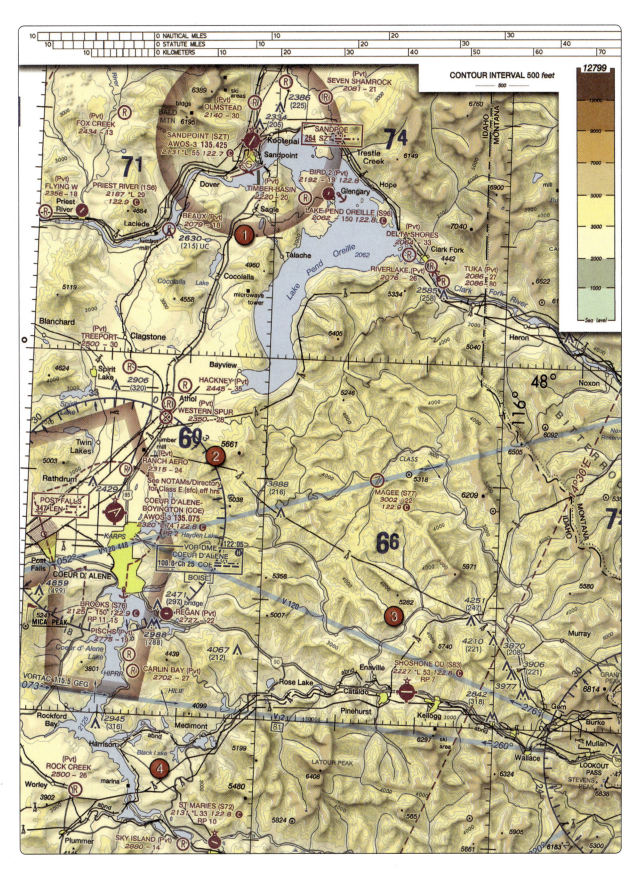

Figure 22. Sectional Chart Excerpt.
NOTE: Chart is not to scale and should not be used for navigation. Use associated scale.

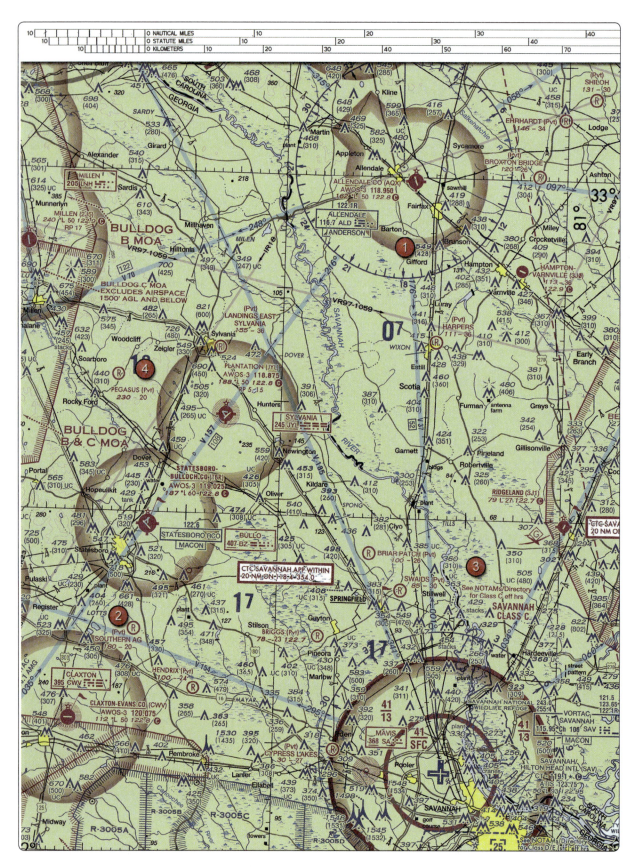

Figure 23. Sectional Chart Excerpt.

NOTE: Chart is not to scale and should not be used for navigation. Use associated scale.

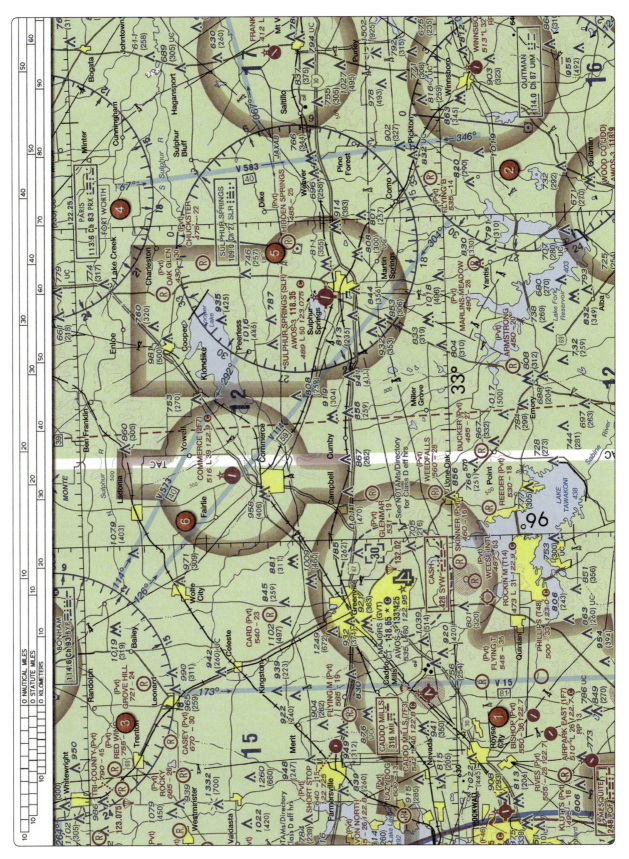

Figure 24. Sectional Chart Excerpt.
NOTE: Chart is not to scale and should not be used for navigation. Use associated scale.

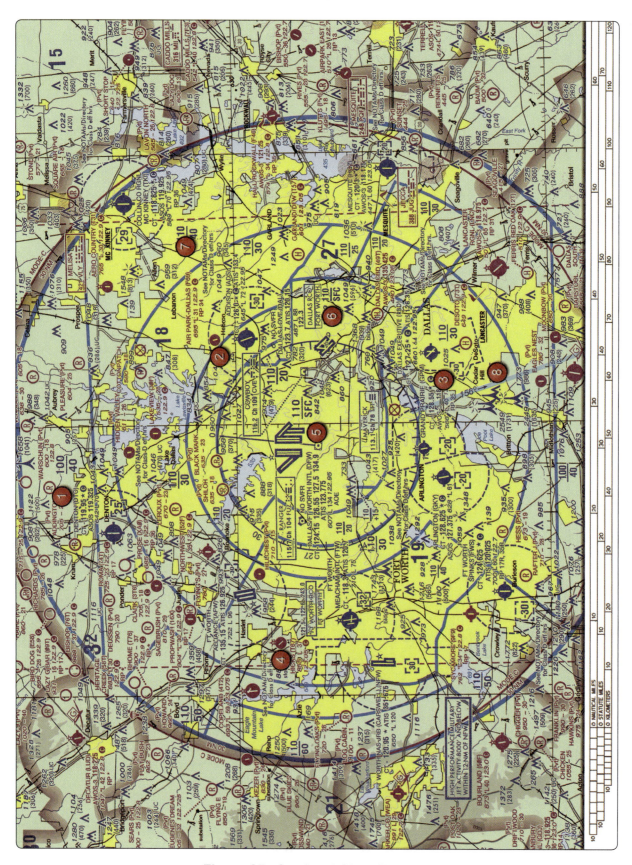

Figure 25. Sectional Chart Excerpt.

NOTE: Chart is not to scale and should not be used for navigation. Use associated scale.

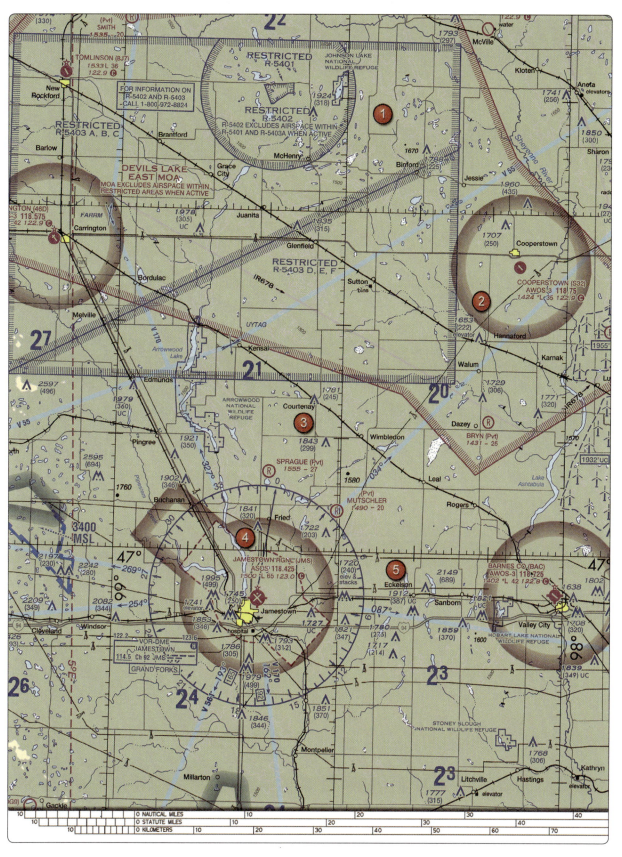

Figure 26. Sectional Chart Excerpt.
NOTE: Chart is not to scale and should not be used for navigation. Use associated scale.

APPENDIX 2 ■ FAA Figures

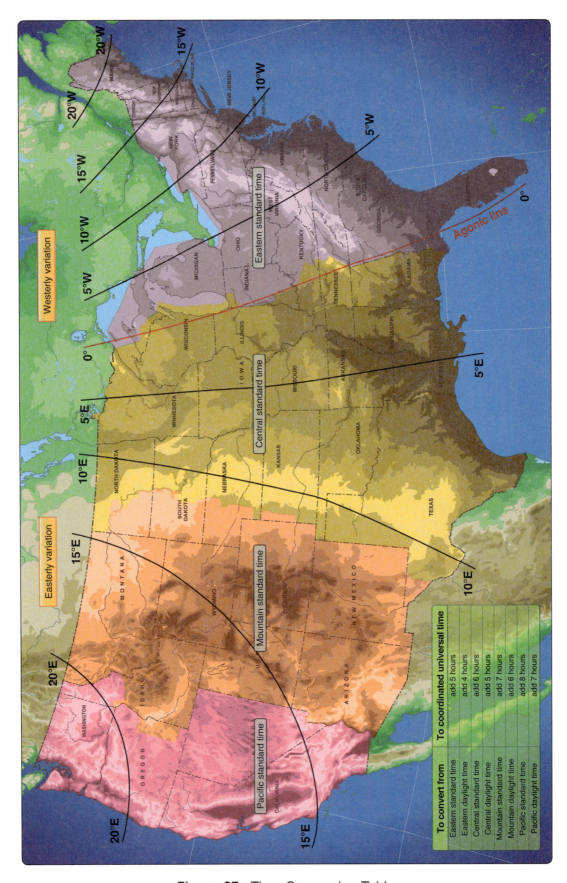

Figure 27. Time Conversion Table.

Figure 28. VOR.

IDAHO   31

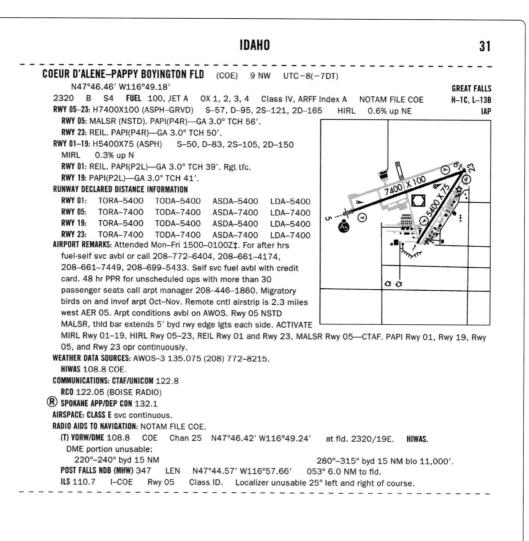

COEUR D'ALENE–PAPPY BOYINGTON FLD (COE)   9 NW   UTC–8(–7DT)

    N47°46.46′ W116°49.18′   **GREAT FALLS**

2320   B   S4   **FUEL** 100, JET A   OX 1, 2, 3, 4   Class IV, ARFF Index A   NOTAM FILE COE   **H–1C, L–13B**

RWY 05–23: H7400X100 (ASPH–GRVD)   S–57, D–95, 2S–121, 2D–165   HIRL   0.6% up NE   **IAP**

  RWY 05: MALSR (NSTD). PAPI(P4R)—GA 3.0° TCH 56′.

  RWY 23: REIL. PAPI(P4R)—GA 3.0° TCH 50′.

RWY 01–19: H5400X75 (ASPH)   S–50, D–83, 2S–105, 2D–150

  MIRL   0.3% up N

  RWY 01: REIL. PAPI(P2L)—GA 3.0° TCH 39′. Rgt tfc.

  RWY 19: PAPI(P2L)—GA 3.0° TCH 41′.

RUNWAY DECLARED DISTANCE INFORMATION

RWY 01:	TORA–5400	TODA–5400	ASDA–5400	LDA–5400
RWY 05:	TORA–7400	TODA–7400	ASDA–7400	LDA–7400
RWY 19:	TORA–5400	TODA–5400	ASDA–5400	LDA–5400
RWY 23:	TORA–7400	TODA–7400	ASDA–7400	LDA–7400

AIRPORT REMARKS: Attended Mon–Fri 1500–0100Z‡. For after hrs fuel-self svc avbl or call 208–772–6404, 208–661–4174, 208–661–7449, 208–699–5433. Self svc fuel avbl with credit card. 48 hr PPR for unscheduled ops with more than 30 passenger seats call arpt manager 208–446–1860. Migratory birds on and invof arpt Oct–Nov. Remote cntl airstrip is 2.3 miles west AER 05. Arpt conditions avbl on AWOS. Rwy 05 NSTD MALSR, thld bar extends 5′ byd rwy edge lgts each side. ACTIVATE MIRL Rwy 01–19, HIRL Rwy 05–23, REIL Rwy 01 and Rwy 23, MALSR Rwy 05—CTAF. PAPI Rwy 01, Rwy 19, Rwy 05, and Rwy 23 opr continuously.

WEATHER DATA SOURCES: AWOS–3 135.075 (208) 772–8215.

  HIWAS 108.8 COE.

COMMUNICATIONS: CTAF/UNICOM 122.8

  RCO 122.05 (BOISE RADIO)

Ⓡ **SPOKANE APP/DEP CON** 132.1

AIRSPACE: CLASS E svc continuous.

RADIO AIDS TO NAVIGATION: NOTAM FILE COE.

  (T) VORW/DME 108.8   COE   Chan 25   N47°46.42′ W116°49.24′   at fld. 2320/19E.   **HIWAS.**

    DME portion unusable:

      220°–240° byd 15 NM   280°–315° byd 15 NM blo 11,000′.

  POST FALLS NDB (MHW) 347   LEN   N47°44.57′ W116°57.66′   053° 6.0 NM to fld.

  ILS 110.7   I–COE   Rwy 05   Class ID.   Localizer unusable 25° left and right of course.

Figure 31. Chart Supplement.

Useful load weights and moments

Baggage or 5th seat occupant

ARM 140

Weight	Moment / 100
10	14
20	28
30	42
40	56
50	70
60	84
70	98
80	112
90	126
100	140
110	154
120	168
130	182
140	196
150	210
160	224
170	238
180	252
190	266
200	280
210	294
220	308
230	322
240	336
250	350
260	364
270	378

Occupants

Front seats ARM 85		Rear seats ARM 121	
Weight	Moment / 100	Weight	Moment / 100
120	102	120	145
130	110	130	157
140	119	140	169
150	128	150	182
160	136	160	194
170	144	170	206
180	153	180	218
190	162	190	230
200	170	200	242

Usable fuel

Main wing tanks ARM 75

Gallons	Weight	Moment / 100
5	30	22
10	60	45
15	90	68
20	120	90
25	150	112
30	180	135
35	210	158
40	240	180
44	264	198

Auxiliary wing tanks ARM 94

Gallons	Weight	Moment / 100
5	30	28
10	60	56
15	90	85
19	114	107

*Oil

Quarts	Weight	Moment / 100
10	19	5

*Included in basic empty weight.

Empty weight~2,015

MOM/100~1,554

Moment limits vs weight
Moment limits are based on the following weight and center of gravity limit data (landing gear down).

Weight condition	Forward CG limit	AFT CG limit
2,950 lb (takeoff or landing)	82.1	84.7
2,525 lb	77.5	85.7
2,475 lb or less	77.0	85.7

Figure 32. Airplane Weight and Balance Tables.

Moment limits vs weight (continued)						
Weight	Minimum Moment 100	Maximum Moment 100	Weight	Minimum Moment 100	Maximum Moment 100
2,100	1,617	1,800	2,500	1,932	2,143
2,110	1,625	1,808	2,510	1,942	2,151
2,120	1,632	1,817	2,520	1,953	2,160
2,130	1,640	1,825	2,530	1,963	2,168
2,140	1,648	1,834	2,540	1,974	2,176
2,150	1,656	1,843	2,550	1,984	2,184
2,160	1,663	1,851	2,560	1,995	2,192
2,170	1,671	1,860	2,570	2,005	2,200
2,180	1,679	1,868	2,580	2,016	2,208
2,190	1,686	1,877	2,590	2,026	2,216
2,200	1,694	1,885	2,600	2,037	2,224
2,210	1,702	1,894	2,610	2,048	2,232
2,220	1,709	1,903	2,620	2,058	2,239
2,230	1,717	1,911	2,630	2,069	2,247
2,240	1,725	1,920	2,640	2,080	2,255
2,250	1,733	1,928	2,650	2,090	2,263
2,260	1,740	1,937	2,660	2,101	2,271
2,270	1,748	1,945	2,670	2,112	2,279
2,280	1,756	1,954	2,680	2,123	2,287
2,290	1,763	1,963	2,690	2,133	2,295
2,300	1,771	1,971	2,700	2,144	2,303
2,310	1,779	1,980	2,710	2,155	2,311
2,320	1,786	1,988	2,720	2,166	2,319
2,330	1,794	1,997	2,730	2,177	2,326
2,340	1,802	2,005	2,740	2,188	2,334
2,350	1,810	2,014	2,750	2,199	2,342
2,360	1,817	2,023	2,760	2,210	2,350
2,370	1,825	2,031	2,770	2,221	2,358
2,380	1,833	2,040	2,780	2,232	2,366
2,390	1,840	2,048	2,790	2,243	2,374
2,400	1,848	2,057	2,800	2,254	2,381
2,410	1,856	2,065	2,810	2,265	2,389
2,420	1,863	2,074	2,820	2,276	2,397
2,430	1,871	2,083	2,830	2,287	2,405
2,440	1,879	2,091	2,840	2,298	2,413
2,450	1,887	2,100	2,850	2,309	2,421
2,460	1,894	2,108	2,860	2,320	2,428
2,470	1,902	2,117	2,870	2,332	2,436
2,480	1,911	2,125	2,880	2,343	2,444
2,490	1,921	2,134	2,890	2,354	2,452
			2,900	2,365	2,460
			2,910	2,377	2,468
			2,920	2,388	2,475
			2,930	2,399	2,483
			2,940	2,411	2,491
			2,950	2,422	2,499

Figure 33. Airplane Weight and Balance Tables.

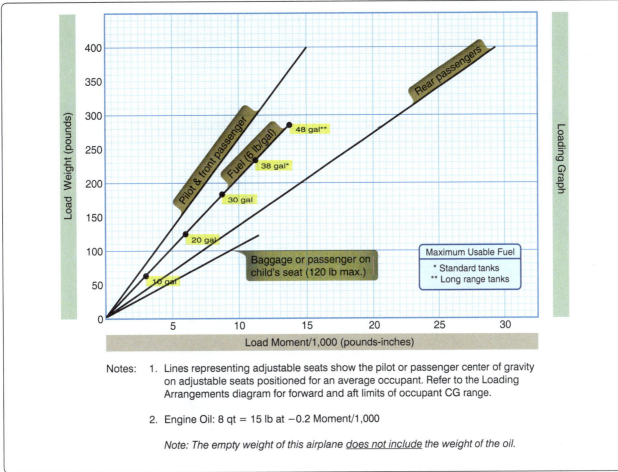

Notes: 1. Lines representing adjustable seats show the pilot or passenger center of gravity on adjustable seats positioned for an average occupant. Refer to the Loading Arrangements diagram for forward and aft limits of occupant CG range.

2. Engine Oil: 8 qt = 15 lb at −0.2 Moment/1,000

Note: The empty weight of this airplane does not include the weight of the oil.

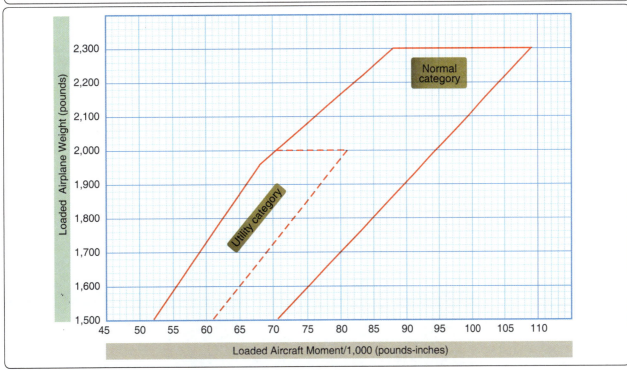

Figure 34. Airplane Weight and Balance Graphs.

APPENDIX 2 ■ FAA Figures

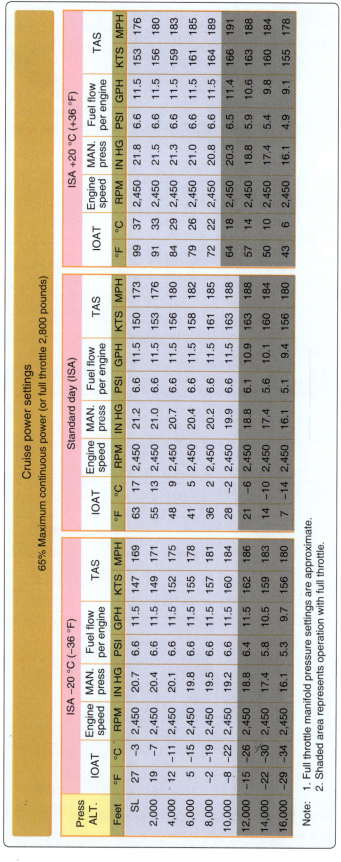

Cruise power settings
65% Maximum continuous power (or full throttle 2,800 pounds)

Note: 1. Full throttle manifold pressure settings are approximate.
2. Shaded area represents operation with full throttle.

Figure 35. Airplane Power Setting Table.

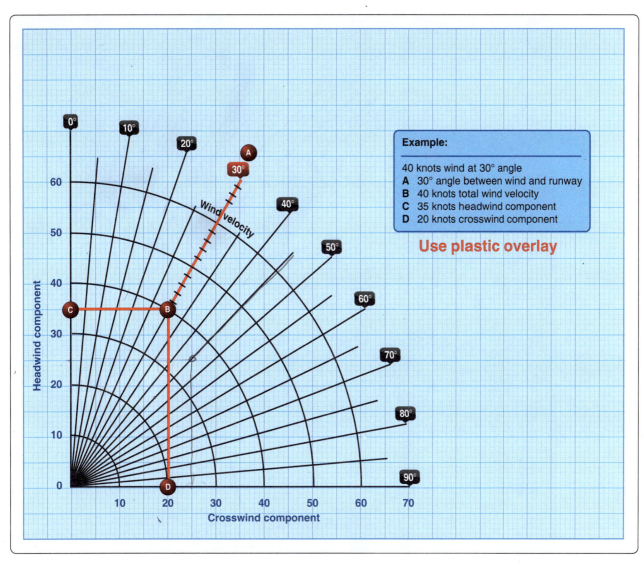

Figure 36. Crosswind Component Graph.

APPENDIX 2 ■ FAA Figures

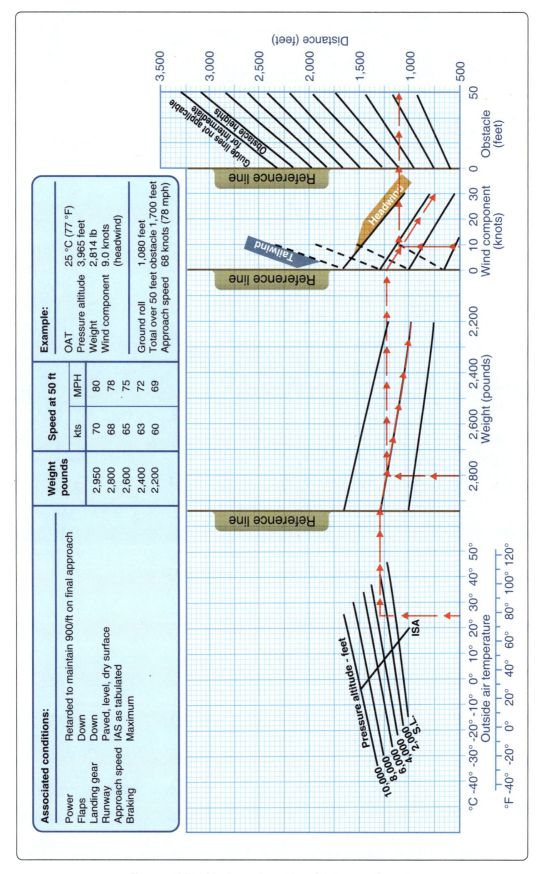

Figure 37. Airplane Landing Distance Graph.

Gross weight lb	Approach speed, IAS, MPH	Landing distance									
		Flaps lowered to 40° – Power off Hard surface runway – Zero wind									
		At sea level & 59 °F		At 2,500 feet & 50 °F		At 5,000 feet & 41 °F		At 7,500 feet & 32 °F			
		Ground roll	Total to clear 50 feet OBS	Ground roll	Total to clear 50 feet OBS	Ground roll	Total to clear 50 feet OBS	Ground roll	Total to clear 50 feet OBS		
1,600	60	445	1,075	470	1,135	495	1,195	520	1,255		

NOTE:
1. Decrease the distances shown by 10% for each 4 knots of headwind.
2. Increase the distance by 10% for each 60 °F temperature increase above standard.
3. For operation on a dry, grass runway, increase distance (both "ground roll" and "total to clear 50 feet obstacle") by 20% of the "total to clear 50 feet obstacle" figure.

Figure 38. Airplane Landing Distance Table.

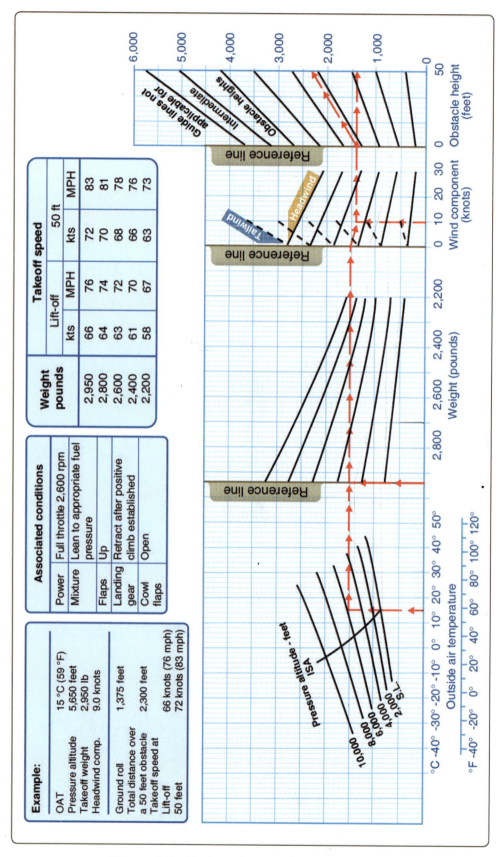

Figure 40. Airplane Takeoff Distance Graph.

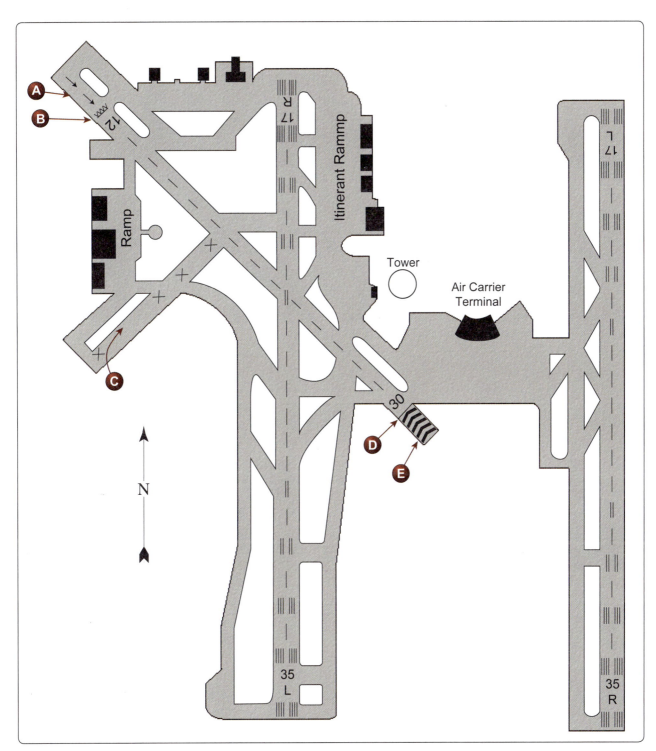

Figure 48. Airport Diagram.

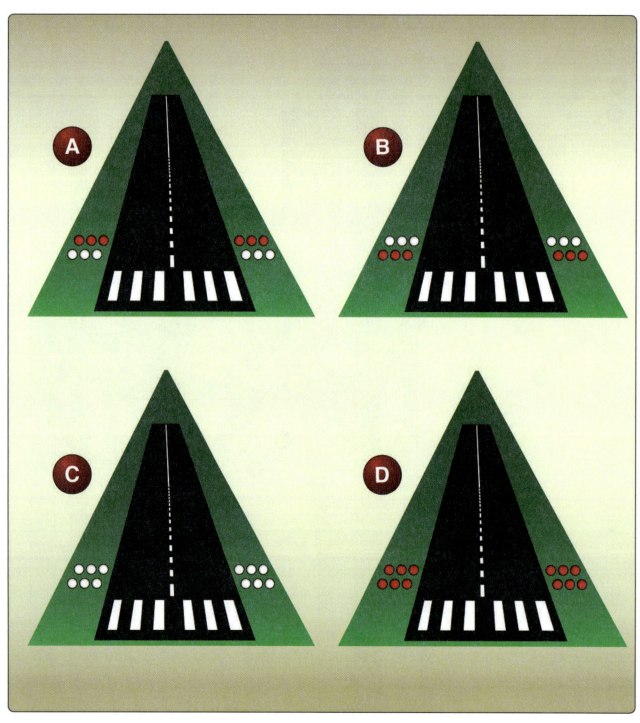

Figure 47. VASI Illustrations.

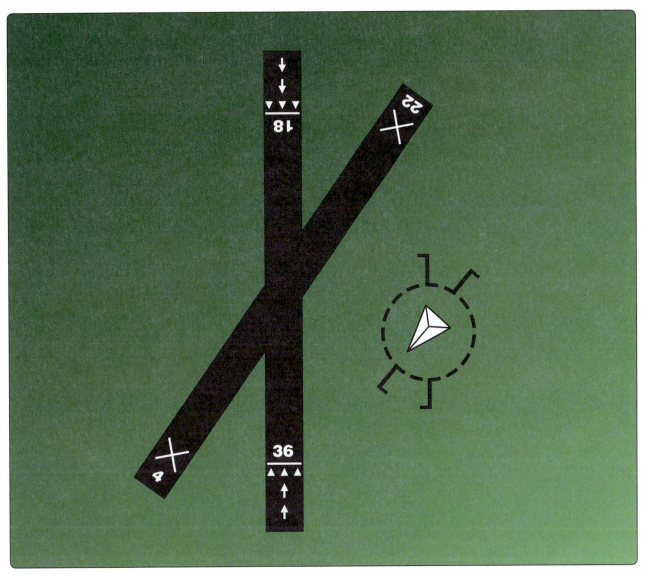

Figure 49. Airport Diagram.

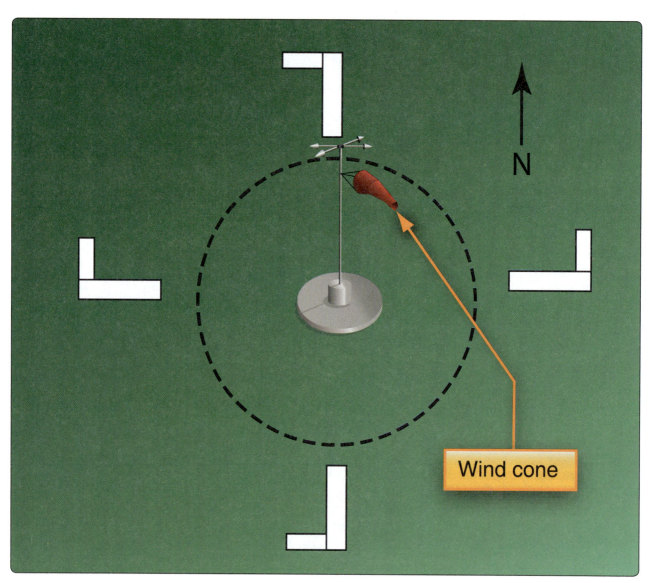

Figure 50. Wind Sock Airport Landing Indicator.

Approved OMB No. 2120-0026
Exp. 5/31/2017

U S Department of Transportation
Federal Aviation Administration

International Flight Plan

PRIORITY   ADDRESSEE(S)

<=FF

<=

FILING TIME   ORIGINATOR

<=

SPECIFIC IDENTIFICATION OF ADDRESSEE(S) AND / OR ORIGINATOR

3 MESSAGE TYPE   7 AIRCRAFT IDENTIFICATION   8 FLIGHT RULES   TYPE OF FLIGHT

<=(FPL   —   —   —   <=

9 NUMBER   TYPE OF AIRCRAFT   WAKE TURBULENCE CAT.   10 EQUIPMENT

—   /   —   /   <=

13 DEPARTURE AERODROME   TIME

—   <=

15 CRUISING SPEED   LEVEL   ROUTE

—

<=

16 DESTINATION AERODROME   TOTAL EET
HR   MIN   ALTN AERODROME   2ND ALTN AERODROME

<=

18 OTHER INFORMATION

—

<=

SUPPLEMENTARY INFORMATION (NOT TO BE TRANSMITTED IN FPL MESSAGES)

19   ENDURANCE
HR   MIN   PERSONS ON BOARD   EMERGENCY RADIO
UHF   VHF   ELT

—**E/**   **P/**   **R/** U   V   E

SURVIVAL EQUIPMENT   JACKETS
POLAR DESERT MARITIME JUNGLE   LIGHT FLUORES UHF   VHF

/ P   D   M   J   / L   F   U   V

DINGHIES
NUMBER CAPACITY COVER   COLOR

D /   C   <=

AIRCRAFT COLOR AND MARKINGS

A/

REMARKS

N /   <=

PILOT-IN-COMMAND

C/   )<=

FILED BY   ACCEPTED BY   ADDITIONAL INFORMATION

FAA Form 7233-4 (7/15)

Figure 51. Flight Plan Form.

NEBRASKA　　　　271

LINCOLN (LNK)　4 NW　UTC–6(–5DT)　N40°51.05′ W96°45.55′　　　**OMAHA**
　　1219　B　S4　**FUEL** 100LL, JET A　TPA—See Remarks　ARFF Index—See Remarks　　H–5C, L–10I
　　NOTAM FILE LNK　　　　　　　　　　　　　　　　　　　　　　　　IAP, AD 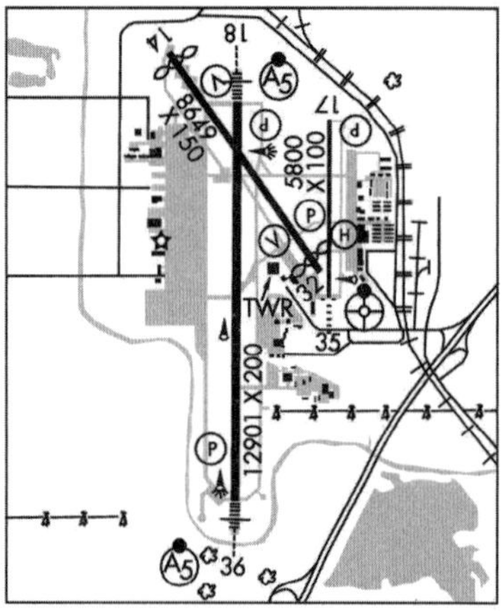
　　RWY 18–36: H12901X200 (ASPH–CONC–GRVD)　S–100, D–200,
　　　2S–175, 2D–400　HIRL
　　　RWY 18: MALSR. PAPI(P4L)—GA 3.0° TCH 55′. Rgt tfc.　0.4%
　　　　down.
　　　RWY 36: MALSR. PAPI(P4L)—GA 3.0° TCH 57′.
　　RWY 14–32: H8649X150 (ASPH–CONC–GRVD)　S–80, D–170,
　　　2S–175, 2D–280　MIRL
　　　RWY 14: REIL. VASI(V4L)—GA 3.0° TCH 48′. Thld dsplcd 363′.
　　　RWY 32: VASI(V4L)—GA 3.0° TCH 50′. Thld dsplcd 470′.
　　　　Pole.　0.3% up.
　　RWY 17–35: H5800X100 (ASPH–CONC–AFSC)　S–49, D–60
　　　　HIRL　0.8% up S
　　　RWY 17: REIL. PAPI(P4L)—GA 3.0° TCH 44′.
　　　RWY 35: ODALS. PAPI(P4L)—GA 3.0° TCH 30′. Rgt tfc.
　　RUNWAY DECLARED DISTANCE INFORMATION
　　　RWY 14: TORA–8649　TODA–8649　ASDA–8649　LDA–8286
　　　RWY 17: TORA–5800　TODA–5800　ASDA–5400　LDA–5400
　　　RWY 18: TORA–12901 TODA–12901 ASDA–12901 LDA–12901
　　　RWY 32: TORA–8649　TODA–8649　ASDA–8286　LDA–7816
　　　RWY 35: TORA–5800　TODA–5800　ASDA–5800　LDA–5800
　　　RWY 36: TORA–12901 TODA–12901 ASDA–12901 LDA–12901
　　AIRPORT REMARKS: Attended continuously. Birds invof arpt. Rwy 18 designated calm wind rwy. Rwy 32 apch holdline
　　　on South A twy. TPA—2219 (1000), heavy military jet 3000 (1781). Class I, ARFF Index B. ARFF Index C level
　　　equipment provided. Rwy 18–36 touchdown and rollout rwy visual range avbl. When twr clsd MIRL Rwy 14–32
　　　preset on low ints, HIRL Rwy 18–36 and Rwy 17–35 preset on med ints, ODALS Rwy 35 operate continuously on
　　　med ints, MALSR Rwy 18 and Rwy 36 operate continuously and REIL Rwy 14 and Rwy 17 operate continuously
　　　on low ints. VASI Rwy 14 and Rwy 32, PAPI Rwy 17, Rwy 35, Rwy 18 and Rwy 36 on continuously.
　　WEATHER DATA SOURCES: ASOS (402) 474–9214. LLWAS
　　COMMUNICATIONS: CTAF 118.5　　**ATIS** 118.05　　**UNICOM** 122.95
　　　RCO 122.65 (COLUMBUS RADIO)
　Ⓡ **APP/DEP CON** 124.0 (180°–359°) 124.8 (360°–179°)
　　　TOWER 118.5 125.7 (1130–0600Z‡)　**GND CON** 121.9　**CLNC DEL** 120.7
　　AIRSPACE: CLASS C svc 1130–0600Z‡ ctc **APP CON** other times CLASS E.
　　RADIO AIDS TO NAVIGATION: NOTAM FILE LNK.
　　　(H) VORTACW 116.1　LNK　Chan 108　N40°55.43′ W96°44.52′　181° 4.4 NM to fld. 1370/9E
　　　POTTS NDB (MHW/LOM) 385　LN　N40°44.83′ W96°45.75′　355° 6.2 NM to fld. Unmonitored when twr clsd.
　　　ILS 111.1　I–OCZ　Rwy 18.　Class IB　OM unmonitored.
　　　ILS 109.9　I–LNK　Rwy 36　Class IA　LOM POTTS NDB. MM unmonitored. LOM unmonitored when twr
　　　　clsd.
　　COMM/NAV/WEATHER REMARKS: Emerg frequency 121.5 not available at twr.

LOUP CITY MUNI　(Ø F4)　1 NW　UTC–6(–5DT)　N41°17.20′ W98°59.41′　　**OMAHA**
　　2071　B　**FUEL** 100LL　NOTAM FILE OLU　　　　　　　　　　　　　L–10H, 12H
　　RWY 16–34: H3200X60 (CONC)　S–12.5　MIRL
　　　RWY 34: Trees.
　　RWY 04–22: 2040X100 (TURF)
　　　RWY 04: Tree.　　**RWY 22:** Road.
　　AIRPORT REMARKS: Unattended. For svc call 308–745–1344/1244/0664.
　　COMMUNICATIONS: CTAF 122.9
　　RADIO AIDS TO NAVIGATION: NOTAM FILE OLU.
　　　WOLBACH (H) VORTAC 114.8　OBH　Chan 95　N41°22.54′ W98°21.22′　253° 29.3 NM to fld. 2010/7E.

MARTIN FLD　(See SO SIOUX CITY)

Figure 52. Chart Supplement.

For	N	30	60	E	120	150
Steer	0	27	56	85	116	148
For	S	210	240	W	300	330
Steer	181	214	244	274	303	332

Figure 58. Compass Card.

Figure 59. Sectional Chart Excerpt.
Note: Chart is not to scale and should not be used for navigation. Use associated scale.

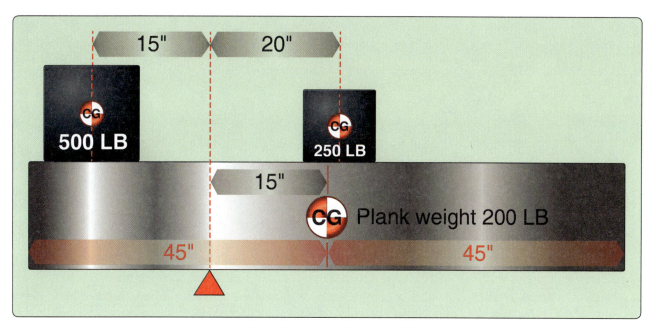

Figure 60. Weight and Balance Diagram.

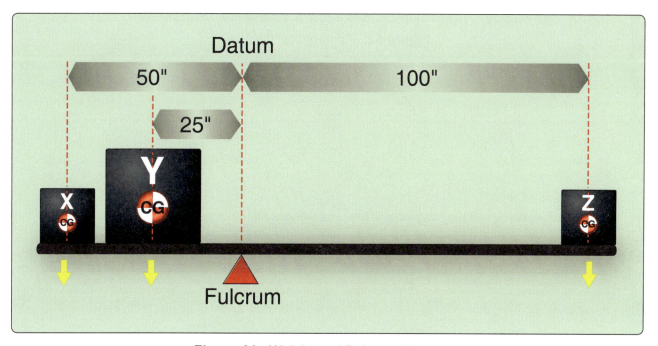

Figure 61. Weight and Balance Diagram.

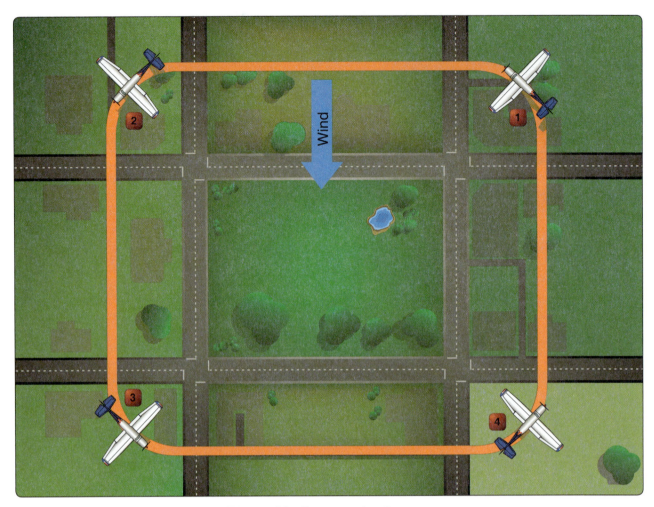

Figure 62. Rectangular Course.

OHIO 263

TOLEDO

TOLEDO EXECUTIVE (TDZ) 6 SE UTC−5(−4DT) N41°33.90′ W83°28.93′

DETROIT
H−10G, L−28J
IAP

623 B S4 **FUEL** 100LL, JET A OX 1, 3 NOTAM FILE TDZ
RWY 14−32: H5829X100 (ASPH−GRVD) S−63, D−85, 2S−107 MIRL
   RWY 14: REIL. PAPI(P4L)—GA 3.0° TCH 34′. Thld dsplcd 225′.
Tower.
   RWY 32: VASI(V4L)—GA 3.0° TCH 43′. Thld dsplcd 351′. Road.
RWY 04−22: H3799X75 (ASPH) S−63, D−85, 2S−107 MIRL
   RWY 04: REIL. PAPI(P4L)—GA 3.5° TCH 35′. Thld dsplcd 100′.
Road.
   RWY 22: REIL. PAPI(P4L)—GA 3.0° TCH 25′. Thld dsplcd 380′.
Railroad.
AIRPORT REMARKS: Attended Mon−Fri continuously, Sat−Sun
   1300−0100Z‡. Parallel twy Rwy 04−22 and Rwy 14−32 35′ wide.
   Seagulls on and invof arpt. Ldg fee. ACTIVATE MIRL Rwy 04−22
   and Rwy 14−32, REIL and PAPI Rwy 04, Rwy 22, Rwy 14 and VASI
   Rwy 32—CTAF.
WEATHER DATA SOURCES: ASOS 121.575 (419) 838−5034.
COMMUNICATIONS: CTAF/UNICOM 123.05
Ⓡ **APP/DEP CON** 126.1 **CLNC DEL** 125.6
RADIO AIDS TO NAVIGATION: NOTAM FILE CLE.
   WATERVILLE (L) VOR/DME 113.1 VVV Chan 78 N41°27.09′
   W83°38.32′ 048° 9.8 NM to fld. 664/2W.

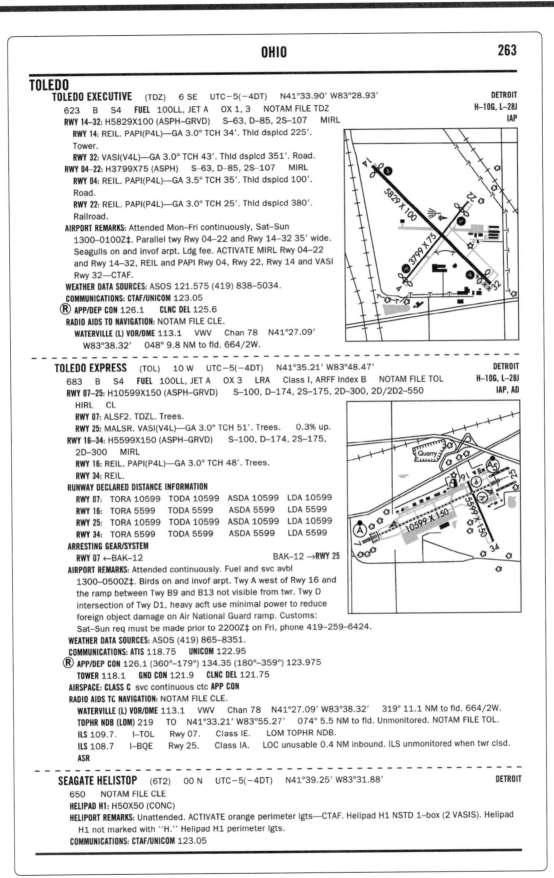

TOLEDO EXPRESS (TOL) 10 W UTC−5(−4DT) N41°35.21′ W83°48.47′

DETROIT
H−10G, L−28J
IAP, AD

683 B S4 **FUEL** 100LL, JET A OX 3 LRA Class I, ARFF Index B NOTAM FILE TOL
RWY 07−25: H10599X150 (ASPH−GRVD) S−100, D−174, 2S−175, 2D−300, 2D/2D2−550
HIRL CL
   RWY 07: ALSF2. TDZL. Trees.
   RWY 25: MALSR. VASI(V4L)—GA 3.0° TCH 51′. Trees. 0.3% up.
RWY 16−34: H5599X150 (ASPH−GRVD) S−100, D−174, 2S−175,
   2D−300 MIRL
   RWY 16: REIL. PAPI(P4L)—GA 3.0° TCH 48′. Trees.
   RWY 34: REIL.
RUNWAY DECLARED DISTANCE INFORMATION
   RWY 07: TORA 10599 TODA 10599 ASDA 10599 LDA 10599
   RWY 16: TORA 5599 TODA 5599 ASDA 5599 LDA 5599
   RWY 25: TORA 10599 TODA 10599 ASDA 10599 LDA 10599
   RWY 34: TORA 5599 TODA 5599 ASDA 5599 LDA 5599
ARRESTING GEAR/SYSTEM
   RWY 07 ←BAK−12 BAK−12 →RWY 25
AIRPORT REMARKS: Attended continuously. Fuel and svc avbl
   1300−0500Z‡. Birds on and invof arpt. Twy A west of Rwy 16 and
   the ramp between Twy B9 and B13 not visible from twr. Twy D
   intersection of Twy D1, heavy acft use minimal power to reduce
   foreign object damage on Air National Guard ramp. Customs:
   Sat−Sun req must be made prior to 2200Z‡ on Fri, phone 419−259−6424.
WEATHER DATA SOURCES: ASOS (419) 865−8351.
COMMUNICATIONS: ATIS 118.75 UNICOM 122.95
Ⓡ **APP/DEP CON** 126.1 (360°−179°) 134.35 (180°−359°) 123.975
   TOWER 118.1 **GND CON** 121.9 **CLNC DEL** 121.75
AIRSPACE: CLASS C svc continuous ctc **APP CON**
RADIO AIDS TO NAVIGATION: NOTAM FILE CLE.
   WATERVILLE (L) VOR/DME 113.1 VVV Chan 78 N41°27.09′ W83°38.32′ 319° 11.1 NM to fld. 664/2W.
   TOPHR NDB (LOM) 219 TO N41°33.21′ W83°55.27′ 074° 5.5 NM to fld. Unmonitored. NOTAM FILE TOL.
   ILS 109.7. I−TOL Rwy 07. Class IE. LOM TOPHR NDB.
   ILS 108.7 I−BQE Rwy 25. Class IA. LOC unusable 0.4 NM inbound. ILS unmonitored when twr clsd.
   ASR

SEAGATE HELISTOP (6T2) 00 N UTC−5(−4DT) N41°39.25′ W83°31.88′

DETROIT

650 NOTAM FILE CLE
HELIPAD H1: H50X50 (CONC)
HELIPORT REMARKS: Unattended. ACTIVATE orange perimeter lgts—CTAF. Helipad H1 NSTD 1−box (2 VASIS). Helipad
   H1 not marked with "H." Helipad H1 perimeter lgts.
COMMUNICATIONS: CTAF/UNICOM 123.05

Figure 63. Chart Supplement.

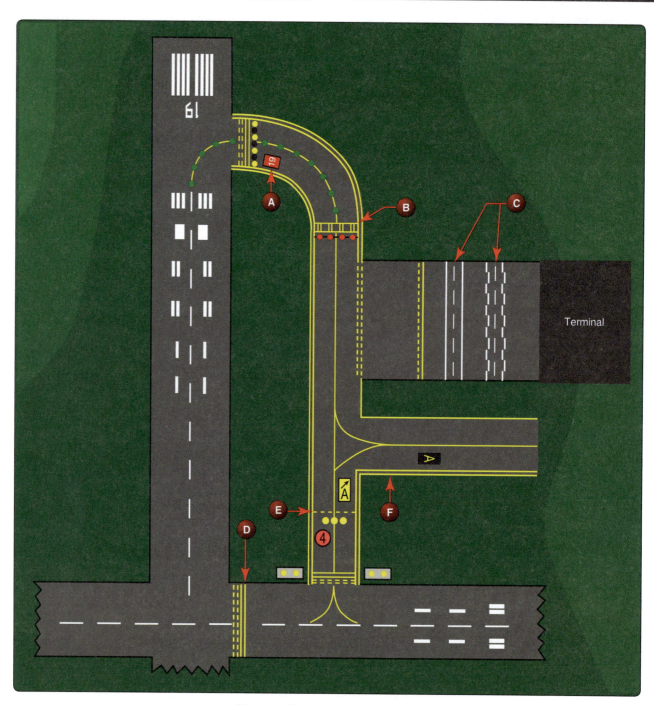

Figure 64. Airport Markings.

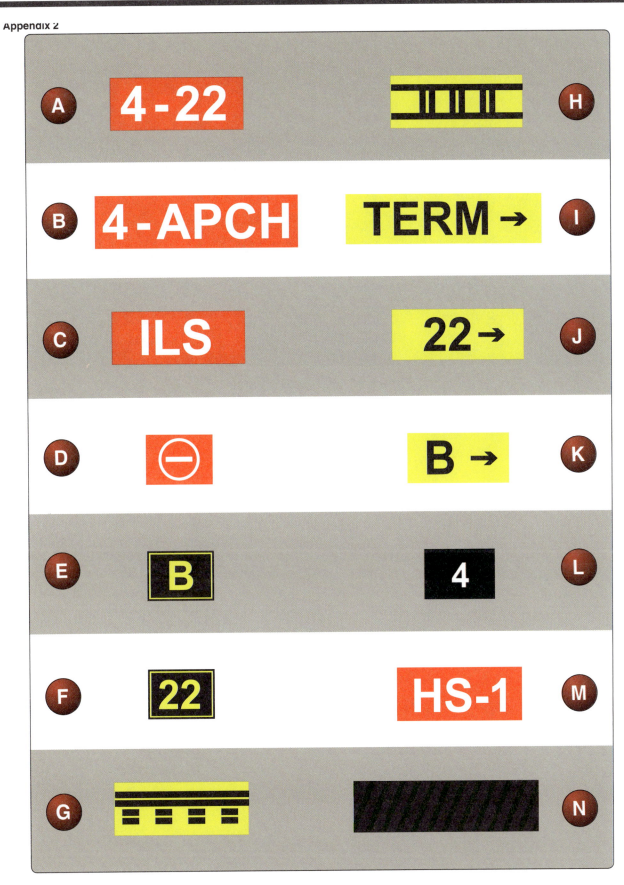

Figure 65. U.S. Airport Signs.

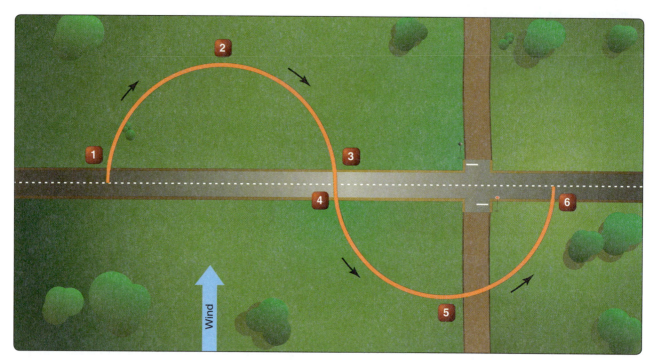

Figure 66. S-Turn Diagram.

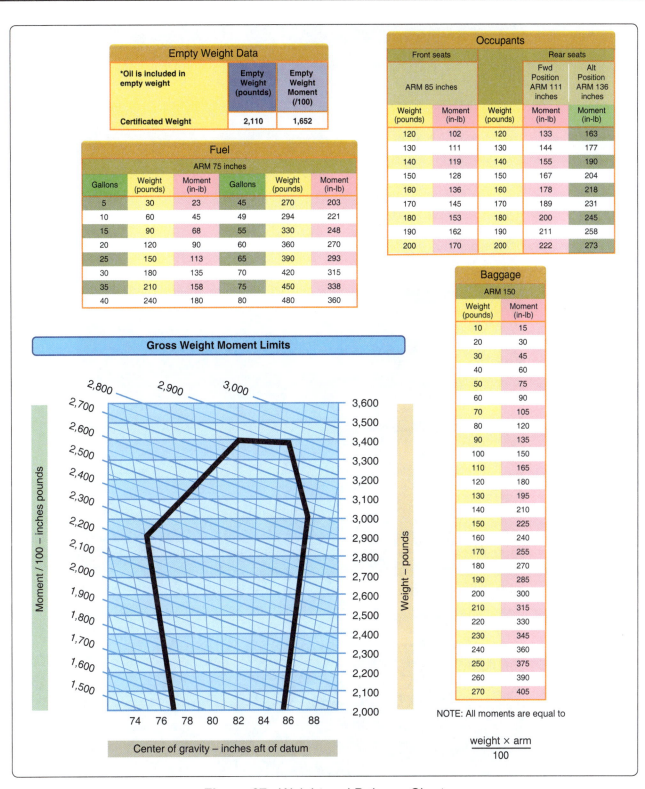

Empty Weight Data

*Oil is included in empty weight	Empty Weight (pountds)	Empty Weight Moment (/100)
Certificated Weight	2,110	1,652

Fuel

		ARM 75 inches			
Gallons	Weight (pounds)	Moment (in-ib)	Gallons	Weight (pounds)	Moment (in-lb)
5	30	23	45	270	203
10	60	45	49	294	221
15	90	68	55	330	248
20	120	90	60	360	270
25	150	113	65	390	293
30	180	135	70	420	315
35	210	158	75	450	338
40	240	180	80	480	360

Occupants

Front seats		Rear seats		
ARM 85 inches			Fwd Position ARM 111 inches	Alt Position ARM 136 inches
Weight (pounds)	Moment (in-lb)	Weight (pounds)	Moment (in-lb)	Moment (in-lb)
120	102	120	133	163
130	111	130	144	177
140	119	140	155	190
150	128	150	167	204
160	136	160	178	218
170	145	170	189	231
180	153	180	200	245
190	162	190	211	258
200	170	200	222	273

Baggage

	ARM 150
Weight (pounds)	Moment (in-lb)
10	15
20	30
30	45
40	60
50	75
60	90
70	105
80	120
90	135
100	150
110	165
120	180
130	195
140	210
150	225
160	240
170	255
180	270
190	285
200	300
210	315
220	330
230	345
240	360
250	375
260	390
270	405

NOTE: All moments are equal to

$$\frac{\text{weight} \times \text{arm}}{100}$$

Gross Weight Moment Limits

Moment / 100 – inches pounds

Weight – pounds

Center of gravity – inches aft of datum

Figure 67. Weight and Balance Chart.

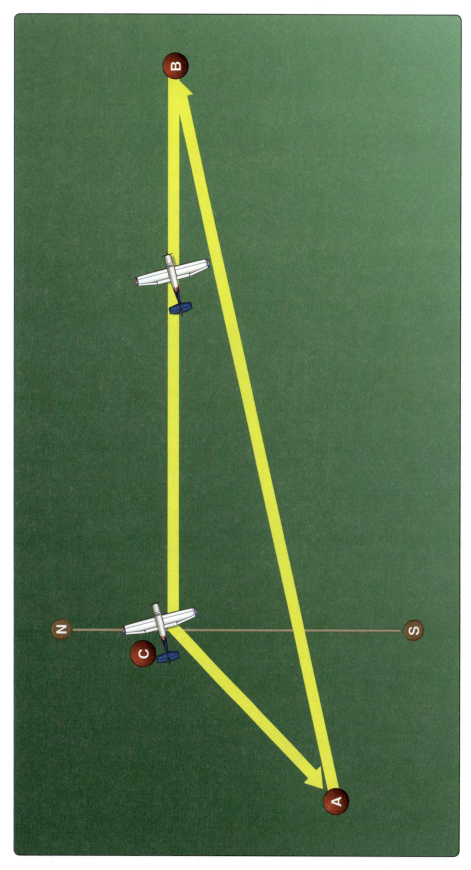

Figure 68. Wind Triangle.

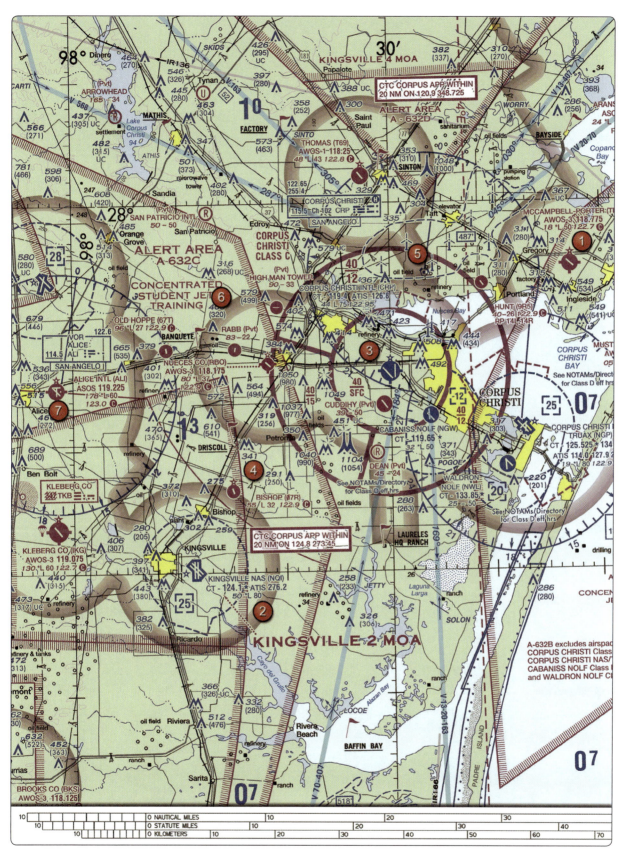

Figure 69. Sectional Chart Excerpt.
NOTE: Chart is not to scale and should not be used for navigation. Use associated scale.

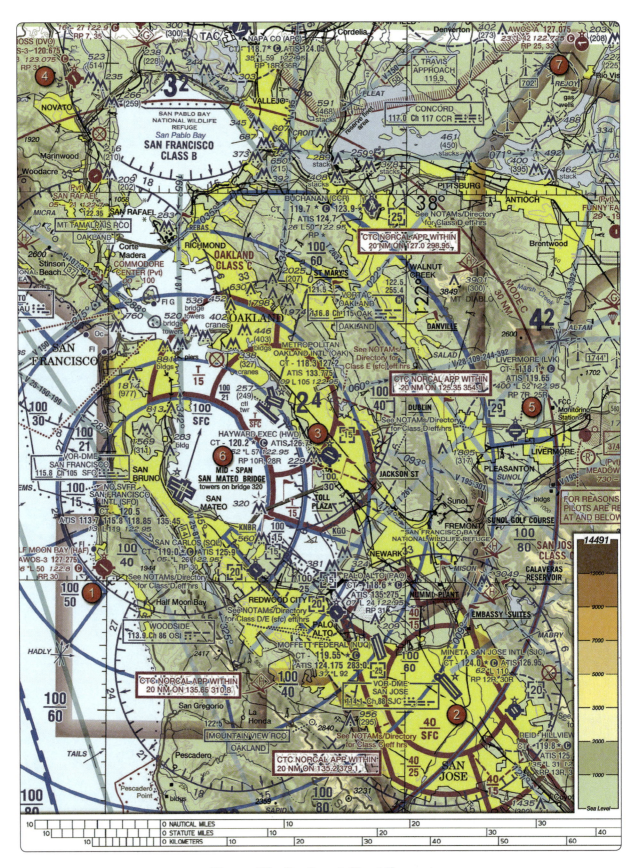

Figure 70. Sectional Chart Excerpt.
NOTE: Chart is not to scale and should not be used for navigation. Use associated scale.

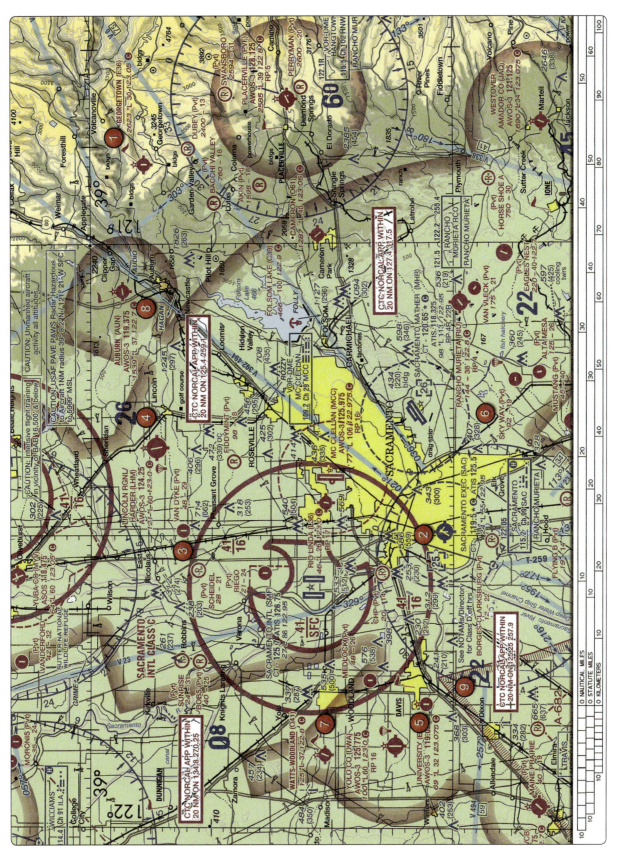

Figure 71. Sectional Chart Excerpt.
NOTE: Chart is not to scale and should not be used for navigation. Use associated scale.

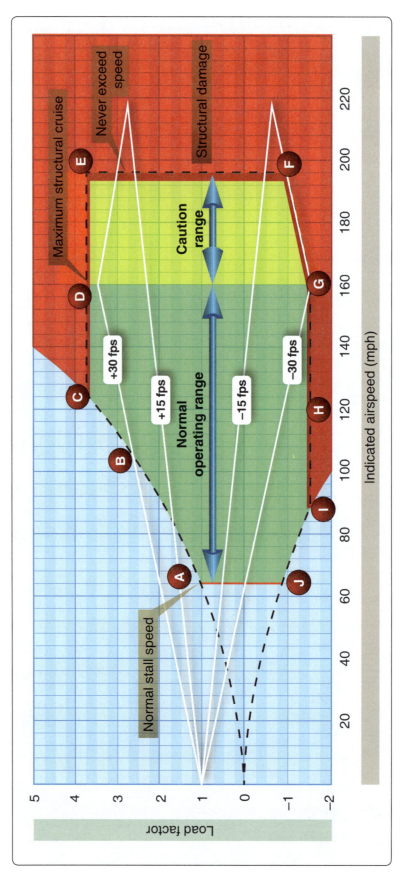

Figure 72. Velocity vs. G-Loads.

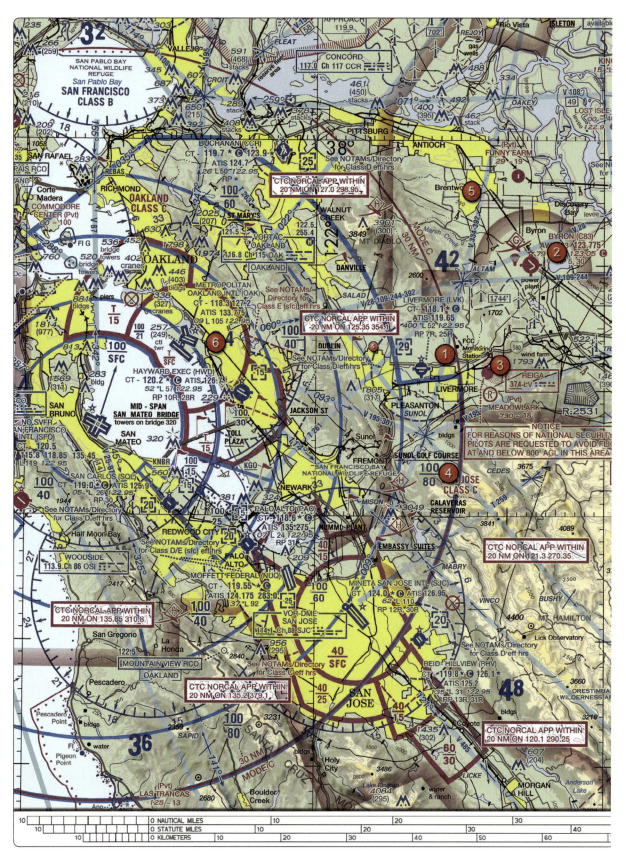

Figure 74. Sectional Chart Excerpt.
NOTE: Chart is not to scale and should not be used for navigation. Use associated scale.

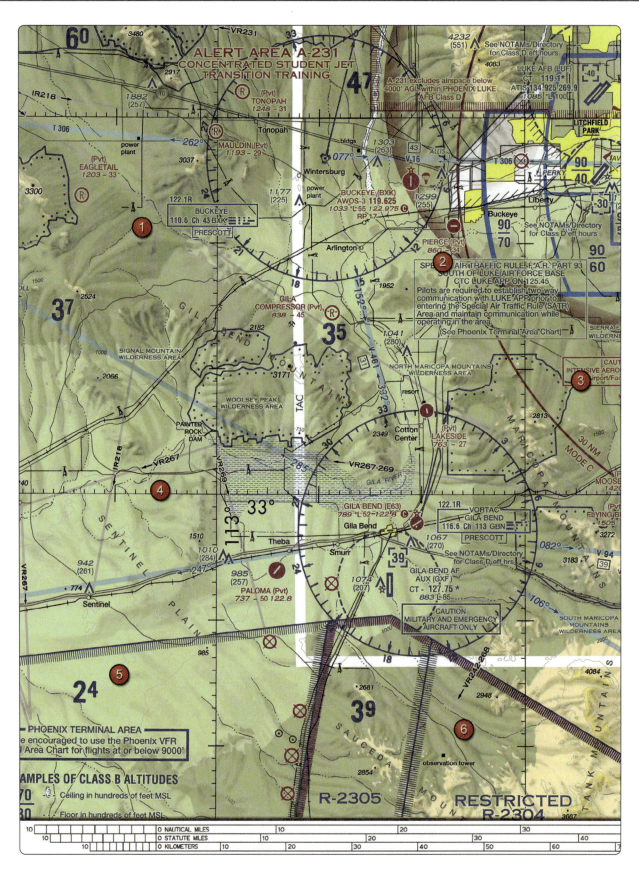

Figure 75. Sectional Chart Excerpt.

NOTE: Chart is not to scale and should not be used for navigation. Use associated scale.

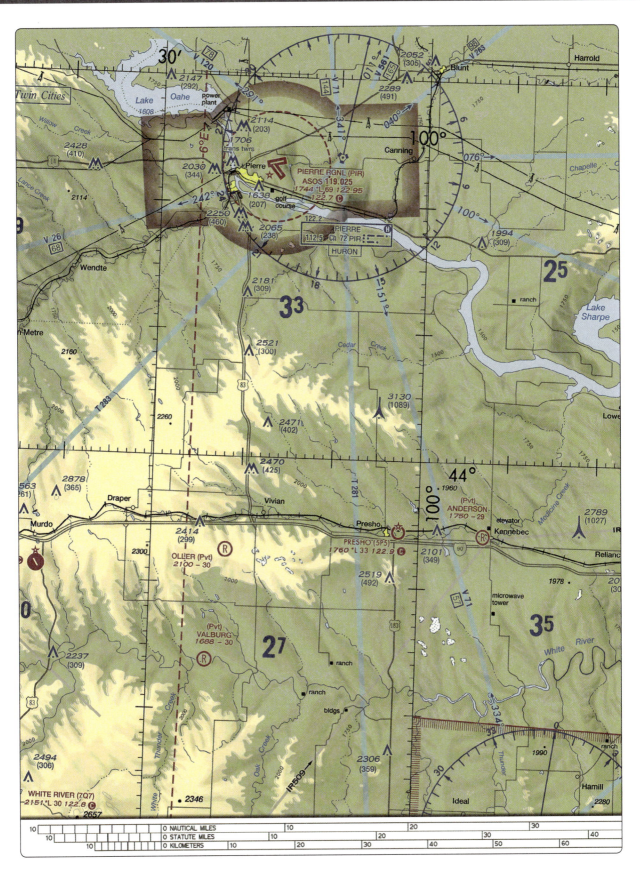

Figure 76. Sectional Chart Excerpt.

NOTE: Chart is not to scale and should not be used for navigation. Use associated scale.

340   SOUTH DAKOTA

PIERRE RGNL (PIR)   3 E   UTC −6(−5DT)   N44°22.96′ W100°17.16′   OMAHA
  1744   B   S4   **FUEL** 100LL, JET A   OX 1, 2, 3, 4   Class I, ARFF Index A   NOTAM FILE PIR   **H−2I, L−12H**
  RWY 13−31: H6900X100 (ASPH−GRVD)   S−91, D−108, 2S−137, 2D−168   HIRL   IAP
    RWY 13: REIL. PAPI(P4L)—GA 3.0 ° TCH 52′.
    RWY 31: MALSR. PAPI(P4L)—GA 3.0 ° TCH 52′.
  RWY 07−25: H6881X150 (ASPH−GRVD)   S−91, D−114, 2S−145,
    2D−180   HIRL   0.6% up W
    RWY 07: REIL. PAPI(P4L)—GA 3.0 ° TCH 47′. Tank.
    RWY 25: REIL. PAPI(P4L)—GA 3.0 ° TCH 54′.
  RUNWAY DECLARED DISTANCE INFORMATION
    RWY 07:   TORA−6881   TODA−6881   ASDA−6830   LDA−6830
    RWY 13:   TORA−6900   TODA−6900   ASDA−6900   LDA−6900
    RWY 25:   TORA−6881   TODA−6881   ASDA−6881   LDA−6881
    RWY 31:   TORA−6900   TODA−6900   ASDA−6900   LDA−6900
  AIRPORT REMARKS: Attended Mon−Fri 1100−0600Z‡, Sat−Sun
    1100−0400Z‡. For attendant other times call
    605−224−9000/8621. Arpt conditions unmonitored during
    0530−1000Z‡. Numerous non−radio acft operating in area. Birds
    on and invof arpt and within a 25 NM radius. No line of sight
    between rwy ends of Rwy 07−25. ARFF provided for part 121 air
    carrier ops only. 48 hr PPR for unscheduled acr ops involving acft
    designed for 31 or more passenger seats call 605−773−7447 or
    605−773−7405. Taxiway C is 50′ wide and restricted to acft 75,000 pounds or less. ACTIVATE HIRL Rwy 13−31
    and Rwy 07−25, MALSR Rwy 31, REIL Rwy 07, Rwy 13 and Rwy 25, PAPI Rwy 07, Rwy 25, Rwy 13 and Rwy
    31−CTAF 122.7. NOTE: See Special Notices Section—
    Aerobatic Practice Areas.
  WEATHER DATA SOURCES: ASOS 119.025 (605) 224−6087.   HIWAS 112.5 PIR.
  COMMUNICATIONS: CTAF 122.7   UNICOM 122.95
    RCO 122.2 (HURON RADIO)
  ® MINNEAPOLIS CENTER APP/DEP CON 125.1
  RADIO AIDS TO NAVIGATION: NOTAM FILE PIR.
    (L) VORTACW 112.5   PIR   Chan 72   N44°23.67′ W100°09.77′   251° 5.3 NM to fld. 1789/11E.   HIWAS.
    ILS/DME 111.9   I−PIR   Chan 56   Rwy 31.   Class IA   ILS GS unusable for coupled apch blo 2,255′. GS
      unusable blo 2135′.

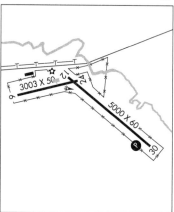

PINE RIDGE (IEN)   2 E   UTC −7(−6DT)   N43°01.35′ W102°30.66′   CHEYENNE
  3333   B   NOTAM FILE IEN   **H−5B, L−12G**
  RWY 12−30: H5000X60 (ASPH)   S−12   MIRL   0.7% up SE   IAP
    RWY 12: P−line.
    RWY 30: PAPI(P2L)—GA 3.0 ° TCH 26′. Fence.
  RWY 06−24: H3003X50 (ASPH)   S−12   0.7% up NE
    RWY 24: Fence.
  AIRPORT REMARKS: Unattended. Rwy 06−24 CLOSED indef. MIRL Rwy
    12−30 and PAPI Rwy 30 opr dusk−0530Z‡, after 0530Z‡
    ACTIVATE—CTAF.
  WEATHER DATA SOURCES: ASOS 126.775 (605) 867−1584.
  COMMUNICATIONS: CTAF 122.9
    DENVER CENTER APP/DEP CON 127.95
  RADIO AIDS TO NAVIGATION: NOTAM FILE RAP.
    RAPID CITY (H) VORTAC 112.3   RAP   Chan 70   N43°58.56′
      W103°00.74′   146° 61.3 NM to fld. 3160/13E.

Figure 77. Chart Supplement.

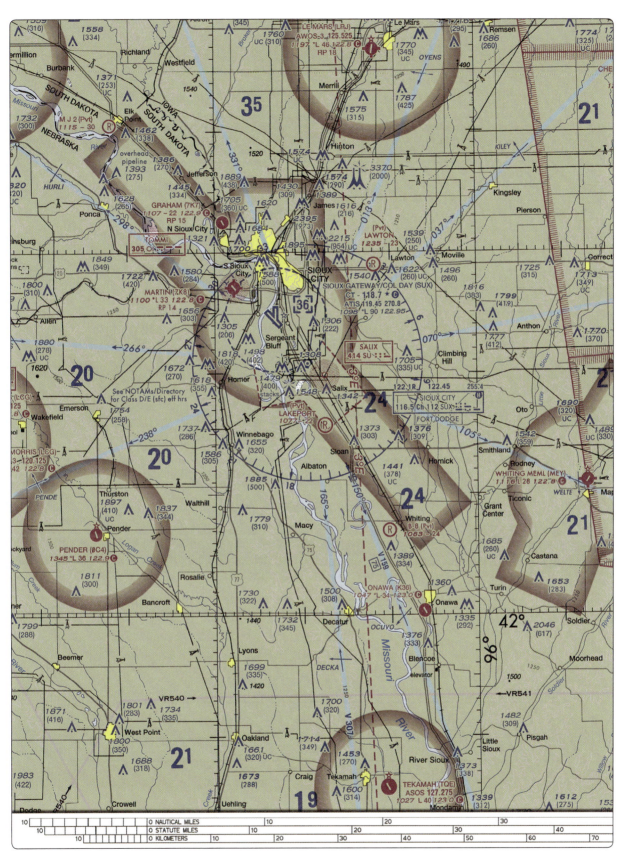

Figure 78. Sectional Chart Excerpt.

NOTE: Chart is not to scale and should not be used for navigation. Use associated scale.

64 **IOWA**

SIOUX CITY N42°20.67' W96°19.42' NOTAM FILE SUX **OMAHA**
(L) **VORTAC** 116.5 SUX Chan 112 313 ° 4.4 NM to Sioux Gateway/Col Bud Day Fld. 1087/9E. **HIWAS.** **L–12I**
VOR unusable:
280°-292° byd 25 NM 306°-350° byd 20 NM blo 3,000'
293°-305° byd 20 NM blo 4,500' 350°-280° byd 30 NM blo 3,000'
293°-305° byd 35 NM
RCO 122.45 122.1R 116.5T (FORT DODGE RADIO)

SIOUX CITY
SIOUX GATEWAY/COL BUD DAY FLD (SUX) 6 S UTC −6(−5DT) N42°24.16' W96°23.06' **OMAHA**
1098 B S4 **FUEL** 100LL, 115, JET A OX 1, 2, 3, 4 Class I, ARFF Index—See Remarks **H–5C, L–12I**
NOTAM FILE SUX **IAP, AD**
RWY 13–31: H9002X150 (CONC-GRVD) S-100, D-120, 2S-152,
2D-220 HIRL
RWY 13: MALS. VASI(V4L)—GA 3.0 ° TCH 49'. Tree.
RWY 31: MALSR. VASI(V4L)—GA 3.0 ° TCH 50'.
RWY 17–35: H6600X150 (ASPH-PFC) S-65, D-80, 2S-102,
2D-130 MIRL
RWY 17: REIL. VASI(V4R)—GA 3.0 ° TCH 50'. Trees.
RWY 35: PAPI(P4L)—GA 3.0 ° TCH 54'. Pole.
LAND AND HOLD SHORT OPERATIONS

LANDING	HOLD SHORT POINT	DIST AVBL
RWY 13	17-35	5400
RWY 17	13-31	5650

ARRESTING GEAR/SYSTEM
RWY 13 ←BAK-14 BAK-12B(B) (1392')
BAK-14 BAK-12B(B) (1492') →RWY 31
AIRPORT REMARKS: Attended continuously. PAEW 0330-1200Z ‡ during
inclement weather Nov-Apr. AER 31-BAK-12/14 located (1492')
from thld. Airfield surface conditions not monitored by arpt
management between 0600-1000Z ‡ daily. Rwy 13-BAK-12/14
located (1392') from thld. All A-gear avbl only during ANG flying ops. Twr has limited visibility southeast of
ramp near ARFF bldg and northeast of Rwy 31 touchdown zone. Rwy 31 is calm wind rwy. Class I, ARFF Index
B. ARFF Index E fire fighting equipment avbl on request. Twy F unlit, retro-reflective markers in place. Portions
of Twy A SE of Twy B not visible by twr and is designated a non-movement area. Rwy 13-31 touchdown and
rollout rwy visual range avbl. When twr clsd, ACTIVATE HIRL Rwy 13-31; MIRL Rwy 17-35; MALS Rwy 13;
MALSR Rwy 31; and REIL Rwy 17—CTAF.
WEATHER DATA SOURCES: ASOS (712) 255-6474. **HIWAS** 116.5 SUX. LAWRS.
COMMUNICATIONS: CTAF 118.7 **ATIS** 119.45 **UNICOM** 122.95
SIOUX CITY **RCO** 122.45 122.1R 116.5T (FORT DODGE RADIO)
Ⓡ **SIOUX CITY APP/DEP CON** 124.6 (1200-0330Z ‡)
Ⓡ **MINNEAPOLIS CENTER APP/DEP CON** 124.1 (0330-1200Z ‡)
SIOUX CITY TOWER 118.7 (1200-0330Z ‡) **GND CON** 121.9
AIRSPACE: CLASS D svc 1200-0330Z ‡ other times CLASS E.
RADIO AIDS TO NAVIGATION: NOTAM FILE SUX.
SIOUX CITY (L) **VORTAC** 116.5 SUX Chan 112 N42 °20.67' W96°19.42' 313° 4.4 NM to fld. 1087/9E.
HIWAS.
NDB (MHW) 233 GAK N42°24.49' W96°23.16' at fld.
SALIX NDB (MHW/LOM) 414 SU N42°19.65' W96°17.43' 311° 6.1 NM to fld. Unmonitored.
TOMMI NDB (MHW/LOM) 305 OI N42°27.61' W96°27.73' 128° 4.9 NM to fld. Unmonitored.
ILS 109.3 I-SUX Rwy 31 Class IT. LOM SALIX NDB. ILS Unmonitored when twr clsd. Glide path
unusable coupled approach (CPD) blo 1805'.
ILS 111.3 I-OIQ Rwy 13 LOM TOMMI NDB. Localizer shutdown when twr clsd.
ASR (1200-0330Z‡)

SNORE N43°13.96' W95°19.66' NOTAM FILE SPW. **OMAHA**
NDB (LOM) 394 SP 121° 6.8 NM to Spencer Muni.

SOUTHEAST IOWA RGNL (See BURLINGTON)

Figure 79. Chart Supplement.

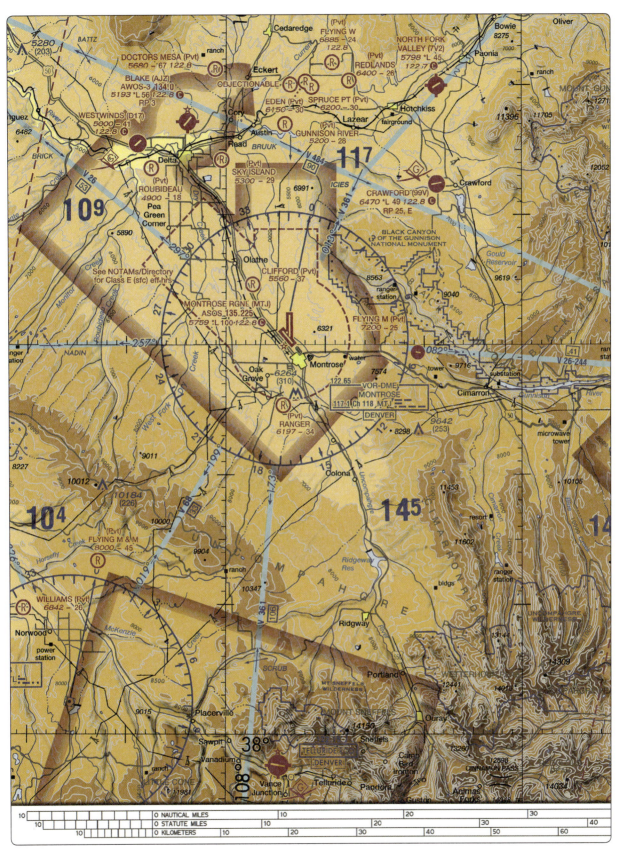

Figure 80. Sectional Chart Excerpt.

NOTE: Chart is not to scale and should not be used for navigation. Use associated scale.

216   **COLORADO**

CRAWFORD (99V)   2 W   UTC −7(−6DT)   N38°42.25′ W107°38.62′   DENVER
6470   S2   OX 4   TPA−7470(1000)   NOTAM FILE DEN   L−9E
RWY 07−25: H4900X20 (ASPH)   LIRL (NSTD)
  RWY 07: VASI (NSTD). Trees.   RWY 25: VASI (NSTD) Tank. Rgt tfc.
RWY E−W: 2500X125 (TURF)
  RWY E: Rgt tfc.   RWY W: Trees.
AIRPORT REMARKS: Attended continuously. Rwy 07−25 west 1300 ′ only 25′ wide. Heavy glider ops at arpt. Land to the
    east tkf to the west winds permitting. 100LL fuel avbl for emergency use only. Pedestrians, motor vehicles, deer
    and wildlife on and invof arpt. Unlimited vehicle use on arpt. Rwy West has +15 ′ building 170′ from thld 30′ left,
    +10′ road 100′ from thld centerline. +45′ tree 100′ L of Rwy 07 extended centerline 414 ′ from rwy end. −8′ to
    −20′ terrain off both sides of first 674 ′ of Rwy 25 end. E−W rwy occasionally has 6 inch diameter irrigation
    pipes crossing rwy width in various places. Rwy 07 has 20 ′ trees and −10′ to 20′ terrain 20′ right of rwy first
    150′. E−W rwy consists of +12 inch alfalfa vegetation during various times of the year. Arpt lgts opr
    dusk-0800Z‡. Rwy 07 1 box VASI left side for local operators only or PPR call 970-921-7700 or
    970-921-3018. Rwy 07−25 LIRL on N side from Rwy 25 end W 3800 ′. Rwy 07 1300 ′ from end E 300′. No thld
    lgts Rwy 07−25 3800′ usable for ngt ops.
COMMUNICATIONS: CTAF/UNICOM 122.8
RADIO AIDS TO NAVIGATION: NOTAM FILE MTJ.
    MONTROSE (H) VORW/DME 117.1   MTJ   Chan 118   N38 °30.39′ W107°53.96′   033° 16.9 NM to fld. 5713/12E.

CREEDE
  MINERAL CO MEM   (C24)   2 E   UTC −7(−6DT)   N37°49.33′ W106°55.79′   DENVER
8680   NOTAM FILE DEN   H−3E, L−9E
RWY 07−25: H6880X60 (ASPH)   S-12.5, D-70, 2D-110
  RWY 07: Thld dsplcd 188 ′.   RWY 25: Road.
AIRPORT REMARKS: Unattended. Elk and deer on and invof arpt. Glider and hang glider activity on and in vicinity of
    arpt. Mountains in all directions. Departure to NE avoid over flight of trailers and resident homes, climb to 200 ′
    above ground level on centerline extended prior to turn. Acft stay to right of valley on apch and/or departure
    route. 2′ cable fence around apron.
COMMUNICATIONS: CTAF 122.9
RADIO AIDS TO NAVIGATION: NOTAM FILE DEN.
    BLUE MESA (H) VORW/DME 114.9   HBU   Chan 96   N38 °27.13′ W107°02.39′   158° 38.1 NM to fld. 8730/14E.

CUCHARA VALLEY AT LA VETA   (See LA VETA)

DEL NORTE
  ASTRONAUT KENT ROMINGER   (8V1)   3 N   UTC −7(−6DT)   N37°42.83′ W106°21.11′   DENVER
7949   NOTAM FILE DEN   H−3E, L−9E
RWY 06−24: 6050X75 (ASPH)   1.1% up SW
RWY 03−21: 4670X60 (TURF-DIRT)
  RWY 21: Mountain.
AIRPORT REMARKS: Unattended. Wildlife on and invof arpt. Unlimited vehicle access on arpt. Mountainous terrain
    surrounds arpt in all directions.
COMMUNICATIONS: CTAF 122.9
RADIO AIDS TO NAVIGATION: NOTAM FILE ALS.
    ALAMOSA (H) VORTACW 113.9   ALS   Chan 86   N37 °20.95′ W105°48.93′   298° 33.7 NM to fld. 7535/13E.

Figure 81. Chart Supplement.

Figure 82. Altimeter.

APPENDIX 2 ■ FAA Figures